ANCIENT TEXTILES SERIES VOL. 4

TEXTILE PRODUCTION IN PRE-ROMAN ITALY

TEXTILE PRODUCTION IN PRE-ROMAN ITALY

Margarita Gleba

ANCIENT TEXTILES SERIES VOL. 4

OXBOW | books
Oxford & Philadelphia

First published in 2008. Reprinted as a paperback in the United Kingdom in 2023 by
OXBOW BOOKS
The Old Music Hall, 106-108 Cowley Road, Oxford, OX4 1JE

and in the United States by
OXBOW BOOKS
1950 Lawrence Road, Havertown, PA 19083

Paperback Edition: ISBN 979-8-88857-059-3
Digital Edition: ISBN 978-1-78297-603-5 (ePub)

A CIP record for this book is available from the British Library

Printed in England by Micropress, Suffolk

For a complete list of Oxbow titles, please contact:

UNITED KINGDOM
Oxbow Books
Telephone (0)1226 734350
Email: oxbow@oxbowbooks.com
www.oxbowbooks.com

UNITED STATES OF AMERICA
Oxbow Books
Telephone (610) 853-9131, Fax (610) 853-9146
Email: queries@casemateacademic.com
www.casemateacademic.com/oxbow

Oxbow Books is part of the Casemate Group

Front cover: Spinning and weaving women, drawing by Marianne Bloch Hansen, inspired by images from Daunian stelae.
Back cover: Women weaving, detail from the Throne of Verucchio (after Stærmose Nielsen 1999, 60 fig. 33).

CONTENTS

LIST OF FIGURES

LIST OF TABLES

Notes on tables

The sites are presented in the chronological order. The data in tables is by no means complete, but are intended to illustrate the overview presented in Parts 3, 4 and 5. The presence of information on each parameter in the tables is contingent on its availability in relevant publications.

Abbreviations within tables

c.	century
Cat. No.	Textile Catalogue Number
D	Diameter
EBA	Early Bronze Age
FBA	Final Bronze Age
fr.	fragmentary
g	gram
MBA	Middle Bronze Age
mm	millimetre
S	Spool

LIST OF MAPS

PREFACE

This work is intended for both textile researchers and archaeologists working on pre-Roman sites in Italy. For the former group, I have included a brief background on chronology and the cultural history of Italy, while for the latter the steps of textile production process are explained in detail. Some aspects were necessarily simplified but the references throughout the text should provide guidance for readers wishing to explore specific topics further.

The material and sites studied were chosen to represent a range appropriate for the chronological and geographical limits of study. The choice of sites has also been heavily conditioned by the quality of published information available for them. The study relies as much as possible on the recently excavated sites where material has been recovered stratigraphically and recorded in detail. Since both excavation and publication of material vary drastically from region to region, evidence from some cultures happens to be better represented than from others. Thus, the prevalence of scholarship on the Villanovan and Etruscan cultures of pre-Roman Italy has dictated a heavier concentration on Central and North Italy in my study. Furthermore, since, apart from recently excavated sites, the data on spinning and weaving implements are poorly published for many Italic excavations, I have supplemented the picture with information gleaned during extensive visits to archaeological sites, collections, and museums of Italy in 2000–2003.

As a matter of consistency, I have chosen to use the current Italian names of tombs, sites and regions, except in the cases where an English equivalent is commonly recognised, such as Sicily (as opposed to Sicilia), Sardinia (as opposed to Sardegna) and so forth. The ancient equivalents are given when relevant.

The standard chronology is presented in the following table (after Guidi 1998, 143, and Pacciarelli 2000, 277):

Neolithic	6000–2800 BCE
Chalcolithic	2800–2300 BCE
Early Bronze Age	2300–1700 BCE
Middle Bronze Age	1700–1350 BCE
Late Bronze Age	1350–1200 BCE
Final Bronze Age	1200–1000 BCE
Iron Age/Villanovan	1000–750/700 BCE
Orientalising	750/700–600 BCE
Archaic	600–475/450 BCE
Classical	475/450–300 BCE
Hellenistic	300–1st century BCE

In the bibliography, references and footnotes, I have followed the style and abbreviations of the *American Journal of Archaeology* (see Abbreviations). Latin and Greek sources are abbreviated according to *The Oxford Classical Dictionary*. Translations of Latin and Ancient Greek texts are from Loeb editions. Translations of secondary sources are my own.

ACKNOWLEDGEMENTS

This book is largely based on my PhD dissertation, written for the Department of Classical and Near Eastern Archaeology at Bryn Mawr College and completed in December 2004. My biggest thanks go to my mentor and dissertation advisor, Professor Jean MacIntosh Turfa, who has been there for me in spirit and deed from the very beginning and continues to provide her unwavering support. I also thank my co-advisor, Professor Richard S. Ellis for all his help with the project.

The research in the libraries of Rome and visits to the museums of Tuscany, Lazio and Umbria were accomplished with the help of the Bryn Mawr Ella Riegel grant in 2001–2002. My trips to the museums and sites in Veneto were funded by the Delmas Krieble Grant in 2002 and my travel in South Italy and Sicily was supported by the Archaeological Institute of America Olivia James Traveling Fellowship in 2002–2003. The writing of the dissertation was supported by the Whiting Fellowship in Humanities in 2003–2004.

I thank my parents who have supported me through every step of my graduate career. Special gratitude also goes to my friends Lesley Lundeen, Thomas Chen and Michelle Domondon, who have kept me sane throughout the writing process. Friends and colleagues who offered their insightful comments in response to the papers I have presented at various conferences while working on this project are too numerous to name – I thank them with all my heart.

During the preparation of the manuscript for publication my colleagues at the Centre for Textile Research have provided invaluable help. Jean M. Turfa, Marie-Louise B. Nosch, Eva B. Andersson, Lise Bender Jørgensen, Françoise Rougemont, Dominique Cardon and Annemarie Stauffer have read the manuscript at its various stages and provided useful comments and suggestions. Claude Albore Livadie, Larisa Bonfante, Hero Granger-Taylor, Christophe Moulherat, Emanuele Papi, Annemarie Stauffer, Simonetta Stopponi and Bouke L. van der Meer provided important references and unpublished information. Any remaining errors are my own.

For illustrations and permissions for their use I thank Carmen Alfaro Giner, Marta Bazzanella, Alberto Cazella, Nancy de Grummond, Eero Jarva, David Morton, Museo Civico di Bologna, Antoinette Rast-Eicher, Lise Ræder Knudsen, Annemarie Stauffer, Judith Swaddling, Anthony Tuck, University of Pennsylvania Museum of Archaeology and Anthropology, Zagreb Archaeological Museum. Many of the illustrations have been carefully re-elaborated by Sissel Tomm-Petersen and Marianne Bloch Hansen. The maps were made by Esbern Holms.

The publication of the book has been generously funded by the Alwyn M. Cotton Foundation Publication Grant and supported by the Danish National Research Foundation's Centre for Textile Research.

I dedicate this work to my grandparents, Šarlota and Martynas Keželis.

Margarita Gleba
Copenhagen, November 2007

ABBREVIATIONS

Abbreviations of ancient sources

Aen. Tac. *Tact.*	Aeneas Tacitus *Tactics*
Athen. *Deipn.*	Athenaeus *Deipnosophistae*
Apul. *Met.*	Apuleus *Metamorphoses*
Cato *Agr.*	Cato *De Agri Cultura*
Catul.	Catullus
Colum. *Rust.*	Columella *Rei Rusticae*
Gal.	Galen *De faciis*
Grat. Fal. *Cyn.*	Grattius Faliscus *Cynegeticon quae supersunt*
Herod.	Herodotos *Historiae*
Homer *Il.*	Homer *Iliad*
Jul. C. *B. Gal.*	Julius Caesar *De Bellum Gallicum*
Liv.	Livy *Ab Urbe Condita*
Lucr.	Lucretius *De Rerum Natura*
Ov. *Met.*	Ovid *Metamorphoses*
Paus.	Pausanias *Description of Greece*
Philostratos *Apol.*	Philostratos *The Life of Apollonius of Tyana*
Plin. *NH*	Pliny the Elder *Naturalis Historia*
Plut. *Q. R.*	Plutarch *Quaestiones Romanae*
Pollux	Pollux
Servius *ad Aen.*	Servius *ad Aeneida*
Sil. *Pun.*	Silius Italicus *Punica*
Tert. *De Pallio*	Tertullian *De Pallio*
Theophrastus	Theophrastus
Thuc.	Thucydides *History of the Peloponnesian War*
Varro *Rust.*	Varro *Rerum Rusticarum*
Virg. *G.*	Virgil *Georgics*
Virg. *Ecl.*	Virgil *Eclogues*
Vitruvius	Vitruvius *De Architectura*
Xen. *Anab.*	Xenophon *Anabasis*
Xen. *Cyr.*	Xenophon *Cyropaedia*

Abbreviations of periodicals, series, books

ActaArch	*Acta Archaeologica*
ActaHyp	*Acta Hyperborea*
AION	*Annali di Archeologia e Storia Antica. Annali dell'Istituto Orientale di Napoli*
AIONQuad	*Quaderni di Archeologia e Storia Antica. Annali dell'Istituto Orientale di Napoli*

AJA	*American Journal of Archaeology*
AmerAnt	*American Antiquity*
AmerAnthr	*American Anthropologist*
AnatSt	*Anatolian Studies*
AnnBari	*Annali della Facoltà di Lettere e Filosofia, Università di Bari*
AnnFaina	*Annali della Fondazione per il Museo "Claudio Faina"*
AnnInst	*Annales Institutorum*
AntK	*Antike Kunst*
AntW	*Antike Welt*
AR	*Archaeological Reports*
ArchAustr	*Archaeologia Austriaca*
ArchCl	*Archeologia Classica*
ArchKorrBl	*Archäologisches Korrespondenzblatt*
ArchLaz	*Archeologia Laziale*
ArchNews	*Archaeological News*
ASAtene	*Annuario della Scuola archeologica di Atene e delle Missioni italiane in Oriente*
ATN	*Archaeological Textiles Newsletter*
AttiMGrecia	*Atti e memorie della Società Magna Grecia*
BaBesch	*Bulletin Antieke Beschaving*
BAR	*British Archaeological Reports*
BAR-IS	*British Archaeological Reports, International Series*
BClevMus	*The Bulletin of the Cleveland Museum of Art*
BerRGK	*Bericht der Römisch-Germanischen Kommission*
BMCR	*Bryn Mawr Classical Review*
BolArch	*Bollettino di archeologia*
BPI	*Bollettino di paletnologia italiana*
BSA	*Annual of the British School at Athens*
BTCGI	*Bibliografia topografica della colonizzazione greca in Italia e nelle isole tirreniche. Pisa-Roma. Vols. I–XVII (1977–2000)*
BullCom	*Bullettino della Commissione Archeologica Comunale di Roma*
CAH	*Cambridge Ancient History*
CAJ	*Cambridge Archaeological Journal*
CIETA	*Centre International d'Etude des Textiles Anciens*
CIL	*Corpus Inscriptionum Latinarum*
CR	*Classical Review*
CronCatania	*Cronache di Archeologia e Storia dell'Arte, Università di Catania*
CSE	*Corpus Speculorum Etruscorum*
CSVI	*Corpus delle Stipi Votive in Italia*
CurrAnthr	*Current Anthropology*
CVA	*Corpus Vasorum Antiquorum*
CW	*Classical World*
DialArch	*Dialoghi di Archeologia*
EJA	*European Journal of Archaeology*
EtrStud	*Etruscan Studies*
IFAR Journal	*International Foundation for Art Research*
IJNA	*International Journal of Nautical Archaeology*

JAS	*Journal of Archaeological Science*
JdI	*Jahrbuch des Deutschen Archäologischen Instituts*
JFA	*Journal of Field Archaeology*
JHS	*Journal of Hellenic Studies*
JNES	*Journal of Near Eastern Studies*
JRGZM	*Jahrbuch des Römisch-Germanischen Zentralmuseums Mainz*
MeditArch	*Mediterranean Archaeology. Australian and New Zealand Journal for the Archaeology of the Mediterranean World*
MEFRA	*Mélanges d'archéologie et d'histoire de l'École française de Rome, Antiquité*
MM	*Madrider Mitteilungen*
MonAnt	*Monumenti Antichi*
NESAT	*North European Symposium for Archaeological Textiles*
NSc	*Notizie degli scavi di antichità*
OJA	*Oxford Journal of Archaeology*
OpArch	*Opuscula Archaeologica*
OpRom	*Opuscula Romana*
PBF	*Prähistorische Bronzefunde*
PBSR	*Papers of the British School at Rome*
PCIA	*Popoli e civiltà dell'Italia antica*
PP	*La Parola del Passato*
QArchEtr	*Quaderni del centro di studio per l'archeologia etrusco-italica*
QdAV	*Quaderni di Archeologia Veneta*
RA	*Revue Archéologique*
RdA	*Rivista di Archeologia*
RendAcadNazLinc	*Rendiconti di Accademia Nazionale dei Lincei*
RendIstLomb	*Rendiconti dell'Istituto Lombardo di Scienze e Lettere*
RendNap	*Rendiconti dell'Accademia di archeologia, lettere e belle arti, Napoli*
RivArchCom	*Rivista Archeologica Comense*
RivStorAnt	*Rivista storica dell'antichità*
ScAnt	*Scienze dell'Antichità*
SicArch	*Sicilia Archeologica*
SlovArch	*Slovenská Archeológia*
StEtr	*Studi Etruschi*
StMat	*Studi e Materiali. Soprintendenza ai beni arceologici per la Toscana*
StSard	*Studi Sardi*
VAMZ	*Vjesnik Arheološog Muzeja u Zagrebu*
WorldArch	*World Archaeology*
ZSchwArch	*Zeitschrift für Schweizerische Archäologie und Kunstgeschichte*

INTRODUCTION

Textile production is older than other technologies such as ceramic and metallurgy and it has always been a fundamental part of economy and trade. It began with a realisation that parts of plants and animals can be made into threads, which in turn can be woven into a two-dimensional structure of fabric. The tools required were simple and remained fundamentally unchanged until the Industrial Revolution. Textile technology was conservative and at the same time widespread: the same types of tools were used over long time periods and large geographical areas. In order to see differences and changes in textile technology and production organisation over time and space, it is therefore necessary to consider long chronological periods and relatively large geographical areas. Cultural differences observed in other archaeological material are not always present in textiles and textile tools.

This work examines evidence for textile production from the entire Apennine peninsula and Sicily focusing on the period between 1000 and 400 BCE, although material dated to earlier and later periods is included when relevant to the breadth of study. The upper end of this chronological period is the transition from the Bronze Age to the Early Iron Age in Italy. The lower chronological boundary is the time immediately preceding the Roman expansion. The period from the 10th through the 6th century is particularly important as it was the time of development from small villages of mostly egalitarian type to large urban centres with social stratification and specialised crafts. Organised production intensified significantly during these centuries, as did commercial exchange throughout and beyond the Italic sphere.

Publications that deal with the transitions in Italy during the 1st millennium BCE often mention technological development as a fact (Bartoloni 1989; Cornell 1995), but our understanding of the underlying steps is incomplete. Studies discussing a specific technique are few and usually based on a description of archeological evidence (Formigli 1993). A notable exception is a work published by Nijboer (1998) on the crafts of pottery and metalworking using industrial structures and waste products as primary sources, but textiles, as often, are not treated in the same depth. An intensive discussion of textile production will help to show the technological development of the protohistoric Italy in a more balanced fashion.

Issues connected to the question of production are those of the specialisation and of the social position of craftsmen in ancient Italy (Nijboer 1997). It is important to consider the case of textiles here because they entered the sphere of specialised production only in the 8th century BCE, after a long existence as a part of the subsistence economy. The workshop mode of textile manufacture, as defined by Peacock (1982, 8–11), seems to be in existence in Italy by the end of 7th century BCE. This is suggested by the appearance of an inordinate quantity of textile implements at certain sites, such as Poggio Civitate (Gleba 1999). The existence of such specialist production sites also raises a question of sedentary craft specialisation (Bonghi Jovino 1990; Nijboer 1997). Here again, it is important to look at textile production in isolation, since it was an activity specifically associated with women (Bartoloni 1989; Bietti Sestieri 1992; Barber 1991; Barber 1994). In small-scale societies, gender is one of the primary variables of the labor process and thus, must have an important role in studies of technology (Sassaman 1992). Women's role in Iron Age society has been a focus of studies (Rallo 1989; Nielsen 1998b), and their economic role as spinners and weavers has been definitively shown on the basis of burial evidence (Bietti Sestieri 1992) and iconographic material. Because of this association, textile equipment can present a way of tracking movements of women (Barber 1991), while diffusion of technology in some cases may be explained to be a consequence of intermarriage.

The primary evidence examined in the following pages consists of the surviving textiles and the implements associated with textile production. The data collected and presented here is by no means intended to be definitive; rather the aim of the present study is to set the stage for future studies of textile production in ancient Italy. A comprehensive analysis of textiles is needed before one can speak more concretely of developmental trends in ancient Italian technology. The variety and complexity of ancient Italian textiles is already apparent from the preliminary analyses of surviving specimens. Further studies will permit us to trace the development of new techniques and their relationship to textile technology in other European regions. An in-depth analysis of the iconographic material, such as the painted tombs of Tarquinia, is needed to correlate the patterns of the surviving textiles with those depicted in art. Research into ancient dyeing is still in its infancy, and one may hope that it will soon yield important results for textile studies. Here, paleoethnobotanic and geological studies are crucial to determining the possible sources of ancient dyes. Furthermore, different classes of tools must be studied separately in their varied contexts. Thus, implements found in burial contexts must be correlated with other burial goods. Votive contexts must be examined more carefully and sanctuary textile production has to be taken into account. Implement distribution studies are necessary for the settlements. Most importantly, physical parameters for various implements, such as weight, shape and size, have to be examined more carefully because they provide important information about textiles that could have been produced with their help. Furthermore, comparative work with other regions of the Mediterranean is needed to set the Italian material in a larger context. It is my hope that this study will draw the attention not only of textile specialists but also of excavators to this frequently overlooked class of archaeological objects, and lead towards a better understanding of ancient textile technology and the broader issues associated with it.

The following chapters have been organised with regard to the stated aims of this study. Part 1 provides a chronological and cultural background for this analysis. The sources for the study of textile production in Italy are discussed in Part 2. Extant Italian archaeological textiles are considered in Part 3. Tools used in various stages of textile production are at the core of Part 4. In Part 5, a discussion of archaeological contexts in which textile implements have been found aims to recover information about the role of this craft in ancient Italian society. Part 6 considers textile production process and trade in pre-Roman Italy and its development throughout time. Part 7 places textile technology into the social context of the period under consideration.

PART 1

Geographical and Chronological Context

Through textiles we express who we are – our gender, status, age, profession, family, religion and ethnic affiliation. Textiles, thus, bring us very close to the individual. At the same time, textile production is tightly bound to the social structure of any culture. Archaeologically, characteristics of an individual are especially evident in a burial context, while characteristics of a society can be recognised in a variety of contexts. The chronology of textiles and textile tools is largely dependent on the context.

Before we can proceed to the discussion of textile production in Italy, a short introduction into Italian cultural history is in order. The large territory of the Apennine peninsula, Sicily and Sardinia (Map 1), and their varied topography and environment account for ancient Italy's cultural diversity. The literature on the archaeology of pre-Roman Italy is vast, and it is not my intention here to provide a comprehensive overview. Rather, the following pages will set the chronological and cultural stage for discussing textile production in the Apennine peninsula in ancient times.

Chronology

The chronological framework for the present study encompasses the period between 1000 and about 300 BCE; however, material from earlier and later periods will be considered in selected cases. The chronologies of ancient Italy are specific for each region, with certain type sites providing the sequences on the basis of the typologies developed for the study of burials and hoards (cf. Table 1). In this volume, however, absolute dates are used instead of names of periods developed for cultures or sites, and it should be emphasised that dates present ranges and not specific years. Regional sequences are constructed on the basis of all available materials (metal, ceramic, lithic, *etc.*), which are often of local character. Metal artefact types are most reliable for inter-regional chronologies since they have a wider distribution than non-metal objects (Pare 1996, 99). Typologies of bronze objects are particularly useful for the Bronze Age, when inter-regional contacts were especially widespread (Carancini, Peroni and Peroni 1996; Carancini and Peroni 1999). The Bronze Age/Iron Age transition was marked by the collapse of this exchange and the cessation of earlier traditions of metal production and hoard deposition. The start of the Iron Age in Italy had traditionally been set at 900 BCE. Recently, however, Peroni raised the beginning of the Iron Age at Este, Golasecca and Bologna to 1020 BCE (Peroni 1994; 1996; Nijboer 1998, 24–26); the dates of 975/950, however, have gained wider acceptance (Pare 1996, 100; Guidi 1998, 142; Pacciarelli 2000, 277).

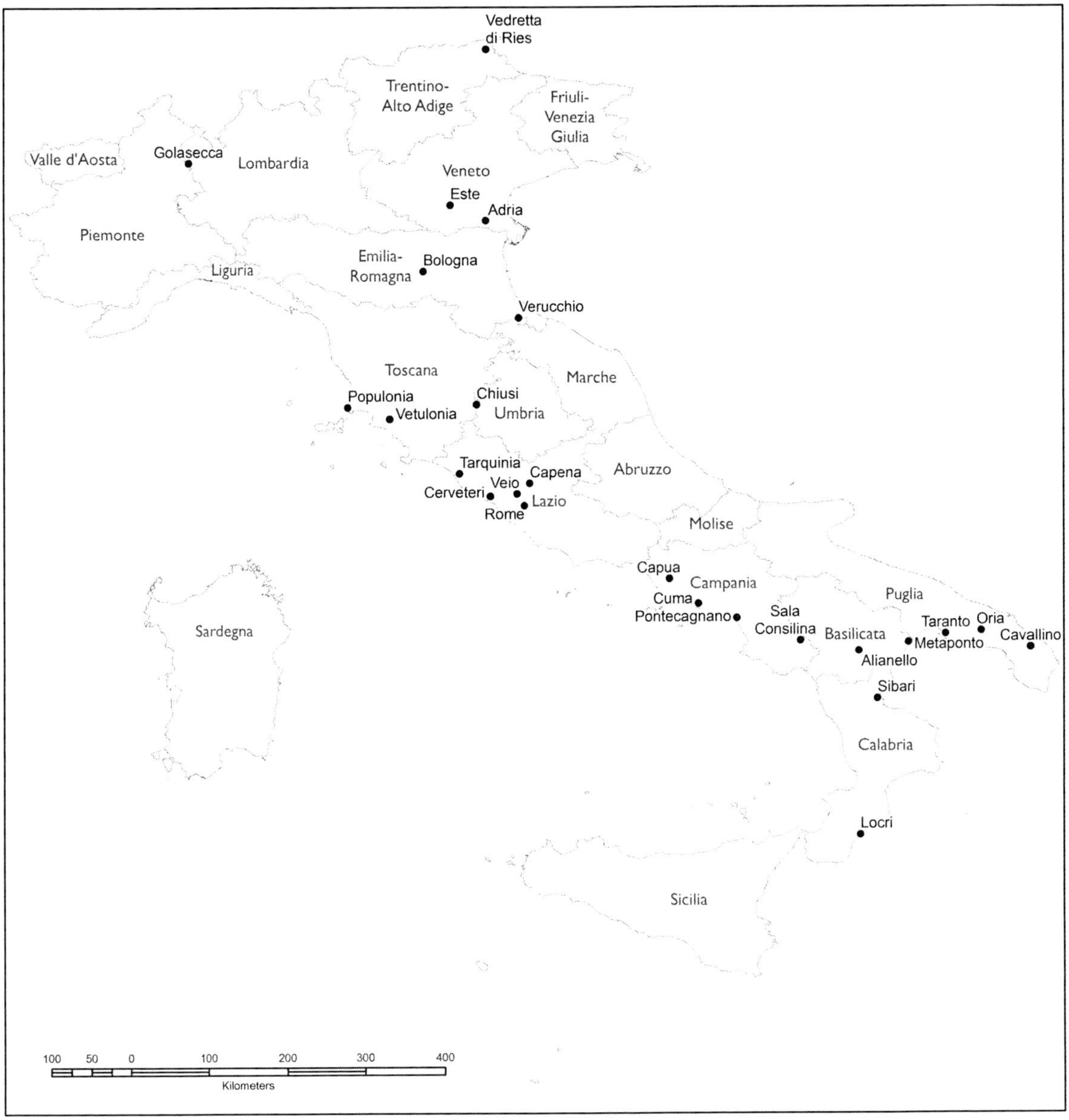

Map 1. General map of Italy with main sites mentioned in Part 1.

In the 8th century BCE, Italian chronology becomes more reliable thanks to the appearance of well-dated Greek ceramic imports. In fact, for a long time comparisons with Greek and Near Eastern material have been the basis for the chronology of the 1st millennium BCE in Italy. At present, the prehistoric chronology of the Mediterranean is undergoing a process of redefinition using radiocarbon and dendrochronological dating systems, which have allowed for a readjustment of relative chronologies and for an establishment of independent absolute dating systems (Guidi 1998, 140–142). Thus, the calibrated radiocarbon dating, applied to material which is dated by relative chronology, gives earlier absolute dates than the traditional ones (Randsborg 1991). These calibrated radiocarbon datings agree with dendrochronological sequences based on the materials from the pile-dwellings in Switzerland, south-western Germany and North Italy (Randsborg 1991, 94; Pare 1996, 102; Nijboer 1998, 22–26). Dates previously established are thus being pushed back.[1] Where possible, the dates provided in the following discussion are taken from this new absolute framework. The regional chronological sequences and their correlations, as well as absolute dates are provided in Table 1.

Bronze Age

The cultural processes that culminate in the Early Iron Age have their roots in the Bronze Age (Peroni 1971, 1989; Ridgway 1988a, 624–628; Guidi and Piperno 1992; Cocchi Genick 1996). The Early and Middle Bronze Ages in Italy are still poorly defined archaeologically. The Early Bronze Age was characterised by a multiplicity of cultural facies (Peroni 1971), while in the Middle Bronze Age a relatively uniform Apennine civilisation covered Central Italy. The better-investigated Late Bronze Age is the period most relevant for the present study, since it immediately precedes the Early Iron Age, when the material culture of the different ethnic groups became well-defined. In Italy, this period has been subdivided into the Recent and Final phases. The Recent Bronze Age is also known as the Subapennine, being contemporary with the culture that characterised the greater part of Italy from the Po valley to southern Calabria (Pacciarelli 2000, 36). The Final Bronze Age corresponds to the appearance of the Protovillanovan culture, so called because of its affinities with Villanovan cultural traits of the following period (Bartoloni 2002, 85). The term 'Villanovan' was coined in the middle of the 19th century after the site of Villanova di Castenaso excavated near Bologna. It is important to emphasize that the term is used in an exclusively cultural sense since an ethnic entity that could be recognised in the 'Villanovan' funerary rite and associated ceramic decorative style never existed (see Ridgway 1988a, 628–630; Bartoloni 2002, 101–102). The essential characteristics of Protovillanovan culture include an economy based on agricultural and metallurgical activities, a socio-political structure of aristocratic type marked by warrior ideology, the rite of cremation with the appearance of large necropoleis, a distinctive type of ceramics with varied decorative schemes, and a differentiated use of territory, the occupation of which is tied to its control.

Italy maintained continuous and systematic ties with the eastern Mediterranean since the Middle Bronze Age, and Mycenaean Greece played an important role in these connections (Vagnetti 1985; Peroni and Trucco 1994; Vagnetti 2000; Bettelli 2002; Ridgway 2002). The significance of the Mycenaean presence in Italy is still debated, scholars ascribing varied degrees of influence that Aegean goods, technologies and institutions had on the development of South Italian native communities. Whatever the outcome of this debate, it is clear that the destruction of the Mycenaean kingdoms around the year 1200 BCE did not produce the same repercussions in Italy that it had in the Aegean area. There are no major breaks in the archaeological record of the Italian Late Bronze Age sites. Thus, the 'Dark Age' concept commonly used to describe the time between the 12th and 10th centuries BCE is not applicable in Italy or, in fact, anywhere outside

Date scale: 900 – 800 – 700 – 600 – 500 – 400

HALLSTATT	GOLASECCA	ESTE	BOLOGNA	VERUCCHIO	PICENUM	VEIO	TARQUINIA	LATIUM	PONTECAGNANO	SALA CONSILINA	CAPUA
B2	I A1	I	I	I	I	I A	I A	I A	I A	I A	I A
B3	I A2	II	II A	II	II	I B	I B	II B	I B	I B	I B
C	I	II B	II B	III	III	II A	I C	III	II A	I C	II
D	I C	III C	III		IV A	II B	II A	IV A	II B	II A	III A
LA TÈNE	II A	III A	IV A		IV B	II C	II B	IV B	III	II B	III B
	II B	III B	IV B		II	III A	III A	ARCHAIC		III	IV
	III	III C	IV C		IV	III B	III B				
		III D	CERTOSA			IV	IV				
		IV				ARCHAIC	ARCHAIC				

Table 1. Chronological concordance of Hallstatt and Iron Age culture of Italy (After Nijboer 1998; Rittatore Vonwiller 1975; Manfredi and Malnati 2003; Naso 2001a; Toms 1986; Hencken 1968a; Bietti Sestieri 1992a; d'Agostino and Gastaldi 1988; Ruby 1995).

of Greece (Ridgway 1994). In Italy, this period, corresponding to the Final Bronze Age, was marked by significant transformations that culminated in the development of proto-urban centres in many central and southern areas. The passage between the Bronze Age and the Iron Age was marked in certain areas (Etruria, northern Calabria) by a switch from collective land ownership to individual or family ownership and its progressive concentration in the hands of dominant elites, which by the time of Greek colonisation, had developed into a true aristocracy (Peroni and Trucco 1994, 877).

Early Iron Age

The Early Iron Age is generally recognised as the time of the origins of most Italic peoples. From the break-up of a Protovillanovan unity emerged several cultures, which later developed into the major ethnic and linguistic entities of pre-Roman Italy: Golasecca in the north-west; Venetic culture in the north-east; Villanovan culture in the Central Tyrrhenian with Latium Vetus immediately south of it; Picene culture in the Central Adriatic; Iapygian culture in Puglia; *Fossakultur* in Campania and Calabria; and Enotrian culture in Basilicata.

The type of social organisation that in the Early Iron Age was common throughout the peninsula was characterised by a division into broad family groups (*gens*). Each such group was dominated by an elite headed by a male figure (*princeps gentis*). These head figures adopted funeral customs in which weaponry and other valuable goods played a key role, thus promoting the circulation of these prestige items. The control of this economic circulation was kept in the hands of the elite groups. This economic inequality led to a further concentration and accumulation of wealth by a few families that eventually developed into an aristocracy (Cerchiai 1995, 62–64). By the end of the Iron Age, many areas of Italy had a social organisation of gentilitial character in which elite groups controlled economic activities and mediated the relations with Greeks and other groups. Their power was expressed through the great richness of their funerary goods and the adoption of the banquet as an institution. The richness of some children's burials indicates that rank was hereditary.

The Orientalising phenomenon

In the period generally dated from 720 to 580 BCE, Italy was flooded by a massive wave of luxury and prestige objects, many of which were imported from the eastern Mediterranean and the Near East. Brought for the most part by Levantine traders, such goods became indispensable elements that defined and legitimised aristocrats (Bartoloni 2000a). These goods in turn inspired local craftsmen to adopt new techniques and produce works in similar styles. The term 'Orientalising' is thus used for both the new style and the period. Competition among the elite resulted in conspicuous consumption as illustrated by numerous high status tombs, often characterised by monumental architecture. Some of the objects included in these 'princely tombs' also indicate the adoption of practices such as 'Homeric' burial, sympotic banqueting or hoplite warfare (Barker and Rasmussen 1998, 124–125). The extraordinary richness of the Orientalising tombs was never to be reached again in later periods.

Burial customs

The majority of textiles and many of the textile tools examined in this study were found in burials. Funerary data have been considered a primary source for the reconstruction of the social makeup of ancient communities (Bartoloni 2002, 32). In fact, for a long time, the bulk of archaeological material pertaining to pre-Roman Italy was derived from burial contexts – it still is to a significant extent. The typology of the materials and their association within burials, the examination of funerary ritual and structure, and the topographical analysis of the necropoleis permit the identification of the sex, age and status of an individual, as well as the social structure and chronology of the community to which this individual belonged (Bartoloni 2002, 32–35; 2003).

Italian burial practices during the 1st millennium BCE include inhumation with deposition in a trench grave, *fossa*, and cremation, with remains deposited in an urn, which was placed in a well-like pit, *pozzo*. Cremation appeared in Italy in the 12th century BCE, coinciding with similar practices in the Urnfield culture of continental Europe. Over the next two centuries, it spread from the Alps to Sicily. However, while in North Italy and Etruria cremation became a predominant burial rite, in the eastern and southern parts of the peninsula, inhumation was retained as the main way of disposal of the dead (Briquel 2000, 54–55). In some necropoleis, one rite is found exclusively, while in others, both occur in varying degrees. Children's inhumations are occasionally deposited in large containers, such as amphorae, and are called *enchytrismoi*. Both inhumations and cremations could be placed in a chamber tomb (*tomba a camera*), which simulated a dwelling and sometimes was covered with a mound of earth, a *tumulus*. The shape, the dimensions, the cover and the fill of individual tombs could vary. In the case of cremations, the urn shape varied and it could, in turn, be placed in another container, like a stone cist (*ziro*). Sometimes, urns were placed lying prone, as if they were bodies, or dressed in garments and jewellery, demonstrating that cremated remains and their containers were anthropomorphised.

The burial was often accompanied by objects, referred to as grave goods, assemblages of which are known in Italian as *corredi*. They included objects in ceramic and metal, as well as more luxurious materials. *Corredi* attributed to women are often characterised by *parures*, the entirety of personal decorations, such as fibulae, belts, bead necklaces, pendants and earrings. The degree of variation in the quantity and quality of grave goods in burials is an important indicator of social differentiation within a community. Richer burials have a higher quantity and quality of offerings. Burials of the Final Bronze Age show clear signs of this differentiation in the Central Tyrrhenian area, whereas the Villanovan and Latial graves of the Early Iron Age display a more egalitarian character. This 'isonomy' may be a consequence of concentration of people into a few big centres with an "ideological masking of social differences" (Guidi 1998, 146–147). The progressive differentiation of grave-good assemblages that followed provides a clear archaeological correlate for the existence of aristocratic groups. The adoption of heroic ideology and its display was a way of elite representation during the period of power consolidation. These trends were especially prominent throughout Italy during the Orientalising period.

Settlement structure

Settlement archaeology plays a fundamental role in our understanding of economic and social phenomena in ancient societies. Spatial distribution studies, developed in the last twenty years, have been of great importance in this field. Methodologies such as site catchment analysis (Potter 1976; Bartoloni 2002, 26–27), Thiessen's polygon model (Di Gennaro 1982; Bartoloni 2002, 31–32), central place theory (Bartoloni 2002, 28–29), or rank size rule (Guidi 1985; Bartoloni 2002, 29–30), have recently been applied to protohistoric societies in Italy and have yielded important results.

Urbanisation

As outlined by Guidi (1998, 156), during the first half the 1st millennium BCE, many parts of Italy underwent a transformation from chiefdom-type village communities to societies with state characteristics, including urban structures, social stratification, the formation of market systems and private property, the appearance of writing and figurative art, and organised cult activities.

The Tyrrhenian area experienced this urban formation between the 10th and 8th centuries BCE. In the surrounding areas, this process occurred on a smaller scale during the course of the 7th and 6th centuries BCE. The reasons for the delay of this process in North Italy may be connected with economic structures that had been in place there since the Final Bronze Age. The North was "still characterized by a redistributive economy, ruled by elites not able to grasp with their growing complexity" (Guidi 1998, 156). In contrast, the Eastern Adriatic and Southern regions never achieved the level of urbanisation evident in the Central Tyrrhenian area.

As recently as twenty years ago, the most common explanation for the urbanisation of Italian settlements was contact with the founders of Greek colonies. Thanks to recent surveys and settlement excavations this attitude is starting to change. Archaeological evidence has demonstrated that in Etruria and South Italy proto-urban entities already existed in the 9th century BCE, and that early cities arose in the course of autochthonous development. Protourban centres are defined in terms of "the radical change in size with respect to the settlements of the preceding phases with a corresponding growth of their functions and, at the same time, their transformation in[to] true urban sites in the following historical period" (Guidi 1998, 142, citing F. di Gennaro). The process of syncretism of numerous villages of the Final Bronze Age into large Iron Age centres thus was not determined by any external factors (Guidi 1998, 143). The later Greek impact occurred rather on an ideological level.

Exchange and trade

Exchange played an important social and economic role in ancient societies, and has become the focus of one of the growing areas of archaeological research (Parker 1992, 2ff). The movements of people, goods, raw materials, and entire exchange systems can be reconstructed through studies of the distribution and/or source of certain goods. Textiles, textile raw materials and people with the 'know-how' in textile production were certainly involved in the exchange, even if the archaeological evidence for it is often less obvious than in the case of other materials and technologies. While itinerant craftsmen were important in transfer of technologies such as metalwork and glass production, other factors may have played a more important role in the case of textile production. One vector of textile technology transfer was intermarriage, another was gift-giving. This latter form of exchange had more than a purely economic significance, because it reinforced the relationship between the trading parties and took place within a framework of reciprocity. Despite the difficulty of distinguishing archaeologically between gift-giving and commercial exchange, the presence of prestige objects in princely tombs throughout Italy during the Orientalising period has made it possible to reconstruct the elite exchange circuits. This system was particularly important and lasted from the Bronze Age until the middle of the 6th century BCE when a monetary economy gradually began to develop.

Colonisation

The study of trade is inevitably tied to the issue of colonisation. In Italy, the term 'colonisation' is usually associated specifically with Greek foundations (Graham 1982b; Buckley 1996; Pugliese Carratelli 1996; Boardman 1999). In recent years, however, the Etruscan presence in Campania and North Italy has been viewed as a colonising movement (Cerchiai 1995, 11). More importantly, one has to consider Phoenician settlements, which were established soon after the trading contacts with the west were made. They were designed to consolidate these early trade relationships, which were later threatened by the Greeks (Niemeyer 1990, 488). Thus, today we know of several types of colonisation in Italy.

Pre-colonial contacts certainly preceded the establishment of Phoenician and Greek colonies and the evidence for such contacts has been increasing over the last two decades (Coldstream 2003). It is becoming clear, however, that the role of the Euboeans in this pre-colonisation has been overestimated, while the Phoenicians played a more important role than is generally accepted (Aubet 1993). Sardinia is also slowly emerging from its earlier archaeological obscurity to be recognised as one of the crossroads of the Mediterranean.

For now, however, Greek colonial foundations remain better studied and documented than Phoenician settlements. The standard view has been that Greek colonies were organised foundations of the Greek *poleis*, necessitated by crises in the metropoleis. Now many scholars accept that colonisation was not a state enterprise because the *polis* did not yet exist as a fully formed institution (Malkin 1987; Maaskant-Kleibrink 1997). Furthermore, it has been suggested that colonisation was a private rather than a state enterprise and that the process was instituted at random and is not well described by the term 'colonisation' (Osborne 1998). Beginning with the 8th-century Pithekoussai, colonies were, in fact, ethnically diverse and included other groups such as Phoenicians and the indigenous peoples of Italy and Sicily, so that "ethnic identities were not divisive factors in the process of Greek settlement in southern Italy" (Wilson 2000, 41). It is with this 'multiculturalism' in mind that this study has approached Greek and other settlements.

Cultures and *facies*

For the purpose of the present study, the term 'culture' is to be understood as "a constantly recurring assemblage of artefacts assumed to be representative of a particular set of behavioural activities carried out at a particular time and place" (Renfrew and Bahn 1996, 539). In Italy, cultures with large territorial extension have been subdivided into smaller, local units, termed cultural *facies* (Pallottino 1939; Pacciarelli 2000), which are often defined in terms of ceramics excavated at settlement sites and in burials. Archaeological cultures and their *facies* can occasionally be equated with the ethnic groups documented in historic sources; the emergence of such groups with strong identities seems to begin in the Early Iron Age (Pacciarelli 2000, 19). Equating archaeological cultures to ethnicities has not always been done prudently, so ethnic designations are used here only when they are generally accepted and correspond closely to specific assemblages of archaeological material (cf. Shennan 1989).

The nature of the evidence for each particular culture and region is unequal and heterogeneous and necessitates different approaches to their description. North and Central Italy will be reviewed in terms of cultures, which were present there during the first half of the 1st millennium BCE, since the current regional boundaries do not correspond to cultural borders. South Italy, on the other hand, will be considered region by region. The islands of Sicily and Sardinia, having natural boundaries, will be reviewed separately.

North Italy

North Italy comprises the modern regions of Piemonte, Lombardia, Liguria, Veneto, Trentino Alto Adige and Friuli Venezia Giulia.[2] The cultures that developed in these regions in the course of the Early Iron Age received influences from both the Central Italian area and continental Europe. The Middle and Late Bronze Ages in North Italy are characterised by lake-dwellings (*palafitte*) in the Alpine and pre-Alpine regions and by the *Terremare* culture of the ditched sites in the lowlands (Bernabò Brea and Mutti 1994, Bernabò Brea, Cardarelli and Cremaschi 1997). The *palafitte-terremare* settlements formed a complex system with an advanced level of socio-political organisation, craft production and exchange, which, owing to external and internal causes (Cardarelli 1992, 380), suddenly collapsed in the Late Bronze Age, giving rise to the development of several distinct cultures: so-called Golasecca culture in the north-west, Liguri in the south-west and Veneti in the east. All of these cultures would later be eclipsed by the Celtic invasions of the 4th century BCE (Santoro 1979; Grassi 1991; Vitali 1991a; Naso 2000a, 251–255).

Golasecca

The area of the Golasecca culture, named after the type-site, comprises the north-western part of the Po valley including Piemonte, Lombardia, and inland Liguria (Rittatore Vonwiller 1975; De Marinis 1991; Gambari 1998). Unlike other parts of North Italy, this area did not experience any notable discontinuity in the Late Bronze Age. In fact, the local Late Bronze Age Canegrate culture gave rise to the Proto-Golasecca type Ascona complex during the 12th–11th centuries BCE, which, in turn, developed into the proto-Golasecca type Ca' Morta-Malpensa culture. The 10th century BCE marked the beginning of the Golasecca culture, with different *facies* observed on the level of funerary rites since little is known about the settlements (De Marinis 1986, 26–27).

The role played by the Golasecca culture as a bridge between the Mediterranean and Central Europe, starting in the 8th century BCE, may have been facilitated not only by its control of the routes to important Alpine passes, but also by its "ethnic affinities with the Transalpine Celts" (De Marinis 1991, 97). The reciprocal cultural influences are thus not surprising. Furthermore, both cultures were strongly influenced by Etruscan civilisation. In the 5th century BCE, Golaseccan trade routes underwent important changes, and the new main sites located along the trade routes (Como, Milano, Lodi, Bergamo, and Brescia) were soon transformed into Celtic *oppida* and, later still, became important Roman cities.

Liguri

The Liguri were one of the oldest and largest Italic ethnic groups, inhabiting the north-western part of Italy, especially the coastal area of Liguria (Bernardini 1977; Salmon 1988, 715–717). Unfortunately, there are few archaeological data on their culture, and the problems of Ligurian ethnic identity and their connection with the Celts remain unsolved. The necropolis of Chiavari, occupied during the 8th–7th centuries BCE, demonstrates that, at the beginning of the Iron Age, the region already had its own independent culture, which differed from the contemporary Golasecca and Villanovan cultures (De Marinis 1988d, 253). Funerary finds also suggest that the settlement site was a rich commercial and maritime centre. Another important site, Genova, reached its acme in the 5th–4th century BCE, becoming a major maritime emporium and the first proto-urban centre in the area. It is believed to have been an Etruscan foundation that incorporated indigenous settlements (De Marinis 1988d, 255).

Veneti

Veneti is the ancient name used for the peoples who inhabited the territory from Lago Costanza to the northern Adriatic, comprising the modern region of Veneto and parts of Friuli Venezia Giulia and Trentino Alto Adige (Fogolari 1975; Chieco Bianchi and Tombolani 1988; Chieco Bianchi 1988a; Fogolari and Prosdocimi 1988. On Friuli, see Càssola Guida 1995; Càssola Guida 1999). They were the carriers of the Venetic (Paleovenetic) culture, otherwise known as Atestine, after the most important site, Este. As in the case of other Italian Iron Age cultures, Venetic origins go back to the Final Bronze Age. The majority of the Bronze Age *terremare-palafitte* settlements in the area were abandoned in the 12th century BCE, followed by radical changes in the settlement system (Bellintani *et al.* 1986; Bietti Sestieri 1980; 1984a; 1999, 36; Bianchin Citton 1999, 45; Càssola Guida 1999, 54; Guidi and Piperno 1992, 394–396). At the same time, new sites appeared which demonstrate no major changes in the local economy and funerary ritual. This continuity was probably due to the foundation of these new settlements by the populations from the abandoned sites, who brought with them not only their old habits and traditions but also their technological know-how. Among these early sites, a special place is occupied by Frattesina, a site which existed between the 13th and 10th centuries BCE in the delta of the Po and was one of the most important craft production and trade centres in the region, reaching the size of over 30 hectares (Càssola Guida 1999, 63–64). The new Iron Age culture was thus based on the area's favourable socio-economic conditions during the preceding centuries (Chieco Bianchi 1988a, 13). The new settlements were of Protovillanovan origin, characterised by cremation burial rite and the deposition of grave goods inside a cinerary urn. A few settlements, however, demonstrate continuity from the Final Bronze into the Early Iron Age. Some of these sites disappeared during the 9th century BCE due to environmental changes in the Po delta, while others developed into important centres of Venetic culture.

The 9th and 8th centuries BCE marked the beginning of the principal Venetic centres. The most important sites were Este and Padova. Social differentiation is evident from the burial rites from the beginning of the Early Iron Age. Starting in the 8th century BCE, demographic growth and a progressive enrichment in the quality and quantity of burial goods is observed (Chieco Bianchi 1988a, 32). Contacts with Central Italy and continental Europe are reflected in local bronze production, which has affinities with both Villanovan and Central European bronze work. During the 7th century BCE, the consolidation of systematic contact with the Etruscan area took place; the Venetic ties were especially close with the Bologna area. As in other regions of Italy, the Orientalizing period brought imports from Greece and the Near East. During the 7th and 6th centuries BCE, more contacts were established with the Golasecca culture, Alpine and Hallstatt territories, as well as with the Adriatic east (Chieco Bianchi 1988a, 61). Being at the southernmost end of the amber route, Veneto played a crucial role in passing this highly sought material to Central and South Italy. The 7th century BCE also saw the creation of an original, elite art form (Situla art), a local modification of Etruscan and Western Greek iconography that soon spread to Bologna, Slovenia and the Hallstatt areas.

The 6th century BCE marked the cultural and economic peak of the Venetic culture, with the emergence of urban centres (Padova and Este) and the first appearance of the alphabet adopted from Etruria (Chieco Bianchi 1988a, 64). The foundation of the Etruscan-Greek emporium of Adria in the Po delta further intensified the movement of long-distance goods in the region. The 4th century BCE saw the mass appearance of Celts in North Italy. During the subsequent two centuries there is evidence for the peaceful coexistence of the Venetic populations with the Celtic invaders, who became integrated into local communities (Chieco Bianchi 1988a, 90).

Archaeologically, the Venetic culture is better known than its neighbours to the west, Golasecca and Liguri, even though extensive excavation of habitation sites has begun only recently. The predominant Venetic burial rite was cremation, with the deposition of cinerary urns (sometimes multiple) in stone boxes. Numerous necropoleis have been excavated throughout the Venetic area, but the most extensive work has been done at Este (Alfonsi 1900; Chieco Bianchi and Calzavara Capuis 1985; Chieco Bianchi 1987; Bianchin Citton, Gambacurta and Serafini 1998; Calzavara Capuis and Chieco Bianchi 2006). Este was one of the largest and most important Venetian sites, reaching the size of 100 hectares at the time of its maximum expansion (Tosi 1992).

Central Adriatic Italy

Picene culture

The Central Adriatic area is not as well investigated as many other regions of the Apennine peninsula, mainly because it never experienced a true proto-urban phase. During the Late Bronze Age, the area was in the sphere of the Ripatransone culture, characterised by the metalwork types and a cremation burial rite similar to the Subapennine and Protovillanovan traditions. The political and cultural consolidation and development of this local *facies* were influenced by exchanges with the Central Tyrrhenian area and the Po valley (Bietti Sestieri 1999, 42; Naso 2000a, 38–44). Like the other areas of the Apennine peninsula, the Central Adriatic regions of modern Italy (Marche, Abruzzo, Molise and parts of Emilia-Romagna) saw the emergence of a homogeneous cultural group during the Early Iron Age, the Picene culture (Franchi Dell'Orto 1999; 2001; Naso 2000a). One of the reasons why Picene communities never achieved the same level of urbanisation as their neighbours on the Tyrrhenian side of Italy was their economic structure, based on animal husbandry, which resulted in a tribal social organisation, in contrast to the Tyrrhenian economy of agriculture and the exploitation of mineral resources (Naso 2000a, 94).

Umbrian culture

The Umbrians were an Italic people who occupied the modern Italian regions of Romagna, the eastern part of Umbria, and parts of Marche. Archaeological research on the Umbrians has been conducted only in the last thirty years (Salmon 1988, 699ff; Roncalli and Bonfante 1991). In the Middle Bronze Age, the Apennine culture was present over the whole territory that was later inhabited by the Umbrians. During the Late Bronze Age this relative homogeneity was interrupted by a series of external and internal factors, such as climatic changes and the influences of the contemporary Terremare culture (Bonomi Ponzi 1991, 64–65). Between the end of the Late Bronze Age and the beginning of the Early Iron Age, an almost total abandonment of settlements took place and the new sites that emerged did not correspond to the old ones.

Central Tyrrhenian Italy

Central Tyrrhenian Italy played a key role in the economic and historical development of the peninsula, and assumed a central role during the Final Bronze Age, due to the emergence of relatively complex socio-political entities that united the communities of Etruria, Latium and the *Agro Falisco-Capenate.*

Etruria

Archaeologically, Etruscans probably constitute the most extensive and best investigated culture of the Apennine peninsula during pre-Roman times. The literature on Etruria and the Etruscans is hence vast, making it quite an impossible task to cover everything in a short overview (cf. Barker and Rasmussen 1998; Haynes 2000; Camporeale 2000; Torelli 2000; Gleba 2003). The sphere of Etruscan culture extended over the large territory of Central Tyrrhenian Italy, encompassing the modern regions of Emilia Romagna, Tuscany, parts of Lazio and Umbria, and parts of Campania. The past twenty-five years especially have witnessed a dramatic increase in our understanding of the Etruscans and their neighbours in and around Italy. It has become evident that the Etruscans were not only the recipients of influences from other cultures but that they also propagated their own cultural influences throughout the Apennine peninsula and the Mediterranean (*e.g. Etruschi e l'Europa* 1992; Torelli 2000; Camporeale 2001).

Although the debate on the origins of the Etruscans still continues in some circles, the Protovillanovan and Villanovan cultural phenomena in Central Italy are understood by most scholars as representing the Bronze Age and the Early Iron Age developmental stages of Etruscan culture (Moser 1996; Barker and Rasmussen 1998, 83–84).

Late Bronze Age and Protovillanovan Culture

The Late Bronze Age was the first phase in the transition between the village and urbanised society in Etruria (Ridgway 1988a, 630–633; Pacciarelli 2000). It was characterised by an intensive and systematic use of mineral resources and the reorganisation of settlement patterns to facilitate Etruscan control of territory and trade routes. During the Final Bronze Age, a significant demographic growth took place, followed by economic intensification, which is indicated by an expansion in arable cultivation, a growth in secondary animal products, an increase in metal circulation and a greater importance of regional exchange networks evident in the archaeological record (Barker and Rasmussen 1998, 58). Recent finds indicate that the foundation of some of the Etruscan cities dates to the Final Bronze Age, rather than the Early Iron Age as previously thought. There is also strong indication based on burial evidence for a hierarchical structure of this society. The settlements, spread out throughout the landscape during the preceding period, now were concentrated on hilltops controlling rivers and surrounding territories, with larger sites acting as foci for the hinterland settlements (Barker and Rasmussen 1998, 53–56). Such settlement nucleation and hierarchy imply the existence of ranked societies, sometimes defined as chiefdoms, which rapidly developed into Etruscan city-states in the following centuries. Metal production was stable and trade in raw materials and finished goods was an important component of the local economy during the Late Bronze Age (Bietti Sestieri 1999, 36).

The Final Bronze Age is characterised in Etruria, as in many other regions of Italy, by the Protovillanovan culture, recognised by the deposition of cremated remains in biconical urns (Moser 1996; Pacciarelli 2000, 36–46). The Protovillanovan settlements were typically small hut villages about 4.5 hectares in extent with populations of several hundred (Di Gennaro 1988, 69, 76). These excavated huts of circular, elliptical, or rectangular plan often have foundations carved into the bedrock. Many sites have larger single structures, built in the same technique; these are interpreted as the chieftain's residences, but they were also used for religious and public functions.

Early Iron Age and Villanovan Culture

During the 10th century BCE, the Villanovan culture developed in the territory of the future Etruria (Bartoloni 2002). Like its Protovillanovan predecessor, it was characterised by cemeteries of cremation burials placed in biconical urns. Continuity between these two cultures is ascertained by the archaeological evidence from the main Etruscan centers, such as Tarquinia and Veio, where settlement occupation started in the 11th century BCE (Guidi 1998; Moser 1996). Peroni (1994, 47) defines Villanovan culture as designating two cultural 'subsystems' present in all Villanovan *facies*: first is the ritual subsystem reflected in the cremation burial rite general for all *facies*; second is the stylistic subsystem that is peculiar to each *facies* and represented by the ceramic decorative style.

The 9th century BCE witnessed profound developments in the settlement structure of Etruria, with a nucleation process that led to the creation of proto-urban sites. In the last two decades, new data have helped scholars to understand this process of urbanisation. In southern Etruria, the number of sites shrank from about 50 to "perhaps a dozen" while their size increased dramatically from 1–5 hectares to between 100 and 200 hectares (Guidi 1985). These new centres controlled large territories (Di Gennaro 1982). By the second half of the 8th century BCE, they reached their full urban physiognomy as witnessed by the foundation rites and location of their necropoleis. Although less is known about the contemporary settlements in northern Etruria, there is evidence of settlement nucleation around the coastal sites (Populonia, Vetulonia) as well, albeit on a smaller scale. The new Villanovan proto-cities consisted of a series of habitation zones, which, in turn, were made up of clusters of huts ranging in shape and size (Barker and Rasmussen 1998, 67–68).

These settlement changes were paralleled by transformations in the size and structure of the necropoleis. Large centres were now surrounded by large cemeteries, containing hundreds of burials and often spanning centuries in occupation time. The burial rite differed regionally and even locally, with each centre presenting its own peculiarities. The differences in the morphology of the burials and the quantity and the quality of the burial goods have been interpreted as evidence of the various communities' differentiation "into extended families or clans, which were in turn stratified vertically into different social levels mostly according to age and sex but including adult males of particularly high status, assumed to be the clan chiefs" (Barker and Rasmussen 1998, 72). These changes were probably a result of the intensification of agriculture and the exploitation of local mineral resources.

The Orientalising phenomenon in Etruria

Beginning in the late 8th century BCE, a visible transformation in the material record took place; it was especially noticeable in the richness of Etruscan tombs and their contents (Bartoloni 2000a). The monumental necropoleis of this period and objects found there to a great extent continue to define the culture of the Etruscans. During the Orientalising period, Etruria emerges as a major player on the Mediterranean stage, with its large-scale exploitation of natural resources, wide-ranging international contacts, and powerful urban centers.

Metals

The exploitation of mineral resources played a role of particular importance in the development of Etruscan civilisation. It is commonly assumed that the search for metal resources was one of the major reasons for Greek and Phoenician presence in Italy (Camporeale 1992a, 42). Control over local metal ores and/ or access to them was in the hands of Etruscan elites who may have acted as intermediaries for the trade (Paoletti and Tamagno Perna 2002). The Orientalising phenomenon was largely an expression of wealth accumulated by the Etruscan elites through metal trade.

Etruscan maritime supremacy

Many Etruscan centers (Vetulonia, Populonia, Roselle, Vulci, Tarquinia, Cerveteri) were located in the immediate vicinity of the Tyrrhenian coast, with roads or rivers connecting them to the sea or, later, to ports (*e.g.* Gravisca for Tarquinia, Pyrgi for Cerveteri, Regae for Vulci), indicating the importance of maritime routes in Etruscan economy. Ancient literary tradition presents the Etruscans as a formidable maritime power (Cristofani 1983; Gras 1985; Ridgway 1988b, 653–655; Camporeale 1992b; Long, Pomey and Sourisseau 2002). Indications of Etruscan maritime supremacy appear as early as the Villanovan period, as evinced by contacts with Sardinia and Greece (Haynes 2000, 17–20). Over the following centuries, Etruria developed and controlled a vast network of maritime routes between France, Spain, South Italy and Sicily, North Africa and, probably, Greece. Shipwrecks at Isola del Giglio, Bon Porté, Cap d'Antibes, Cap d'Agde and other locations provide a great deal of information on Etruscan trade organisation and the types of goods that were being exchanged (Parker 1992; Long, Pomey, and Sourisseau 2002). The loss of maritime control of Tyrrhenian waters to colonial Greeks during the 5th century BCE, in fact, led to the decline of the Etruscans.

Trade

Trade was directly connected to Etruscan maritime supremacy (Turfa 1986). During the Early Iron Age, a regional, 'dendritic', system of trade between Etruria, Latium and Campania is evident (Smith 1998). Over the next few centuries, however, a long-distance trade network developed, reaching its acme in the 6th century BCE, when ports like Gravisca and Pyrgi became important *emporia* with international sanctuaries (Haynes 2000, 171–184). At the same time, the newly founded Adria and Spina started to function as *emporia* in the Po delta. Etruscan objects have been found in Central Europe (Aigner Foresti 1992; Adam *et al.* 1992), Spain (Almagro-Gorbea 1992), Greece (Naso 2000b), and

France (Boulomié 1992; Landes 2003; *Gli Etruschi da Genova ad Ampurias* 2006).

'Colonisation'

The creation of trade outposts is characteristic of early states from the beginning of their formation (Algaze 1993; Stein 1999, 46–55). In the case of Etruria, this is seen in the 'colonies' of Campania (Pontecagnano, Capua and Sala Consilina) and their expansion into North and Adriatic Italy (Bologna, Verucchio, Fermo and, later, Adria and Spina), most of which took place as early as the 9th century BCE. Both Campania and the Po valley were agriculturally rich areas and were at the crossroads of land and marine trade routes. Verucchio and Fermo (and later Adria and Spina) were crucial in providing access to the Adriatic.

Regional differences

It is common in the scholarly literature to divide northern and southern Etruria. Northern, especially inland sites have often been seen as lagging behind in terms of cultural development, a phenomenon known as *attardamento culturale* (Torelli 1986, 52; Ridgway 1988a, 631). To a large extent, this division is due to the lack of research on the northern Etruscan centres, most of which are buried under the modern towns. Recent archaeological investigations have started to fill these lacunae and the emerging picture is complex and particular to each region. For instance, Chiusi, previously cited as one of the latest Etruscan centres to develop, has yielded the earliest representation of a figure reclining in a symposiast pose (Haynes 2001, 10).

The Archaic period

The 6th century BCE is regarded as the zenith of the Etruscans. The military aristocracy gradually lost its pre-eminence and the social, economic and political power became concentrated in the hands of a mercantile elite. The cities of Etruria flourished, as did Rome, which was transformed into a truly urban centre by its Etruscan kings (Cornell 1995; Hall 1996b). Etruscan art and architecture thrived during the 6th century BCE.

The 5th century crisis and its aftermath

The prosperity of Etruscan cities was tightly bound to their maritime and commercial power. It is not surprising, then, that the loss of control of the trade routes to the Greeks of Magna Graecia, punctuated by the Etruscan defeat in the battle of Cuma in 474 BCE, triggered a period of crisis aggravated by both external and internal factors. The 5th century BCE, thus, marks the beginning of the end of the Etruscan world. The loss of the control of Rome and Latium in the end of the 6th century BCE, which resulted in the cessation of the permanent contacts with Campania, had already weakened the Etruscan hegemony. Rome, growing more powerful, soon turned its expansionist policy against Etruria, conquering southern Etruscan cities in the course of the 4th century BCE. Meanwhile, the north was being weakened by Celtic invasions starting in the late 5th century BCE. Despite a cultural and economic recovery during the 4th century BCE, by the time of the first Roman emperors, Etruria underwent complete Romanisation (Barker and Rasmussen 1998, 262–296).

Agro Falisco-Capenate

The territory between Etruria and the middle Tiber was occupied by the peoples of *ager Faliscus* and Capena (Potter 1976; Carlucci and De Lucia 1998). Although closely related to the Etruscans, they had their own distinct culture and also differed from each other linguistically: the Faliscans spoke a language related to Latin, while the small Capena enclave spoke a Sabine dialect.

Latium Vetus

The area immediately south of the Etruscan territory was inhabited by Latin-speaking peoples and encompassed the area of southern Lazio and northern Campania, defined by the Tiber in the north, the Garigliano to the south, the Valle Latina to the east and the Tyrrhenian Sea to the west. A local cultural process began here during the Late Bronze Age and culminated, by the Early Iron Age, in the development of a culture, which, despite many similarities and strong ties to that of the Etruscans, remained quite distinct (Colonna 1988; Holloway 1994; Bietti Sestieri and De Santis 2000).

In the Late Bronze Age, the most important Latial settlements clustered around the Alban Hills. This period displays a relative homogeneity of archaeological record. The 9th century BCE witnessed important changes in the organisation of area settlements, when occupation shifted to the coastal settlements and to the sites on the Tiber, including Rome, which show signs of early urbanisation.

The history of Latium is strongly tied to the history of Rome (*Enea nel Lazio* 1981; Cristofani 1990; Holloway 1994; Carandini 2007). None of the other Latial sites reached the size of Rome or its Etruscan neighbors. Nonetheless, most of the Iron Age sites developed into notable centers, often of religious importance.

South Italy

South Italy comprises the modern regions of Campania, Calabria, Basilicata and Puglia, a situation that originated in the Augustan division of the Roman Empire; this geographical division already existed in the Early Iron Age. The peoples who inhabited these regions were heterogeneous in terms of ethnic origin, language, demographic distribution and material culture.

The archaeological exploration of indigenous South Italy is a relatively recent phenomenon. One of the major reasons for this lag is the Greek bias that still pervades the vast majority of scholarly literature.[3] For a long time, the only archaeological scholarship concerning South Italy and Sicily was that of Magna Graecia, or the Greek colonies. While the autochthonous populations were lumped together under the name 'Italic people' and defined as *barbaroi*, the role of the Greek colonies has been disproportionately emphasised, especially for the Early Iron Age, mainly because of the greater 'visibility' of Greek materials, such as pottery. Furthermore, an absolute and often undisciplined reliance on ancient written sources has lead to overly confident and skewed reconstructions of South Italian history (Whitehouse and Wilkins 1985).

Recent archaeological research has shown that in its early stages, Greek colonisation differed enormously from the situation depicted by the ancient authors. South Italy was already densely occupied and its population socially (and probably ethnically) diverse by the time of the first appearance of the Greeks. Many Early Iron Age centers demonstrate a precocious capacity for territorial planning, much like the contemporary sites of South Etruria and Latium; thus, many scholars have questioned the traditional theory that urbanism was essentially a Greek phenomenon in Italy (Peroni and Trucco 1994; Burgers and Yntema 2000). As early as the 9th century BCE, it is possible to speak of proto-urban societies in Campania, the Sybarite and Metapontine regions, and, possibly, the Salentine peninsula. The process of urbanisation in South Italy was accompanied by important technological transformations, recorded qualitatively and quantitatively by the excavated artefacts. Fibulae were already standardised by the beginning of the Early Iron Age, indicating the existence of specialised craftsmen, and the standardisation of ceramic production had commenced in some areas already in the mid-9th century. It is thus essential to look at the endogenous developments of native Italic societies in order to achieve a better understanding of the complex processes that took place during the Iron Age.

During the Late Bronze Age, South Italy, like the rest of the Apennine peninsula, is characterised by relative cultural homogeneity, generally known as the Subapennine culture (Peroni 1989, 116). In addition to the local cultures, however, there is also a well-documented Aegean presence (Tinè and Vagnetti 1967; Vagnetti 1985).

Archaeological evidence shows that the 12th century BCE crisis on the Greek mainland did not completely interrupt Cypriot and Phoenician traffic to the Western Mediterranean (Ridgway 1992, 22–26; 1996b; Aubet 1993, 25). The knowledge of Bronze Age connections and trade routes did not disappear entirely, and this may have been a significant factor in the renewal of regular westward traffic by the Greeks in the 9th–8th centuries BCE (Ridgway 1992, 1994; Vagnetti 1996, 116). Euboeans, the first Greeks to move westward in the 1st millennium BCE, were inserting themselves into a pre-existing Cypro-Levantine commercial network (Ridgway 1992, 111, *passim*).

These Phoenician and Euboean traders were the precursors of the colonising movement. This precolonial phase is documented by the Greek materials, mainly ceramics, found at sites throughout South Italy. The culmination of this 'pre-colonisation' is seen in the establishment of Pithekoussai *circa* 770–760 BCE (Ridgway 1992). The first 'true' colony was established at Cuma in 750–740 BCE. During the subsequent generation (last decades of the 8th century BCE), many other major colonies were founded, including Syracuse, Sybaris and Taras. The 7th century BCE saw the foundation of many others, two of the most prominent of which are Locri and Metapontum.

The Greeks were not the only colonisers of South Italy, however. Even before their arrival on the coasts of the Apennine peninsula, Etruscans had already settled in Campania, while the Phoenicians established their own outposts in Sicily and Sardinia. Etruscan 'colonisation' manifested itself in a different use of settlement area by making colonies centres of interchange with the surrounding sites. This reorganisation of territory and the creation of centralised settlements that were integrated into the indigenous groups differ significantly from the Greek colonisation phenomenon. The Phoenician colonies followed yet another model. They were 'ports of trade', places of professional economic exchange that were independent from the hinterland and situated on the border between a non-market-oriented society and a market economy (Niemeyer 1990; Aubet 1993, 82–82).

The relations between the colonisers and the native populations show great variety. Although the assimilation and elimination of autochthonous populations have been cited as the two main ways of establishing a colony, a coexistence model in the form of mixed communities may have played a larger role than previously thought. The fact that the Greeks never expanded beyond the coastal plains is a clear indication of the strength and organisation of the indigenous communities, which played an important role in exchange and intermarriage, and as sources of labour and conflict. The result of the Greek/Etruscan/Phoenician interaction with the indigenous world of South Italy differed from place to place, depending on what Cerchiai has defined as 'cultural permeability' (Cerchiai 1995, 15). Whatever its degree in a particular case, the relationship between the Greeks and the indigenous populations may be summed up as an 'acculturation process' for both sides (Bottini 1996, 541). Intermarriage certainly played an important role in this development (Coldstream 1993; Shepherd 1994; Hodos 1999). The rate of this acculturation and its extent also varied. Thus, Daunia succeeded in retaining its autonomous nature well into the 5th century BCE. Despite the fact that only Greek influence on Italic society was studied until recently, it is becoming more evident that the process was not unidirectional. For instance, the work of Barbara Barletta has demonstrated that the Greek architecture of Archaic and later Magna Graecia was strongly influenced by Italic traditions (Barletta 2001). Although the colonies strove to reproduce the institutions of their homeland, the differences in climate and natural resources, their relations with indigenous people, and the frequency of mixed marriages inevitably led to changes in the habits of the colonists (Forti and Stazio 1986, 643).

Campania

Protohistoric Campania is characterised by the coexistence of very diverse populations (D'Agostino 1988; Cerchiai 1995, 10; De Natale 2000, 14). The data are scarce on the transition from the Final Bronze Age to the Early Iron Age due to uneven distribution of archaeological material and the methodology adopted in the early years of exploration, as well as a general reduction in the number of settlements during the period of transition. Hill settlements, which characterised the Bronze Age, were abandoned, and new sites were chosen in the fertile river plains near important communication routes.

The Early Iron Age in Campania is mainly documented through burial complexes, which present two main *facies*, distinguished by funerary ideology and morphology. The first is the *Fossakultur*, characterised by the ritual of inhumation. The second is the cremating Villanovan culture.

The *Fossakultur* shows signs of continuity from the Bronze Age and extends throughout southern Italy. In Campania, it is further subdivided into two culturally distinct areas (Cerchiai 1995, 10; De Natale 2000, 15): the northern coastal area with numerous sites that were more open to external contacts, including Cuma and the Sarno valley (d'Agostino 1970); and the southern inland region, also known as the Oliveto Citra-Cairano *facies*, which manifests in a more conservative material culture (d'Agostino 1964; Gastaldi 1998b; Bailo Modesti, d'Agostino and Gastaldi 1974, 113–131). The settlements of the *Fossakultur* in Campania are not well known. Most likely they reached the size of small villages and had an agricultural economy (d'Agostino 1996a, 533). Scholars have associated the Campanian *Fossakultur* with the people called Opici by early historians (Cerchiai 1995, 22ff; d'Agostino 1996a, 533). The carriers of the Oliveto Citra-Cairano *facies* exercised considerable control over the communication routes linking internal Campania and Daunia, a role that led to differentiation and cultural development in their society.

The other major cultural manifestation of protohistoric Campania, the Villanovan culture, does not seem to predate the 10th century BCE. It is also subdivided into two distinct areas: the region south of the Sorrento peninsula with its main centres of Pontecagnano and Sala Consilina, where typically Villanovan biconical urns were used for the deposition of cremated remains; and the northern area of Capua, characterised by burials in a different type of vessel, usually termed olla. These variations in material culture have been explained by different regional influences from Central Italy, with Pontecagnano receiving populations from South Etruria, while Capua was settled from the Faliscan/Latial area (Cerchiai 1995, 10, 40). The three main Villanovan centres in Campania – Capua, Pontecagnano and Sala Consilina – were located in particularly favourable positions for agricultural activities and traffic control. The sizes and locations of the necropoleis at these sites indicate that the settlements were formally structured at an early stage: the residential area was large and marked off from the cemeteries at the outset (d'Agostino 1996a, 533). The urbanisation process at these centres was completed by the end of the 7th century BCE. The large Villanovan centres, thus, stand in stark contrast to the small and undifferentiated *Fossakultur* sites.

The battles of Himera (480 BCE) and Cumae (474 BCE) signalled the beginning of the end of Etruscan control in the area (Cerchiai 1995, 127). During the 5th and 4th centuries BCE, the Etruscan colonies in Campania were cut off from Etruria by land and by sea. Weakened and isolated, they were overrun by Oscan-speaking tribes, including Samnites, who took Capua in 423 BCE, effectively ending Etruscan control over southern Campania.

Calabria

The large size of the region predisposes Calabria to the coexistence of different cultures and the central position of the Sibarite area between Basilicata, Campania and South Calabria facilitated a mixture of cultural elements observed in the archaeological record. Archaeologically, Ionian Calabria is better known than its Tyrrhenian coast (Pacciarelli 2000, 116). In terms of material culture Calabria gravitated towards the Protovillanovan style during the Late Bronze Age (Peroni 1987, 98). This has been subdivided into two main groups based on the typology of bronze objects: the first is the Crati group in northern Calabria, which shows connections with Puglia; the second is the Castellace group in South Calabria, more closely connected with Central and North Italy.

Calabria was one of the main Italian regions to engage in Bronze Age trade with the Aegean (Peroni 1987, 81ff). Although Mycenaean imports ceased by the Final Bronze Age, Aegean influences in ceramic technical and stylistic traits continued. During the 11th and 10th centuries BCE, some of these influences became permanent, such as the centralisation of production and, presumably, of socio-political organisation, as indicated by the presence of large vessels for the storage of surplus goods. Furthermore, the large number of bronze axe hoards during transition from the Late Bronze Age to the Early Iron Age has been hypothesised as having pre-monetary significance (Peroni 1987, 109).

Several distinct cultural *facies* have been identified in Calabria from the 8th century BCE (Peroni 1987, 132). The north-east (Sibari area: Torre Mordillo, Francavilla Marittima, Amendolara) is characterised by individual *fossa* burials. The South (Locri area: Canale, Janchina, Calanna, Reggio, Trunca, Sala, Gerace, S. Stefano in Grotteria) is distinguished by collective burials of the *grotticella* type and a more varied material culture. The area between these two poles demonstrates a mixture of traditions.

Basilicata

No synoecistic process such as has been observed in Central Tyrrhenian Italy and Calabria is evident in Basilicata, demonstrating a certain degree of cultural delay (*attardamento culturale*) with respect to neighbouring regions (Bianco 1996, 31, 33; Bianco *et al.* 1996; Adamesteanu 1999). The transition between the Final Bronze Age and the Early Iron Age was marked by a strong cultural continuity.

The foundation of Greek colonies constituted an irreversible crisis for these coastal settlements but the consequent opening of paths towards the Etruscan-Tyrrhenian area stimulated the development of inland Enotrian centers (Bianco 1996, 34). Various communities became active in the control and mediation of traffic routes, and important centers developed at Aliano, Alianello, Roccanova, Chiaromonte and Latronico. Information about these sites comes mainly from their necropoleis.

Puglia (Apulia)

During the Bronze Age, Puglia experienced population growth and had numerous settlements, which developed partly thanks to agricultural activities and partly as a result of extensive trade with the Aegean world. Despite the Mycenaean breakdown, Puglia experienced no cultural fracture during the passage to the Early Iron Age and its contacts with Greece were never completely interrupted. Already at the end of the Bronze Age, in the 11th century BCE, a new cultural climate arose with new Protovillanovan traits coexisting with Subapennine survivals, anticipating the cultural traits of the Early Iron Age (De Juliis 1988, 11). The practice of cremation, however, was never generally adopted, and inhumation remained the typical funerary rite during the Iron Age. In general, graves show similarities with Central Adriatic Italy and Dalmatia (Salmon 1988, 682).

After the first cultural stimuli of Hellenic type at the end of 2nd millennium BCE, the next contact with the Aegean is indicated by the appearance of Corinthian, Euboean and Cycladic pottery in the end of the 9th – first half of the 8th century BCE in Salento (De Juliis 1988, 35). This pre-colonial phase culminated at the end of the 8th century BCE with the foundation of Taranto (Greek Taras and Roman Tarentum) in 706–705 BCE (De Juliis 2000). The reason why no other Greek colonies were founded in Puglia is the presence of an indigenous population with a high level of civilisation that controlled the inland territories and the seas. It is thus not surprising that the level of Hellenic influences in Puglia varied greatly and, apart from a few cases, was not strong until later periods (Salmon 1988, 684). When external influences did start to take over, the process was slow and not all of Puglia received them to the same extent, at the same time, and in the same ways. Over the course of the Early Iron Age, separate tribes developed from the Iapygian substratum, and from the beginning of the 7th century BCE it is possible to distinguish three different cultures in Puglia: Messapii, Peucetii and Daunii. Archaeologically, they have been differentiated by their distinctive pottery traditions (De Juliis 1997). At the same time, the emergence of elites is indicated by rich burials (Bottini 1982). Although urbanisation happened in Puglia later than in most Tyrrhenian areas, native sites like Cavallino and Oria were proto-urban by the 6th century BCE, with massive fortification walls and regular residential architecture (D'Andria 1988; D'Andria and Mannino 1996). Greek influences grew stronger after 600 BCE, becoming particularly dominant among the Peucetii and Mesapii (Salmon 1988, 688–689). Although largely Hellenised by the 4th century BCE, Puglia retained much of its distinctive character.

Sicily

The island of Sicily (Map 1), despite its proximity to the Italian mainland, always had considerable cultural autonomy (Leighton 1999; Tusa 1999; Albanese Procelli 2003). As in the case of South Italy, however, for the period of our interest, past research has concentrated on Greek settlements (Pugliese Carratelli 1985b; Holloway 1991), resulting in a limited amount of information on the indigenous sites (cf. Leighton 1999; Tusa 1999; De Angelis 2001; Albanese Procelli 2003).

During the Bronze Age, two cultural zones can be distinguished on the island: the first comprises the Milazzese sites of northern Sicily, the Aeolian islands and Ustica, which had closer links with Tyrrhenian Italy; the second encompasses the Thapsos sites in eastern and southern Sicily, which had substantial Mycenaean presence (Leighton 1999, 147). Contacts with the Aegean during the Bronze Age contributed to the emergence of complex social structures and the formation of important proto-urban centres, leaving influences that may have lasted throughout the 1st millennium BCE (*e.g.* De Miro 1991; La Rosa 1996; Tusa 1999). Towards the end of the Bronze Age, three cultural *facies* can be distinguished on the island: the Lipari-Milazzo, which may be pertinent to populations which immigrated from the Italian peninsula; the autochthonous Pantalica *facies*; and the mixed Mulino della Badia group (Albanese Procelli 2003, 30).

The Early Iron Age in Sicily is conventionally dated from about 900 BCE to the foundation of first Greek colonies (734–650 BCE) and is named after the Pantalica III complex. The break in chronology is only conventional since many sites continue from the Final Bronze Age into the Iron Age. Sicilian early chronology relies largely on internal sequences based on fibulae and pottery (Leighton 1993b; 1996; 1999, 187–188; Tusa 1999, 552). Iron Age communities seem less hierarchical and more diverse than their Bronze Age predecessors, "possibly with more egalitarian social structures, less concerned with prestige goods and status-enhancing luxuries from distant sources than with local craft production and general consumption" (Leighton 1999, 188). This social shift has been connected to the decline of contacts with the Aegean at the end of the Late Bronze Age, as well as to the new Italian influences appearing in the Final Bronze Age (Tusa 1999, 551; Leighton 1999, 189). The latter is evident in the appearance of the Urnfield type cemeteries at Milazzo and other sites (Tusa 1999, 563).

Ancient authors (cf. Albanese Procelli 2003, 20–22) mention three distinct peoples occupying Sicily at the time of Greek arrival in the 8th century BCE: the Sicels in the east, the Sicans in the southwest, and the Elymians in the extreme north-western area of the island (Tusa 1999, 553; Leighton 1999, 215–217; Albanese Procelli 2003, 18–27). While there is a wide variety of Iron Age craft traditions and burial customs, it is not yet easy to distinguish between these different cultures on the basis of strictly archaeological evidence.

The settling of Sicily by the Phoenicians was likely due to the island's strategic position in the Mediterranean (Aubet 1993, 200). The arrival of the Greeks in the late 8th century BCE forced the Phoenicians to migrate to the western part of the island, where they founded their three main colonies: Motya, Panormos (Palermo) and Soloa (Solunto).

Sardinia

Sardinia will be treated only briefly here, since little material from the island is included in the present study (Map 1). The reason for this is that, apart from the numerous booklets and small guides to museums and archaeological monuments, there is very little definitive publication of the relevant major finds which have come to light only in the last thirty years (Ridgway 2002, 119–121). Still, over the last decades, the importance of Sardinia in the interaction with the mainland Italy over the course of millennia is starting to emerge in light of new archaeological finds (Gastaldi 1994; Lo Schiavo 1994; Paoletti and Tamagno Perna 2002). The presence of Mycenaean pottery at sites like Nuraghe Antigori, indicates early contacts with the Aegean.

The culture that developed in Sardinia over the course of the Bronze and Iron Ages is generally known as Nuragic, named for the characteristic type of stone architectural monument, the *nuraghe*. (Webster 1996; Lilliu 1982; *La civiltà nuragica* 1990). Over the course of the Late Bronze Age, the island saw the emergence of hierarchically organised societies, reflected in the architecture of large centres. The settlement pattern during the following Iron Age was complex and characterised by regional variation (Webster 1996, 153). The period saw social processes similar to those of Central Tyrrhenian Italy and many sites engaged in active contacts with Phoenician colonists.

Conclusions

This quick and necessarily uneven overview should establish the cultural and chronological context for the discussion of textile production in pre-Roman Italy. Between 1000 and 400 BCE, most parts of Italy witnessed profound social, demographic and cultural transformations, which were anchored in the preceding period and had a significant impact on the following development. The great dynamism of interactions created by the mobility of people and ideas during this period is a backdrop against which textile craft shall be considered in the following chapters.

PART 2

Sources

The sources for the study of ancient textile production are diverse, ranging from texts and iconography to archaeological material. While they provide important information about ancient textile manufacture, each source is imbued with certain problems and requires careful consideration before it can be used to reconstruct the social and economic history of the textile craft. This chapter provides a brief overview of the sources used in this study and some of the problems encountered in their investigation.

Written sources

An array of ancient texts contains information on textile materials, technologies, uses and trade. John Peter Wild has divided the written evidence into three categories: 'conventional classical literature,' 'technical handbooks,' and 'public and private documents' consisting of tablet archives, epigraphic sources and papyri (Wild 2000, 210). Each document type provides useful information on ancient textiles, albeit of different kind and with differing degrees of accuracy. These diverse sources have been used with varied success by modern scholars to reconstruct different aspects of textile production in ancient times. Epigraphic evidence has been a useful source for the identification of textile professions (Larsson Lovén 1998; Kudlien 2001), while mythological characters such as Penelope, Kirke, Arachne and the Fates have provided social and religious context for ancient textile activities (Scheid and Svenbro 1996).

The earliest evidence of writing in Italy consists of an inscription on a flask found in burial 482 of the Osteria dell'Osa necropolis in Latium (Figure 1), which has been dated *c.* 775 BCE (Bietti Sestieri 1992b, 687, fig. 3a.270). The excavator has proposed the reading *eulin* for the inscription, which may be an abbreviation of Greek *eulinos*, meaning 'good at spinning' (Bietti Sestieri, De Santis and La Regina 1991, 83–88; Ridgway 1996a; Ampolo 1997). Although other readings have been proposed (Peruzzi 1992; Colonna 2005), it is tempting to see the importance of textile craft reflected in the first manifestation of writing in Italy. It is surely also significant that some of the earliest graffiti in Italy appear on textile implements, such as spindle whorls and

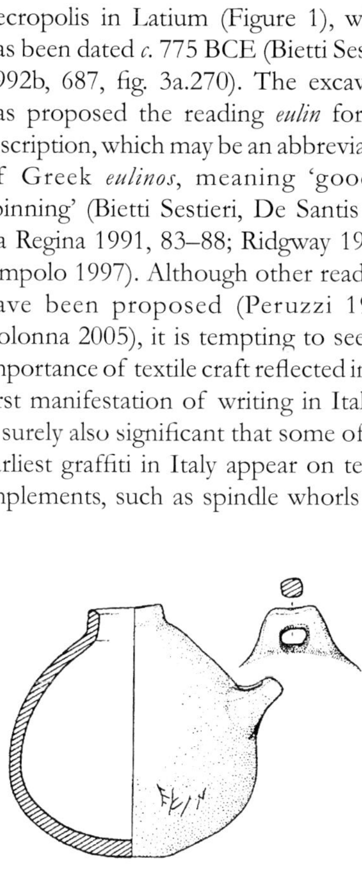

Figure 1. Jug with inscription, Tomb 482, Osteria dell'Osa, 8th century BCE (After Bietti Sestieri 1992b, fig. 3a.270).

spools (see Bagnasco Gianni 1999 and *contra* Benelli 2001).

These first, indirect sources form a backdrop for the Latin literary sources, which provide a great deal of information about textile craft in Italy. For instance, Catullus (64.311–319) gives a very precise description of wool spinning, while Ovid (*Met.* 6.53–60) recounts in detail the weaving process. Although Catullus and Ovid probably never made textiles themselves, they did have a hands-on knowledge of textile production based on daily observation.

More specialised sources also survive. Agricultural manuals, such as *De Re Rustica* by Varro and the writings of Columella, have been of great use in sheding light on Roman flax cultivation and sheep breeding, and may reflect earlier practices. Pliny the Elder's *Naturalis Historia* is an indispensable source of information on types of textiles and their raw materials, as well as on recipes for dyeing. In addition to literary sources, we are well served by epigraphy: funerary inscriptions have been useful in indicating the fibre being worked, as well as the gender and social status of the craftsperson (Larsson Lovén 1998).

While useful to a certain extent, written sources have to be treated with caution. Wild (2000) has summarised some of the main problems with information obtained from literary sources. One of the biggest limitations is 'the semantics of Greek and Latin terminology': the meaning of a particular word is often times unknown or has changed through time (Wild 2000, 209). Furthermore, in the rare instances when descriptions of production processes are preserved in literary accounts, they are not always clear: sometimes the authors themselves did not fully understand the technology or omitted information that seemed obvious to them but is lost to us.[1] Another major problem is chronology. Information provided by Latin sources for the Roman period cannot be projected back in time without reservation. Furthermore, statements such as Lucretius' claim that woven cloth was invented after iron (*De Rerum Natura* 1350–1360) are simply wrong and rather inform us about Lucrecius' view of technology development. For these reasons, the use of ancient literary sources in the present study is limited to instances in which the information they provide is corroborated by other kinds of evidence. Archaeological evidence in particular can shed new light on issues previously investigated only through written sources.

Iconography

After written sources, iconographic material has been the most frequently cited source of information on ancient textiles. Ancient Italian art encompasses one of the most interesting and informative collections of visual material in the ancient Mediterranean. The proliferation of colorful textiles in the painted tombs, statuary and other media has been noted by numerous scholars. This vast corpus of representations of textiles has been used in studies of ancient dress. Work has also been done on inferring textile designs from iconographic material and their connection to other media, for example, transmission of figurative motifs (Ghedini 1995; 1996). The derivation of pottery decoration from textiles has been suggested for Greek Geometric pottery (Barber 1991, 365–372, 374), as well as for the Early Bronze Age pottery of Sicilian Castelluccio culture (Lukesh 1999).

Costume

For pre-Roman Italy, the majority of work on textile iconography has been done in costume studies. A series of exhibitions on *Fashion, costume and beauty in antiquity* throughout Italy a few years ago highlighted the abundance of iconographic evidence on ancient dress (Bottini 2003). Already during the Neolithic, the anthropomorphic stone stelae from the Alpine area show patterns, which have been connected to textiles (Pedrotti 1993). The Iron Age stone stelae of Daunia represent another important group (Nava 1988).

Etruscan clothing depicted on figurines, statues, vases and tomb paintings (Figure 2) has been studied extensively by Larissa Bonfante in her still unsurpassed *Etruscan Dress* (1975 and 2003), as well as in her numerous articles.[2] The late Etruscan statue of Aulus Metellus has been shown to have sufficient details to reconstruct not only the garments but also the techniques of their manufacture (Granger-Taylor 1982).

Costume in iconography has not been studied as extensively for other areas of Italy, although South Italian wall painting (Pontrandolfo and Rouveret 1992), for example, also furnishes extensive information regarding contemporary costume.

Utilitarian textiles

In addition to garments represented on people, there are many depictions of utilitarian textiles, that is those used for purposes other than clothing. In particular, the patterns on the ceilings of the painted Etruscan tombs have attracted the attention of scholars for a long time. In 1952, Massimo Pallottino recognised the possibility that the pattern of the Tomba delle Leonesse in Tarquinia might represent not a solid roof but an awning. Later Ross R. Holloway (1965) and Simonetta Stopponi (1983) elaborated upon this notion by suggesting that within the ceiling patterns that are employed, the checked and flower schemes are unquestionably textile patterns, since these patterns can be matched with representations of textiles on clothing and furniture in the paintings. According to this interpretation, the entire tombs represent tents or pavilions with roofs made of cloth. Furthermore, colorful patterns are shown on utilitarian textiles, such as tablecloths, pillows and bed covers, indicating the variety and richness of textiles used by the Etruscans.

Pattern

Pattern often allows us to distinguish the material of which the textile was made. Thus, in Etruscan art, woollen garments are represented with a plaid or checked design, while linen is usually left white or decorated with small decorative motifs (Bonfante 1975, 12). Such fabric-specific use of pattern in artistic representations is likely to reflect the realities of ancient life. Moreover, the patterns themselves are products of technical capabilities of ancient weaving equipment and its users. For example, geometric patterns on the Daunian stelae have been connected to tablet-woven bands, known from both Italian and Central European textile finds (Nieling 2007).

In some of the 3rd century BCE tombs, the canopy paintings seem to represent imitation rug. Tarquinian tombs of the Cardinal, Anina II and of Thanchvil Anei have ceilings decorated with a particular type of turret border (Serra Ridgway 2000, 304), known also from Macedonian tomb paintings (Miller 1993), mosaics and recalling Near Eastern pile carpet patterns.[3] This suggests that by the later part of the 1st millennium BCE, rug patterns, if not carpets themselves, reached Italy.

Another example is furnished by representations of complex figural scenes on garments: do they represent real textiles? An interesting new study by a  b

Figure 2. Garment depiction in painted tombs of Etruria: a) Vel Sathies from the Tomba François, Vulci, 4th century BCE (After Bonfante 1975, 201 fig. 135); b) Dancer from the Tomba delle Leonesse, Tarquinia c. *520 BCE (AfterTorelli 2000, 354).*

Antonella Pautasso, who has gathered textile representations on Hellenistic female shoulder busts from Sicily, argues that, indeed, they do (Pautasso 2007). The figural freezes on the garment of the so-called Lady of Sibari dated to the 7th century BCE (Figure 3), and the figures on the mantle of Vel Sathies in the 4th century BCE Tomba François (Figure 2a), provide other examples of textiles with complex patterns. Similar opinions have been expressed regarding figural patterns on textiles seen in Greek vase paintings (Barber 1991, 363–365). It is hard to tell in what type of technique such textiles would have been made, but for certain, the work would have been complicated and time-consuming and the resulting textiles precious enough to be worn by gods.

The study of patterns is also important for understanding religious and symbolic meaning of cloth as they are symbols through which textiles transmit information about specific people, places, or events. Thus, certain patterns may have held a well-defined significance in the ancient mortuary practices. The checked motif, for instance, is among the most common patterns in Etruscan painted tombs and may have been associated with a ritual and maybe even specifically with funerary context (Figure 4). Apart from the frequent checked ceilings, in the Tomba Bartoccini of Tarquinia, the pattern is seen around the doors that lead into the funerary chambers (Steingräber 1985, no. 45). Such positioning is possibly intended to evoke the appearance of a curtain hung in order to conceal the entrance to the World of the Dead. In the later Tarquinian Tomba 2327, the use of the checked pattern is even more specific. Here it appears on the ceilings of the loculi, that is, above each individual deceased (Steingräber 1985, no. 143). Parallels for its use can be found in Anatolian tomb painting and on Greek Geometric pottery, where an important iconographic element of the prothesis scene is a rectangular checked area, thought to represent a funerary cloth, suspended above the bier or held above the deceased by attendants. This specific use of funeral cloth in Geometric vase painting may be indicative of a broader tradition of textile use in funerary rituals. Textiles constitute one of the crucial elements of ritual and it is the manifestation of this symbolic meaning that may be seen in the painted ceilings of Etruscan tombs.

Production

While informative about many aspects of ancient cultures, representations of textiles, however, offer little information about the technical details of ancient Italian textiles. Relatively few actual images of the various stages of textile production or tools associated with it exist in ancient Mediterranean art. They are, however, of great importance to our understanding and reconstruction of this ancient technology. Only representations dating to pre-Roman periods will be considered here. There are numerous depictions of textiles and textile manufacture dating to Roman times but these are beyond the scope of this study.

The earliest known depictions of the warp-weighted loom, in fact, come from Italy. They appear on the Great Rock at Naquane, in North Italy, and are dated to Late Bronze Age, that is the second half of the 2nd millennium BCE (Figure 5) (Anati 1961, 138–142; Bevan 2006, 66–74). At least seven images of looms are carved into the rock surface, one of which has a human figure in front of it. The depiction of checked and fringed textiles on the rocks at the Bronze Age site of Borno indicates that

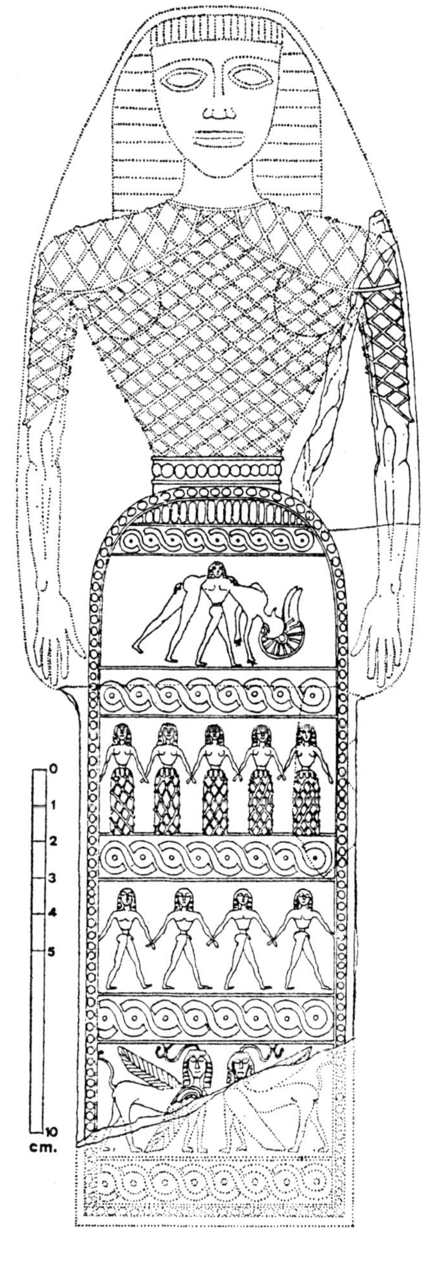

Figure 3. Reconstruction of a terracotta figurine, the so-called Lady of Sibari, with a garment decorated with figural friezes (After Zancani Montuoro 1972, 68 fig. 1).

fabrics produced on such looms may have been polychrome (Figure 6) (Peroni 1971, 100). It further indicates the use of different colours of wool to obtain pattern. Whether the variation in colour was due to natural pigmentation or dyeing is not clear but the latter possibility should be considered.

Iconographic evidence becomes more abundant in the 1st millennium BCE. An important scene related to textile production appears on the wooden cylindrical throne found in Tomb 89 at Verucchio (Figure 7). While interpretations of the intricately carved scenes vary (Gentili 1986, 244; Torelli 1997, 68–69; Kossack 1999, 64–67; Eles 2002), most scholars agree that spinning and weaving are among the activities depicted.

Another important object comes from the area of Bologna and is one of the most important representations of textile production in the ancient world, a bronze *tintinnabulum* found in tomb 5 of Bologna's Arsenale Militare necropolis and dated *c.* 600 BCE (Figure 8) (Morigi Govi 1971; Kossack 1999, 67–68). It is a flat, bell-shaped object, made of two bronze sheets connected at the edges and decorated in a *repoussé* technique. Each side of the *tintinnabulum* is divided into two sections, allowing space for four scenes depicting various stages of textile manufacture. The bottom scene of side A depicts two women seated in throne-like chairs, reminiscent of the Verucchio throne, each holding a distaff in her left hand. With her right, each pulls a strand of fibre from the basket in the centre, that is, they are dressing the distaffs for spinning. The top scene of side A shows a woman at the task of spinning.

Figure 4. The 'tent' of the Tomba del Cacciatore, Tarquinia, 5th century BCE (Drawing by Marianne Bloch Hansen).

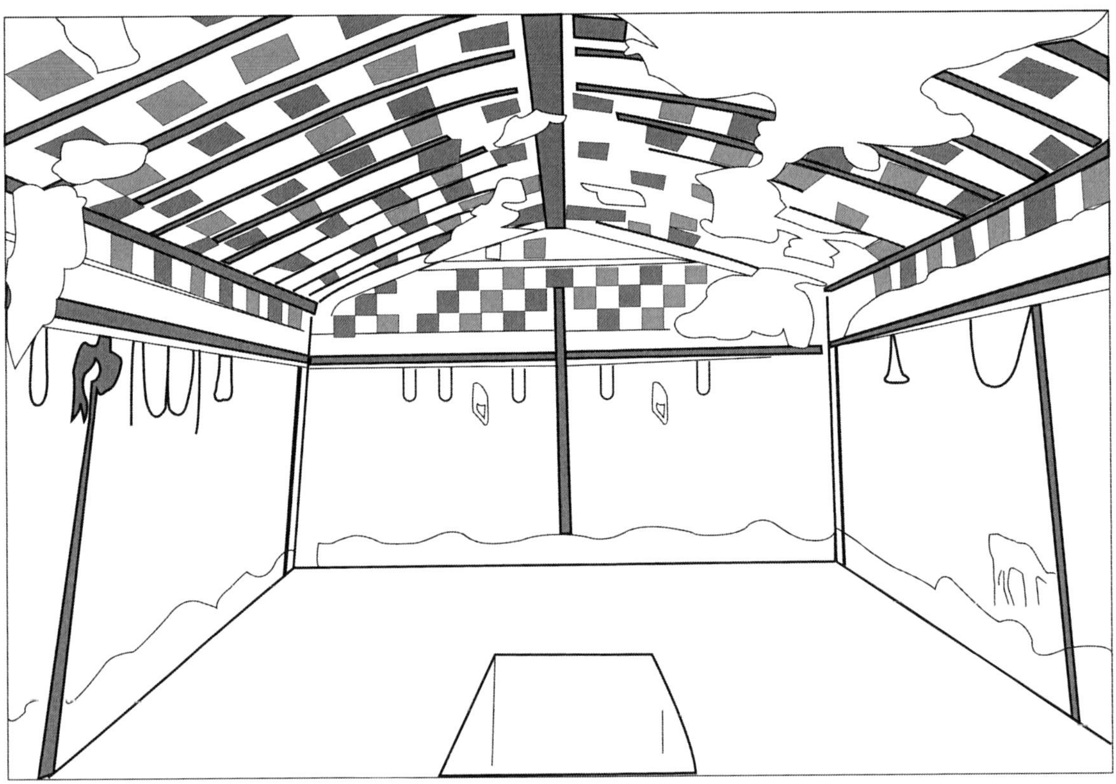

She is standing in profile, extending her left hand to hold a dressed distaff, while a spindle hangs from her right hand, which draws out the yarn. Side B of the *tintinnabulum* shows activities associated with weaving. The bottom scene has not been given a satisfactory explanation. The most likely interpretation has been put forward independently by Stage (1985, 56) and Barber (1991, 116). They suggest that the scene represents the weaving of the starting border necessary for the warp-weighted loom. Such an explanation would fit in with the normal sequence of production, since the last scene provides a rendering of a warp-weighted loom. This representation is unique as it shows what may be the only two-storied loom known from antiquity.

Other relevant finds are a pair of the 5th–4th century BCE stone stelae from the area of Bologna. One such stele shows a woman standing in profile holding a dressed distaff in a fashion similar to the spinner on the *tintinnabulum* (Figure 9) (Zannoni 1876, pl. C.13; Montelius 1895–1910, 465; Ducati 1911, 451 n. 184; Stage 1985, 47). Another stele shows a woman seated in a wagon holding an umbrella in her right hand and a spindle with thread in her left (Zannoni 1876, pl. XXII).

Figure 5. Loom depictions from the Great Rock of Naquane, Middle Bronze Age (After Bazzanella et al. *2003, 93).*

Figure 6. Textile depiction carved into a rock, Borno 1, Bronze Age (After Peroni 1971, 100 fig. 31).

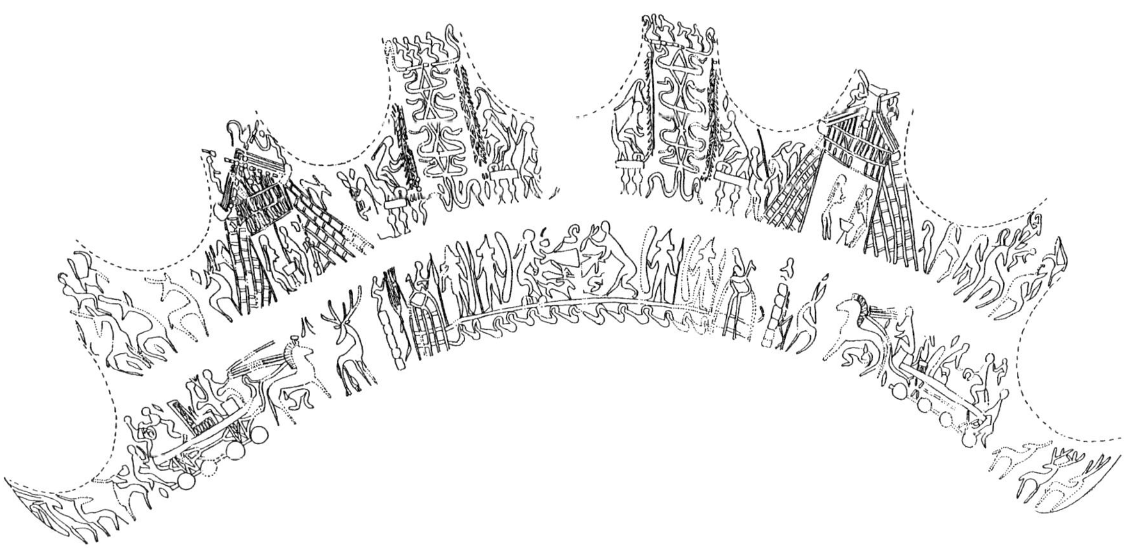

Figure 7. Scenes from the Throne of Verucchio, Tomba del Trono, Rocca Malatestiana, Verucchio, 8th century BCE. Two central scenes in the top register depict women at their looms (After Kossak 1999, fig. 44).

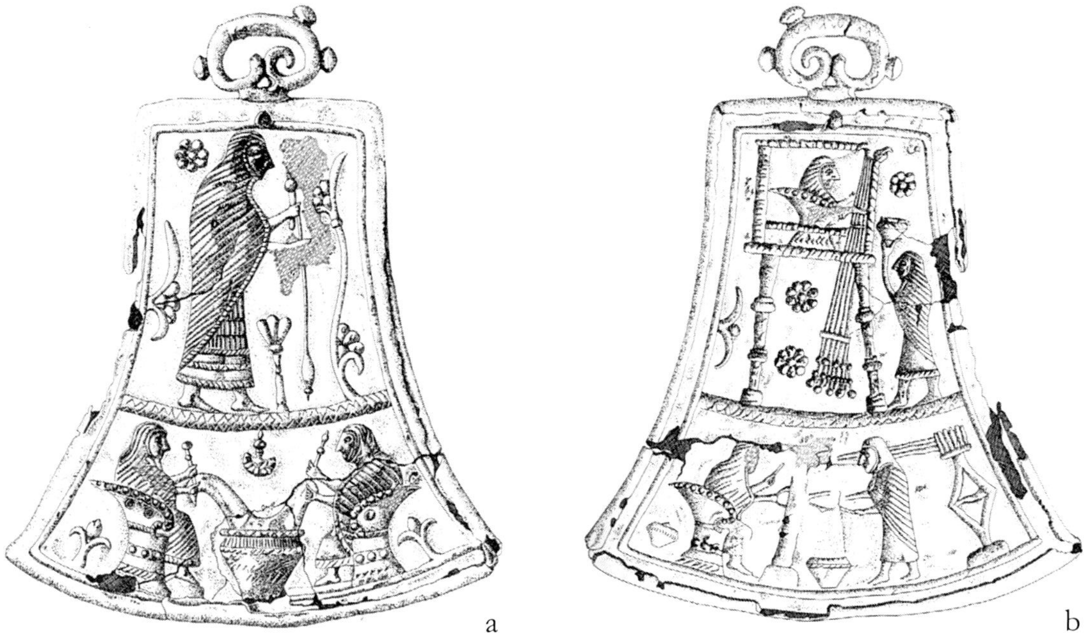

a b

Figure 8 (above). Tintinnabulum *from Bologna, Arsenale Militare, Tomba degli Ori, bronze, late 7th century BCE: a) side A, with scenes of spinning (top) and dressing the distaffs (bottom); b) side B, with scenes of weaving (top) and warping (bottom) (© Bologna Museo Civico Archeologico).*

Figure 9 (right). Stone stele from grave 298, Bologna, showing a woman spinning (© Bologna Museo Civico Archeologico).

Another group of anthropomorphic stone stelae depicting what have been interpreted as weaving scenes are known from Daunia, dated to the 7th–6th centuries BCE (Nava 1980; D'Ercole 2000). The scenes show women seated in front of vertical looms, from which hang two bands with meander pattern (Figure 10).

Several engraved Etruscan mirrors also have depictions of spindles. One, dated to the 4th century BCE, is currently located in Copenhagen (Figure 11). The central design of the mirror shows a young couple sitting and looking at each other while a young boy stands behind the man. The woman holds in her left hand a distaff from which a spindle hangs. A dressed distaff connected to a spindle with the thread wound around its shaft is also depicted on a contemporary mirror from Arezzo (Figure 12). The same object has a depiction of a *situla* with some unidentified objects inside. Given the prominent function of *situlae* on the Bologna *tintinnabulum*, it is tempting to connect them to textile craft.

During the Archaic and Classical periods numerous representations appear on Greek black- and red-figure vases, many of which have been found in Italy. An early 5th century BCE vase from Orvieto shows a girl spinning, holding the thread in her mouth (Barber 1991,

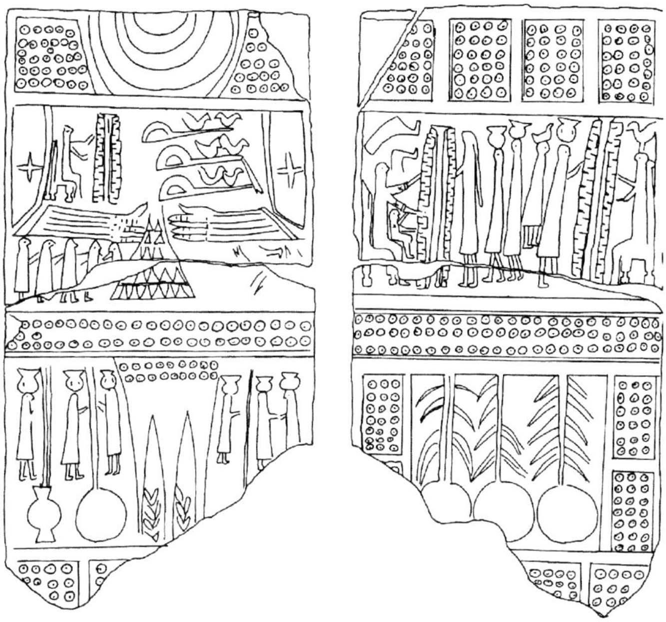

Figure 10. Daunian stele showing seated women at their looms in the upper register of both sides, 7–6th century BCE (After D'Ercole 2000, 330 fig. 1a–b).

70 fig. 2.36). A contemporary vessel from Chiusi represents Penelope at her loom (Figure 13) (Geijer 1977, 52–55; Barber 1991, 108 note 19). This splendidly detailed rendition of a warp-weighted loom is particularly important because it shows the roller-beam and the fabric with figured decoration. A kalyx-krater from Pisticci, in the area of Metaponto, dated 440–420 BCE (Figure 14), shows a lady standing besides the only known representation of a warp-weighted loom in a direct side view showing the deflection of the warp (Quagliati 1904, fig. 4; Carroll 1965, 44–45). Other examples of vases depicting actual or mythological scenes of textile production have been found in Greece or have unknown provenance. Among the latter is the famous 6th century BCE black-figure *lekythos*, now in the Metropolitan Museum of Art in New York, bearing a detailed depiction of the textile production process (Barber 1991, 72 fig. 2.38, 92 fig. 3.13).

Other Greek representations from Italy include a late 5th century BCE terracotta relief plaque with a seated spinning woman from Locri (per. obs. Locri Antiquarium, 2003), and an early 4th century BCE terracotta fragment depicting a spinning Athena Ergane from Scornavacche in Sicily (Figure 15). Greek in style but local in inspiration is the painted depiction of a spinning woman on the east slab of the stone sarcophagus from the 4th century BCE Tomb X of Laghetto necropolis in Paestum (Figure 16) (Pontrandolfo and Rouveret 1992, 356–357; Cipriani and Longo 1996, 131, 182 no. 83.1).

The question of the meaning of the scenes represented on these objects requires a separate investigation. For the purpose of the present discussion, however, the depictions of looms on both the Bologna *tintinnabulum* and the Verucchio throne are considered to be accurate reflections of actual objects, notwithstanding whether the scenes might be mythological or genre. What

Figure 11. Incised mirror showing a seated couple; the lady on the right holds a distaff with a spindle, Copenhagen, 330–320 BCE (From CSE Denmark 1, Fasc. 1*, 122, fig. 26a).*

Figure 12. Incised mirror depicting a distaff with a spindle on the left and a situla on the right, Arezzo, 4th century BCE (After Ciattini 1971, 149).

is to be underlined is the geographical spread of textile making scenes and their importance as status symbols.

At the same time, uneven geographical distribution of extant representations, with heavy emphasis on Etruscan and Greek-influenced areas, precludes meaningful comparisons between different regions. Representations from native Italian sites with strong Greek influences are furthermore culturally ambiguous. Finally, artistic conventions and abbreviations inhibit our ability to read the

Figure 13 (left). Penelope at her loom, Attic red-figure skyphos from Chiusi, c. *460 BCE (After Geijer 1977, 53).*

Figure 14 (below). Side view of a warp-weighted loom, kalyx-krater from Pisticci, 440–420 BCE (After Quagliati 1904, 199 fig. 4).

surviving iconographic sources with confidence.

Archaeological evidence

The most direct but paradoxically least explored type of evidence for ancient textile manufacturing activities consists of archaeological material, that is, the textiles and tools used to produce them.

Textiles

Textiles represent a category of archaeological material little known to general archaeological audience and only recently

Figure 15. Terracotta fragment depicting spinning Athena Ergane from Scornavacche, Sicily, 4th century BCE (From Di Vita 1956, 170 fig. 1).

studied extensively as a part of the cultural record. Over the last few decades, textile studies have developed into an important new field of archaeology (Good 2001).[4] The accumulation of data and the constant development of analytical techniques are permitting more precise fibre and dye identifications. The work on woollen fibres initiated in the 1960s is yielding important results in identifying sheep breeds. This proliferation of technical studies is finally permitting a more synthetic approach to the history of textile technology. Numerous recent publications on the topic, foremost among which is *Prehistoric Textiles* by Elizabeth W. Barber (1991), are demonstrating how much we can learn about the culture, society, technology and economy of the ancient world through textiles.

The study of ancient Italian textiles has progressed much more slowly in comparison to the situation in Central and North Europe. In contrast to the contemporary textiles from Swiss lake-dwellings, textiles of Neolithic and Bronze Age periods recovered in the Terremare district of North Italy since the 19th century have received very little attention over the preceding decades (Rast-Eicher 1997). In fact, the 2004 exhibition at Riva del Garda brought together for the very first time a number of such Italian Bronze Age fragments (Bazzanella *et al.* 2003).

Still, attempts to study pre-Roman archaeological textiles in Italy have been made over the course of the 20th century. In 1921, Lenticchia was the first to examine textiles under the microscope, while a year later, Baserga published one of the earliest analyses of Italian archaeological textiles. After a considerable gap in time, a number of articles on the Ledro material came out in the early 1970s (Perini 1970; 1971). It was not until the 1980s and the 1990s, however, that scientific publications of Italian archaeological textiles started to appear on a more regular basis. This welcome situation is probably partly due to an increase in the recovery of textiles in the field thanks to the improvements in excavation methodology. Material from some sites has been published individually, as, for example, over a hundred fragments recovered from a Late Villanovan boat burial at Sasso di Furbara (Masurel 1977–1982; Mamez and Masurel 1992), while the first attempt to summarise the evidence, at least for the Etruscan period, was made by Jeanine Stage in 1991.

The spectacular finds from Verucchio have added significantly to our knowledge of ancient Italian textile technology. This site boasts the only almost complete

Figure 16. Spinning woman, Tomb X Laghetto, Paestum (After Pontrandolfo and Rouveret 1992, 215).

Italian Iron Age garments. A recent publication of material from one of the Verucchio tombs is already an indispensable source of information for anyone interested in ancient Italian textiles and garments, thanks to the detailed preliminary analysis of two mantles and other fragmentary textiles (Stauffer 2002; Ræder Knudsen 2002). The rest of the material is currently undergoing investigation.

Other recent discoveries have also added significantly to our knowledge about the textiles of ancient Italy. For example, articles of Early Iron Age clothing found frozen at Vedretta di Ries in the Alps may provide a link with Central Europe (Dal Rì 1996; Bazzanella *et al.* 2005). Meanwhile, diverse textiles from the Orientalising cemetery at Casale Marittimo, recently exhibited together with other finds from the site, await analysis (Esposito 1999).

The scarcity and poor preservation of archaeological textiles are two factors usually cited for the absence of interest in textile studies in Italy. Although valid to a certain extent, such reasoning has obscured the fact that textile fragments, while often quite small, are much more abundant in the Italian archaeological record than generally assumed. The vast majority of Italian archaeological textiles, in fact, survive as minute traces on metal objects, such as fibulae and vessels. Until recently, the only such object that has been studied extensively was a bronze bowl, believed to have originated at Veio, and now in the Newark Museum (Carroll 1973). In the recent publications, more attention is being paid to such textile traces (*e.g.* Gandolfo 1981; Buranelli 1983, 129; Terranova and Lo Campo 1999). Examination, documentation and publication of these remains are crucial for a better understanding of ancient Italian textile technology, as will be shown in the following chapter.

Textile tools

The study of textile tools in the investigation of textile production in ancient societies is an even more recent development than textile analysis, John-Peter Wild's *Textile Manufacture in the Northern Roman Provinces* (1970) being the first extensive treatment of the subject. In fact, textile tools have rarely been given attention in archaeological literature beyond general observations or, at best, the publication of a tool catalogue.[5] In the last few years, however, several important studies have been carried out for some prehistoric sites in Switzerland (Bocquet and Berretrot 1989; Médard 2000). Most recently, a groundbreaking study of textile tools from several Swedish Viking Age sites appeared, setting the standard for the future investigations of this kind (Andersson 2003). Currently, a large-scale project on textile tools of Bronze Age Mediterranean is on the way (Andersson and Nosch forthcoming).

Experimental archaeology and ethnographic investigation have been important approaches to the study of ancient textile implements (Peacock 2001, 186). Tools found in archaeological contexts may be tested for their function and suitability by using replicas to reproduce particular types of textiles. Thus, the work of Martha Hoffmann (1964) on the use of the warp-weighted looms in 20th century Scandinavia has been pivotal in understanding the use of ancient looms of the same construction.

Unlike the textiles themselves, textile implements are ubiquitous at Italian archaeological sites. As in the case of textiles however, the major problem with this material is the lack or inadequacy of publication. No study of tools from a functional standpoint exists at present. Furthermore, as with any archaeological source, there are several problems associated with the study of textile tools. Implements made of organic materials are usually not preserved, and the function of some surviving tools is unknown or, at best, ambiguous. Any statistical analysis of the implements recovered at a given site is usually compromised, since it is virtually impossible to ascertain what proportion of tools has survived in the archaeological record. Investigation is further complicated by problematic contexts and the extent to which a site has been excavated. Nevertheless, the great number of implements associated with textile manufacture can be used to study the craft and its technological and economic aspects. Furthermore, many textile tools have been found in burial and votive contexts, providing another interpretative framework.

Spinning and weaving implements constitute the single most important and plentiful type of evidence for the assessment of the scale of production and the technology of the textile industry in ancient Italy. It is the aim of this work to demonstrate the potential of textile and textile tool investigation for archaeology. In the following chapters, textile production in pre-Roman Italy shall be examined from the point of view of both textiles and tools.

PART 3

Fibres and Textiles

Textile is the end product of the textile making and, as such, it contains information about the various stages of the process and the tools used in it. 'Decoding' textiles is thus imperative for our understanding of this process.

Textiles represent a special kind of material, which is rarely considered together with other archaeological artefacts. Like any organic material, they are subject to rapid decomposition in archaeological contexts and their preservation requires special conditions to prohibit their destruction by microorganisms (Wild 1988, 7–13; Peacock 2005; Gillis and Nosch 2007b).

Thus, dry climates have preserved textiles by desiccation in Egypt (Barber 1991, 145ff; Algrove-McDowell 2003) and the Near East, as for instance, in the case of the textiles recovered from the Syrian desert sites of Dura Europos (Pfister and Bellinger 1945) and Palmyra (Pfister 1934, 1937, 1940; Schmidt Colinet, Stauffer and al-As'ad 2000; Stauffer 2000).

Wet environments can also be favourable for preservation of organic materials. The alkaline conditions of the Alpine lakes and acidic environment in the Danish bogs have conserved textiles in Central and North Europe. The pH value of water or soil significantly influences conservation: plant fibres are not preserved in acidic environments, while animal fibres are for the most part destroyed by basic conditions (Rast-Eicher 2003, 47). Thus, in Denmark, only woollens have been preserved (Hald 1980), while Swiss and Italian lakes have yielded predominantly fabrics made of vegetal fibres (Bazzanella *et al.* 2003). This situation even has led to erroneous interpretations of the textile history of Europe (Harding 2000, 254).

Temperatures below 0°C preserve all organic material almost unaltered. A mummy known as the Iceman, the Man of Similaun or Ötzi, dated by radiocarbon to 3350–3100 BCE, was discovered in the Alpine glacier of Italy (Spindler *et al.* 1995). The Iceman's garments were also preserved in permafrost conditions, providing a first glimpse of Bronze Age costume practices. In Eurasia, the contents of Scythian tombs were frozen as well, preserving fabrics almost perfectly (Rudenko 1970; Polosmak and Barkova 2005).

Salt has preserved fabrics in Austrian mines (Bichler *et al.* 2005) and the Tarim desert in north-west China (Trever 1932; Barber 1999; Zhao and Yu 2002). Dry conditions and the presence of salt are particularly favourable for dye preservation while wet environments usually degrade colouring agents (Bazzanella and Mayr 1995b, 127).

Textiles also can be preserved through exposure to fire, which leads to creation of carbonised samples. Vegetal fibres, in fact, are often more stable in carbonised state than in the original shape, especially in wet environments (Rast-Eicher 2003, 47). Animal fibres, on the other hand, usually perish in fire (Rottoli 2005, 66).

In the presence of metal objects, textiles may become pseudomorphs – mineralised formations in which metal corrosion products form casts around fibres retaining their external morphology and size almost unaltered (Jakes and Sibley 1984; Jakes and Ill 1986; Janaway 1987). The conditions conducive to pseudomorph formation depend on pH value, oxidation potential and moisture, as well as the composition of the fibre, and on the elemental composition of the soil and metal (Janaway 1987, 136–142; Gillard *et al.* 1994). Iron and bronze in particular favour formation of easily legible traces. Most of these corrosion-preserved textiles come from inhumation graves. Even when minute, these traces can provide a considerable amount of information about ancient textiles (*e.g.* Bender-Jørgensen 1992).

Textiles can also be preserved in the form of imprints. These are negatives of the original fabrics, created for example when fabric comes in contact with clay objects or surfaces before they are fired. Some of the earliest evidence for woven fabric, in fact, comes in the shape of such imprints, as in the case of some examples from Jarmo dated *c.* 7000 BCE (Lambert 1997, 143).

Last but not least, soil discolouration is another indication of textile presence. At Castel di Decima (Lazio), for example, dark spots of an organic nature were observed at the shoulders of some skeletons, which, excavators believe, represent the traces of bunched cloth where the fibulae held it (Zevi and Bedini 1973, 32).

Whatever the condition of preservation, textile recovery greatly depends on excavation methodology and conservation procedures (cf. Gillis and Nosch 2007b). Thus, the micro-excavation approach, in which entire blocks of earth containing archaeological material are lifted and brought to the laboratory, where they are then carefully excavated by conservators using special equipment and under controlled conditions, has been especially successful in textile recovery and preservation (*e.g.* Esposito 1999). Their prompt consolidation, documentation and conservation, given the usually advanced state of degradation, is then essential. The exact conservation technique depends on the environment, from which the piece was recovered, and the types of damage, whether mechanical, biological or chemical (Bazzanella and Mayr 1995a, 113; Potthast and Riens 2003).

Textiles

To create a textile, raw material has to be transformed in a series of processes. Each stage of this transformation leaves information in the final product, which can be recovered through textile analysis. Thus, preparation of fibres before they become yarn sometimes can be deduced from the appearance of the threads. A smooth, even woollen thread results from the combing of wool, while a softer and flossier yarn is the product of teasing or carding. Fibre analysis may also tell whether the wool was plucked or shorn (Barber 1991, 30).

Textile analysis involves recording of technical information, often using microscopy. Occasionally X-ray is used to see hidden structures, especially relevant in the case of mineralised examples, but the basic examination can be performed by the naked eye, or with little magnification (Walton and Eastwood 1983). The most important features of an archaeological textile to be recorded during such an analysis are its dimensions, condition, colour, fibres, fibre preparation, spin, thread count, weave, edges, applied decoration, and faults (Walton and Eastwood 1983; Masurel 1989; Gillis and Nosch 2007b). Irregularities and faults in the weave can be invaluable evidence for determining the warp and the weft, and the type of loom used to produce the textile. Common faults include missing warps, double wefts, accidental reversing of twill diagonals and crossings of adjacent weft threads. In addition, any evidence of use, such as areas of wear and repair, should be recorded. With the use of a microscope it is even possible to observe the wear and damage of individual fibres (Wild *et al.* 1998).

Yarn

The fibres can be spun or twisted in two directions, producing different structural effects. For convenience, yarn is described as z-twisted if it is spun clockwise, or s-twisted if counter-clockwise (Figure 17). The angle at which fibres lie indicates whether yarn was spun tightly or loosely. The tightness of thread twist may also be measured in twists per centimetre. Fineness or coarseness of the yarn is indicated by its diameter measurement, which can also provide information about the weight of the spindle used to produce it. Two or more threads may be plied together to produce thicker and/or stronger yarn (Figure 17). This is done by twisting two or more threads together, usually in the direction opposite to that of the original spinning. Thus, s-twisted yarn is usually Z-plied and vice versa. Further twisting of plied threads is called cabling. Twist direction is one of the major cultural features (Bender Jørgensen 1992, 13). For example, yarn in ancient Egypt was s-twisted, while in Europe the preferred direction was z.

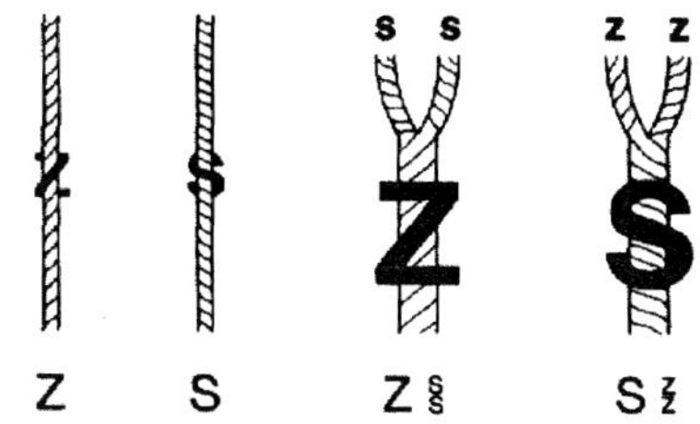

Figure 17. Yan twist directions: z, s, Zs-ply, Sz-ply (After Bender Jørgensen 1992, 15 fig. 2).

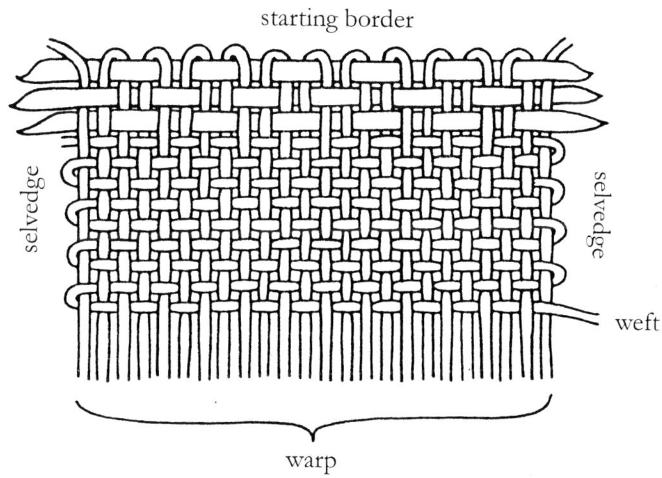

Figure 18. Structural elements of textile (After Walton and Eastwood 1983, fig. 1).

Weave

Once the fibres are spun, they can be woven into a two-dimensional structure (Figure 18).[1] The woven textile is constructed by interlacing two thread systems at right angles. The static system is generally called warp. The perpendicular movable system is called weft. The two systems are not necessarily made of the same yarn (Bender Jørgensen 1992, 13). The textile is then defined in terms of weave (Figure 19). The simplest form of weaving, plain weave or tabby is produced by weft threads passing over and under alternate warps. When the weave is balanced, *i.e.* warp and weft have approximately even amounts of threads, it is called balanced tabby. When one of the systems is denser, the tabby becomes warp-faced or weft-faced. In rep, one of the systems has a considerably higher

Figure 19. Basic weaves used in ancient times (Courtesy of Eva Andersson).

thread count, normally, a ratio of 2:1. Rep can also be warp-faced or weft-faced if warp and weft can be determined. If the warp and weft are paired, the resulting binding system is a basket weave; if only one system is paired, it is a half-basket weave or extended tabby.

In the more complex, twill weave, weft threads pass over and under warps in a regular staggered pattern, each row being stepped to one side of the row above (Figure 19). This creates a diagonal effect. If the binding unit is under two over two, it is described as a 2/2 twill. The variants include a plain diagonal, warp- or weft-chevron, broken and more complex diamond and herringbone twills. If the binding unit is under two over one, the weave is called as a 2/1 twill, and if it is under one and over two, a 1/2 twill. If no intact edge is preserved, it is often impossible to differentiate between the two. Here chevron and diamond variants are also possible. Pattern-making possibilities may be further exploited by 3/1 twills, which produce weft- and warp-faced effects. Compound weaves include multiple sets of wefts and/or warps.

Extra wefts may be worked into the basic weave to produce brocade and, hence, complex pattern. This is a weaving technique well known in Bronze Age Central European textiles (Barber 1991, 139). Soumak is another technique that superficially looks like embroidery and is also known as weft wrapping or 'snare-weave' (Seiler-Baldinger 1994, 57). Soumak textiles can be woven on a warp-weighted loom, as seems to be the case of the Early Iron Age textiles from Gordion (Burke 2007, 66) and Swedish Viking period textiles from Överhogdal (Oscarson 1994).

In another supplementary technique used for decoration, tapestry, several coloured wefts do not pass from selvedge to selvedge but are woven into blocks of colour. Looped textiles are made when threads in one of the systems are not interlaced tightly but allowed to remain loose as protruding loops. Piled textiles, such as carpets, include fabrics of various weaves with extra threads inserted or knotted.

Pattern, however, was not always dependent on a complex weave structure, but could be formed using yarns of different colours, spin directions (spin or shadow pattern), or thickness. As we have seen from iconographic sources, complex patterns, both figural and geometric, were known to the ancient inhabitants of Italy. Depending on the yarn diameter and on how densely the threads are packed in each system, the weave can be more open or more dense. Thread count, that is the number of threads per 1 cm in each system, often indicates the density and fineness of a textile.

Presence of a border or selvedge is very important in textile analysis (Figure 18). Starting borders are characteristic for textiles woven on a warp-weighted loom. Such a border can be a cord, a tabby strip or made by tablet weaving. Selvedges are formed by turning of the weft on itself at the edges of the warp. They can be reinforced with extra threads or cords or a more complex structure may be woven in. The lower edge of a textile may be finished into a fringe or a cord, darned back into the fabric, or left loose. Fringes may be knotted, plaited, twisted, interwoven or finished with tassels. Last, but not least, weaving irregularities are often crucial in determining the weaving direction and other structural aspects of a textile (Stauffer 2005, 205).

Weaving is generally accomplished on a loom but textiles could also be constructed by other means. Tablet weaving is a common technique to make braids but was also popular for making decorative borders (Figure 20) (Collingwood 1982; Seiler-Baldinger 1994, 73–76). Knitting involves the interlooping of continuous thread using two or more long needles, although it probably did not appear as a technique until Roman times (Grass 1955). Sprang is created by twisting adjacent warps without the use of weft yarn (Figure 21); it may be simple or patterned (Hald 1980, 245ff; Barber 1991, 122). Other techniques include lacing, braiding, nalebinding, and netting. Many of these techniques do not require any specialised tools. The fibres can also be entangled without forming a yarn, creating non-woven fabrics such as felts (Forbes 1956, 89–92; Ryder 1962; Barber 1991, 215–222).

Finishing

After a textile has been created, it can be subjected to a variety of finishing processes. Fibres can be matted together by fulling (Forbes 1956, 86–89), or the cloth surface can be brushed and then

Figure 20. Tablet weaving (Courtesy of Lise Ræder Knudsen).

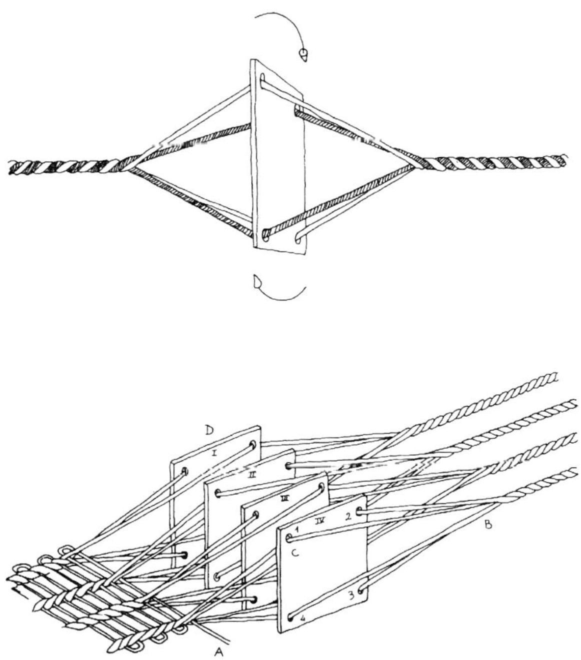

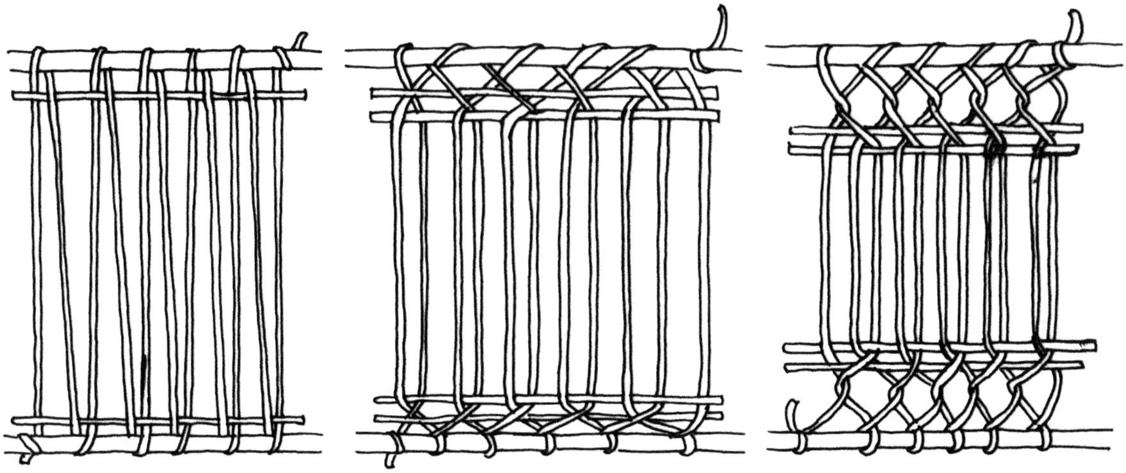

Figure 21. Successive stages of creating sprang (Courtesy of Eva Andersson).

shorn to produce a smooth napped surface. These treatments are generally applied to wool, while linen may be beaten to soften it. Textiles could also be subjected to elaborate pleating. All of these processes will leave traces in archaeological textiles.

A textile can then be sewn into a garment with seams and hems representing the structural elements of stitching. Finally, decoration could be physically applied by a variety of techniques. One of these is embroidery, which includes a variety of stitches. Shells, seeds, beads and metal appliqués could also be attached to the fabric to create a decorative effect.

Effects of use and re-use

Textiles that end up in the archaeological record are not always new or unused, thus providing us with information about their life. For example, wear pattern can indicate the function of a textile, while patching shows re-use and value of textiles.

This brief overview of the technical aspects of textiles demonstrates the importance of textile structure analysis for our understanding of the techniques by which they were created. Some of these techniques and characteristics are chronologically or geographically specific and can be used to trace the date or the origin of an archaeological textile.

Archaeological textiles of Italy

Although almost all of the conditions of preservation described at the beginning of this chapter can be found in Italy, textiles surviving in their original form are rare (Rottoli 2005). The largest corpus consists of linen textile fragments found in various Neolithic and Bronze Age sites in the north of the peninsula, where they have been preserved by the basic conditions of the Alpine lakes. Wool textile fragments, on the other hand, have been recovered from burials, where they were conserved in waterlogged and/or acidic environments; among these, the Verucchio and Sasso di Furbara groups are the most significant. A few Iron Age clothing articles were recovered from the Alpine glacier (Dal Rì 1996; Bazzanella, Dal Rì, Maspero and Tomedi 2005). Carbonised textiles of the Roman period preserved as a result of the eruption of the Vesuvius are also of considerable interest (D'Orazio *et al.* 2000), although chronologically, they are beyond the scope of this study.

The vast majority of textiles in Italy, however, have been preserved in association with metal objects and survive in the shape of minute traces on metal surfaces. Unfortunately, these precious remains often go unnoticed and are only rarely published. In the past, they even were removed deliberately in the process of conservation. Hence, it is particularly important to draw archaeologists' attention to the textile fragments that do survive, as they have great potential in providing a picture of the textile-abundant environment of the ancient world.

No comprehensive review of Italian archaeological textiles exists currently. Neolithic and Bronze Age textiles from North Italy are by far the best published (Bazzanella *et al.* 2003). Textiles related to Etruscan culture have been reviewed collectively by Jeanine Stage in a catalogue, which includes thirty-five sites/fragments (Stage 1991).[2] The following catalogue summarizes the material, sorted both chronologically (from Neolithic to Iron Age) and geographically (from north to south). It is not meant to present a detailed analysis of each fragment; rather, where possible, it records the provenance, location, date, condition, fabric, general technical description and the relevant bibliography. When published, detailed textile analysis is lacking in many cases, often because in the earlier excavations many fragments were discarded and/or not recorded, while others were not conserved properly and are now lost.[3] Furthermore, frequently the illustrations are schematic or of bad quality, hence precluding any observation as to technical structure of a textile (cf. Cat. No. 62). Some textiles listed in the catalogue were observed on objects, which are on display in various museums of Italy. What even briefest of these accounts of textile presence in archaeological contexts of pre-Roman Italy demonstrate, however, is how numerous the material actually is. Map 2 shows locations of sites listed in the Catalogue.

Catalogue

Neolithic period and Bronze Age

1. The oldest textiles in Italy have been found in the submerged Neolithic settlement of La Marmotta in lake Bracciano, dated by radiocarbon to 5480–5260 BCE (Rottoli 1993, 310; Fugazzola Delpino 2001; Rottoli 2003, 68).

Most of the early Italian textiles, however, come from the *terremare* in North Italy and have been dated to the 3rd and 2nd millennia BCE (Bataglia 1943, 51, pl. 50; Barber 1991, 174; Rast-Eicher 1997; Bazzanella *et al.* 2003).

2. Molina di Ledro, Early Bronze Age (Figure 22): numerous textile pieces, which have been cited extensively for their technical complexity and similarity to the Swiss Neolithic textiles (Perini 1970; Peroni 1971, 100; Perini 1971; Barber 1991, 175; Rast-Eicher 1997; Bazzanella 1999; Bazzanella and Mayr 1995, 124–123; Bazzanella *et al.* 2003, 161–174).

   The majority of the surviving fragments are of fine linen tabby woven of S2z yarn, decorated in one case with tiny seeds sewn on like beads, much as the fragment from the Swiss lake-dwelling at Murten (Barber 1991, 140). At least two of the fragments were narrow strips over 2 m long and were found rolled-up in bundles (Bazzanella *et al.* 2003, 161 no. 5 and 162 no. 6). The ground of one of these strips is tabby but the two ends of the cloth had been woven with a pattern of concentric abutting lozenges formed with a technique similar to twill.

3. Lucone di Polpenazze, Early Bronze Age (Figure 23): six carbonised linen tabby fragments woven in S2z yarn and 24

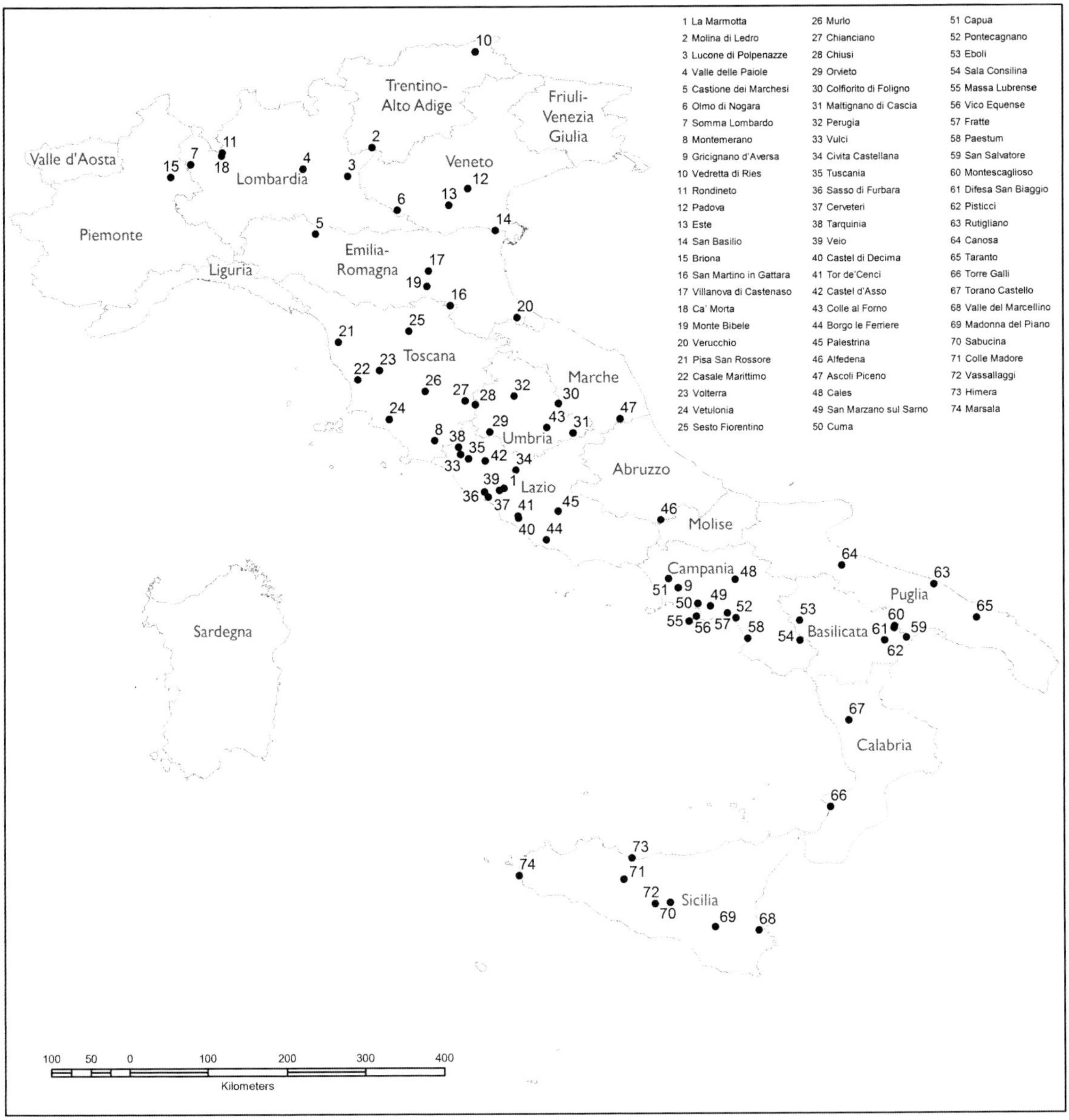

Map 2. Map of textile finds.

cordage examples were found (Bazzanella *et al.* 2003, 186–193). Some of the fragments preserve fringes and one has been mended in antiquity.

4. Valle delle Paiole, Early–Middle Bronze Age: 23 fragments of the same tabby textile made of Z2s wool and tree bast were recovered (Bazzanella *et al.* 2003, 198). The textile was found folded several times.
5. Castione dei Marchesi, Middle–Late Bronze Age (Figure 24): a fragment of tabby woven in Z2s wool (Mutti *et al.* 1988, 274 no. 204; Rast-Eicher 1997, 550, fig. 304; Bazzanella *et al.* 2003, 200; Rast-Eicher 2005, 127).

6. Olmo di Nogara, Bronze Age: pseudomorph of linen textile on a bronze object (Rottoli 2005, 79; Castiglioni, Rettore and Rottoli in press).
7. Somma Lombardo, Late Bronze Age: pseudomorph on a bronze fibula (Castiglioni 1995; Rottoli 2005, 79).
8. Montemerano, Early Bronze Age (Figure 25): tabby pseudomorph on a sword from a hoard (Gandolfo 1981, 362–365).[4]
9. Gricignano d'Aversa, Tomb 6, Eneolithic–Early Bronze Age (Figure 26): partly mineralised traces on a metal blade of a tabby woven in 2Sz hemp yarn (Bazzanella *et al.* 2003, 210).

Iron Age

North Italy

In the 1st millennium BCE, textile finds are more evenly distributed throughout the Apennine peninsula. North Italy, however, yielded a significant share of this material.

10. Vedretta di Ries/Riesenferner, 795–466 BCE (Figure 27): six nearly complete articles of male clothing: two pairs of leggings and a pair of inner shoes, all made of naturally-pigmented goat wool (Dal Rì 1996; Bazzanella *et al.* 2003, 179–182; Bazzanella *et al.* 2005; Rottoli 2005, 73). Different weaving techniques were employed in the making of these textiles: the under-leggings are made in tabby, the over-leggings are in diagonal or herringbone twill, while the shoes are in 2/2 twill. The Vedretta di Ries textiles were sewn together, in some cases, from several pieces of fabric. Numerous mends are also clearly visible, indicating the re-use of textiles that once constituted larger garments.
11. Rondineto (Como), Iron Age: a small sack of very fine 2/2 twill (Baserga 1922, 130, fig. 9).
12. Padova, Largo Europa, Iron Age: sections of woollen cloth which were wrapped around a left foot and tied with two pairs of strings, creating a kind of sock (Maspero 1998, 63, fig. 20).

Figure 22. Textiles from Molina di Ledro, Early Bronze Age (After Bazzanella et al. *2003, 161, 168, 169).*

13. Este, burials: several fragments have been recovered. Tomb 61, dated to the 6th century BCE, contained a fragment of a 2/2 twill that had probably wrapped the urn (Maspero 1998, 64–65). Two types of textile of plain weave were found in the

Figure 23. Textiles from Lucone di Polpenazze, Early Bronze Age (After Bazzanella et al. *2003, 192, 193).*

Figure 24. Textile from Castion dei Marchesi, Middle–Late Bronze Age (After Bazzanella et al. *2003, 200).*

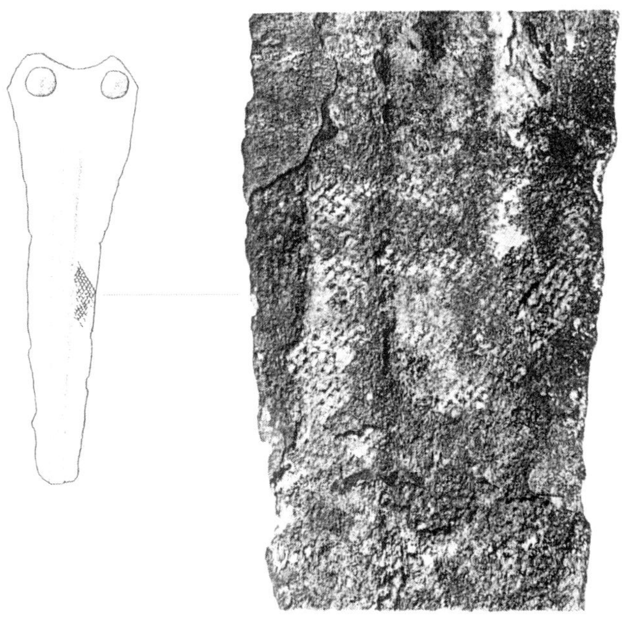

Figure 25. Textile pseudomorph on a bronze weapon from Montemerano, Early Bronze Age (After Gandolfo 1981, 361 pl. 94 and 365 pl. 96).

ossuary from Tomb 126, dated to the end of the 4th century BCE (Maspero 1998, 64). Capodaglio Tomb 38, dated 450–350 BCE, had an iron weapon scabbard with textile remains that appear to be in 2/2 twill. Traces of textiles have also been noted on metal objects from tombs 31 and 37 (Chieco Bianchi 1987, 232), Nazari 161 (Chieco Bianchi 1988a, 76), Casa di Ricovero 32, 230 (Chieco Bianchi and Calzavara Capuis 1985, 255 no. 41, pl. 164), Benvenuti, Tombs 116, 117, 296 (Calzavara Capuis and Ruta Serafini 1987,

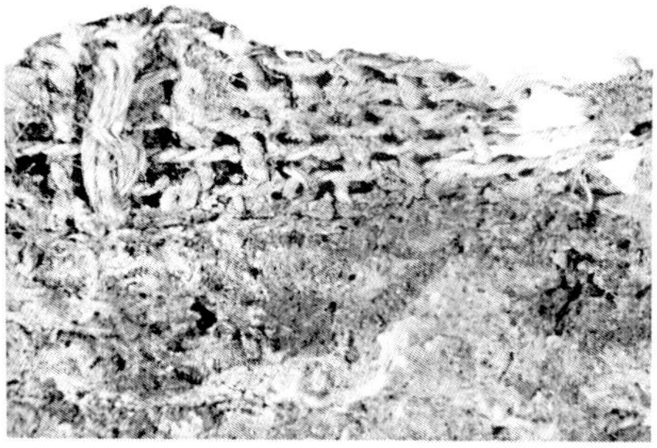

Figure 26. Textile from Gricignano d'Aversa, Eneolithic–Early Bronze Age (After Bazzanella et al. *2003, 211).*

Figure 27. Textiles from Vedretta di Ries: over-legging (top left), under-legging (top right), inner shoe (bottom), 795–499 BCE (After Bazzanella et al. *2003, 180–182).*

301 no. 20 and fig. 21, 303 no. 18c and fig. 23; Calzavara Capuis and Chieco Bianchi 2006, 251 no. 18, 254 nos. 9–10, 385 no. 25).

14. San Basilio, 7th century BCE: z/s twill fragment (pers. com. Christophe Moulherat, 2005).
15. Briona, fraz. S. Bernardino, Tomb 1, tumulo XXXVI, late 6th–early 5th century BCE: bracelet with mineralised traces of twill textile (pers. obs. Torino Museum, 2003).
16. San Martino in Gattara, 5th century BCE: Tomb 6 contained an iron lance head with traces of textile; Tomb 9 yielded a pseudomorph on an iron lance (Bermond Montanari 1969, 14 no. 15 and fig. 17c, 20 no. 11 and fig. 25a).
17. Villanova di Castenaso, Ca' dell'Orbo, Tomb 14, mid-7th century BCE: an iron axe with traces of two different textiles (Tovoli and Vitali 1979, 54 no. 23, 56 fig. 33 no. 23; Forte and Eles 1994, 137 no. 23, 136 pl. IX).
18. Ca' Morta, Iron Age: tabby pseudomorphs on a fibula with thread count of 36 threads/cm (Baserga 1922, 82, 128–131, fig. 8; Rittatore Vonwiller 1966; Castelletti 1978; Casini 1986, 140; Maspero 1999, 119).
19. Monte Bibele, late 5th–early 3rd century BCE (Figure 28): mineralised textile fragments on 26 metal objects from 15 burials (Vitali 1983, 112; 2003; Vellani 2005; Moulherat forthcoming). Textiles include balanced and weft-faced tabbies, 2/2 twills and the fibres are wool, linen and mixed.
20. Verucchio, Early Iron Age: the site boasts the only surviving, almost complete garments of Villanovan period. Mantles and other pieces of clothing were found folded or displayed in the tombs of Verucchio. Traces or actual textiles have been found in Moroni tombs 24 and 26 (Gentili 1985, 80 no. 36, 89 no. 57 and pl. XLIII), La Rocca tombs 16, 17, 18, 31, 47, 54, 85, 89, 92, 96, 111, 113, 116, 118, 119, 122 (Gentili 2003, 24, 26, 28, 187, 189, 191, 216, 234, 237, 283, 286 no. 5, 294, 311, 314, 330, 334, 338, 343, 345, 346 no. 8, 349), Le Pegge tomb 14 (Gentili 2003, 68 no. 10 and pl. 28.10), Lippi tomb 52 (Eles 1998, 60), Strada Marecchiese B/1971 (pers. com. Annemarie Stauffer, 2007). The presence of hundreds of amber and glass beads and other decorative elements indicates that many of these textiles were luxuriously decorated (Forte 1994; Gentili 2003). Some of the Verucchio textiles bear traces of regular folding (Eles 2006a, 14), confirming that artistic representations of garments, like those seen in the painted tombs of Tarquinia, are accurate depictions of contemporary practices.

The recent publication of the textiles found in one of the tombs at Verucchio has given us the first glimpse of this sensational material (Stauffer 2002; 2003; 2004; Ræder Knudsen 2002). The 8th century BCE Tomb 89, also known as the Tomba del Trono, the richly appointed cremation burial of a important male personage, contained over 160 textile fragments that were preserved by the cremation process, as well as two mantles (Figure 29) and what was probably a tunic-like garment, which survived almost intact. The mantles are of very fine wool, woven in 2/2 twill with groups of s and z twisted single yarns in both directions ('spin-pattern') and originally coloured red, with borders dyed blue. The tunic-like garment was woven in 2/2 twill with

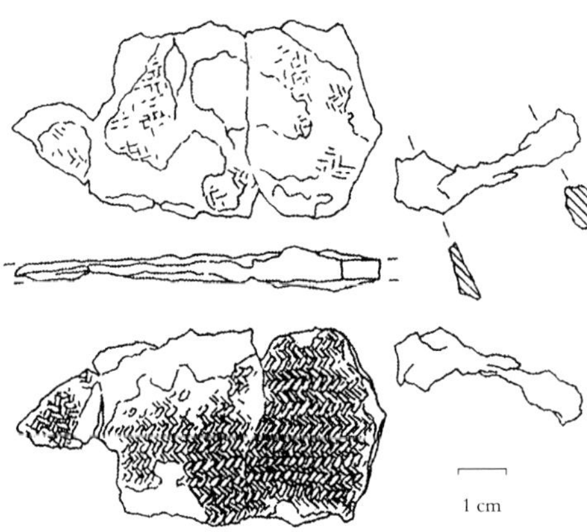

Figure 28. Textile traces on a metal knife from Monte Bibele Tomb 9, 4th century BCE (After Vitali 2003, 17 pl. 9).

'spin-pattern' and was dyed red with red borders. Both mantles have elaborate borders (Figure 30) made by the tablet weaving technique and featuring a triangle motif and 3 horizontal lines (Ræder Knudsen 2002, 222–225). Among the numerous fragments preserved in the cremated remains, most of them twills with varied spin effect, there are some pieces that are very fine and dyed red or blue. One of the fragments is decorated in the soumak technique (Figure 31) and

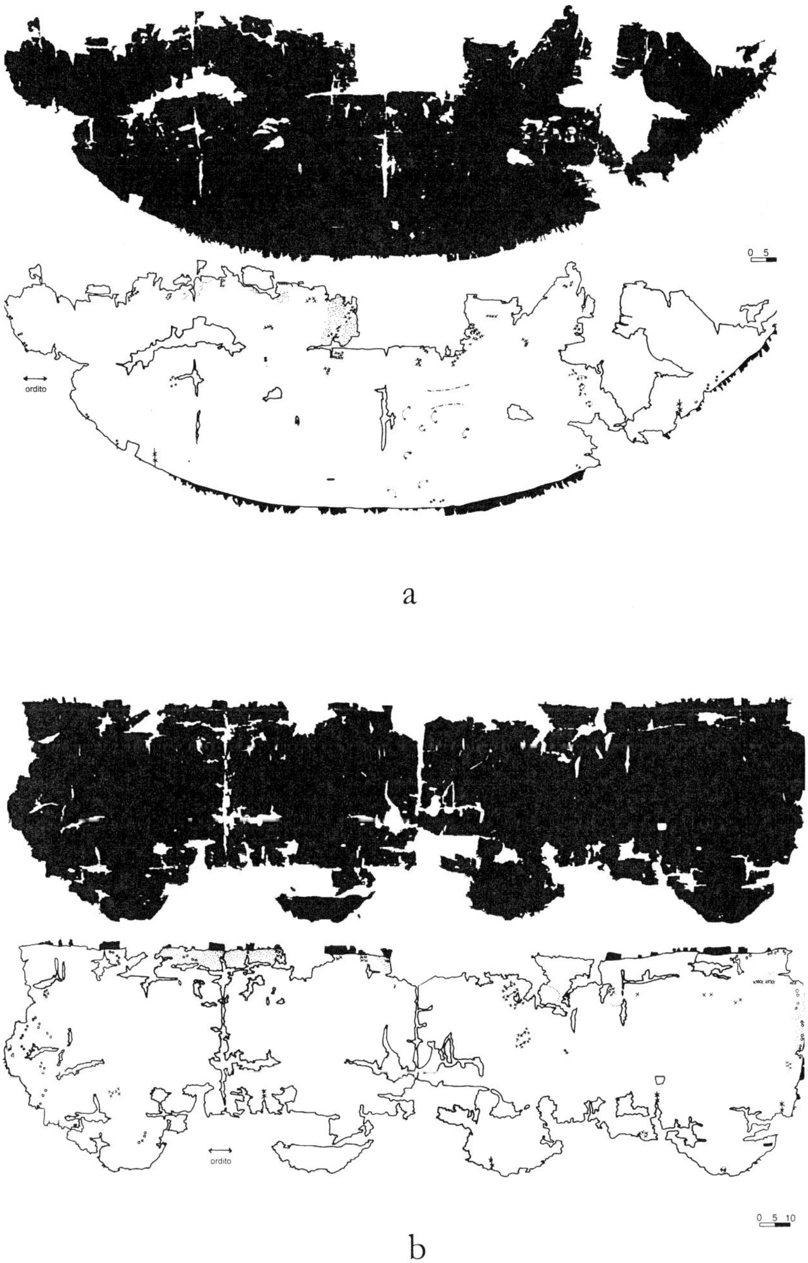

Figure 29. Mantles from the Tomba del Trono, Verucchio, 8th century BCE: a) mantle 1; b) mantle 2 (© Cologne University of Applied Sciences).

Figure 30 (right). Reconstruction of tablet-woven border of a mantle from Tomba del Trono, Verucchio, 8th century BCE (After von Eles 2002, pl. XXII).

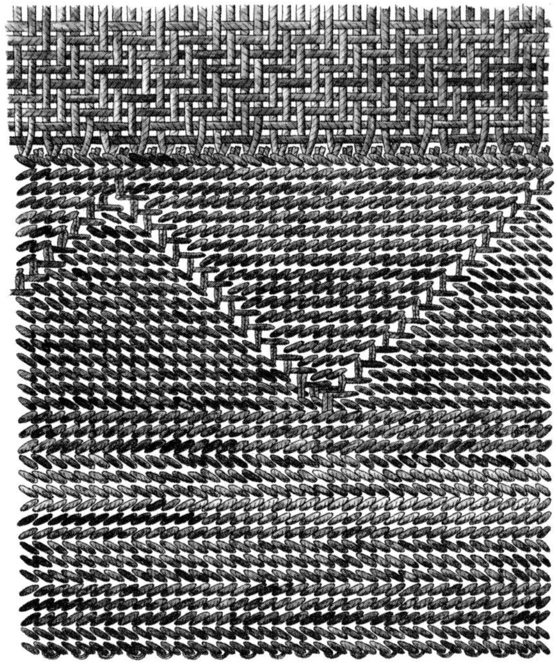

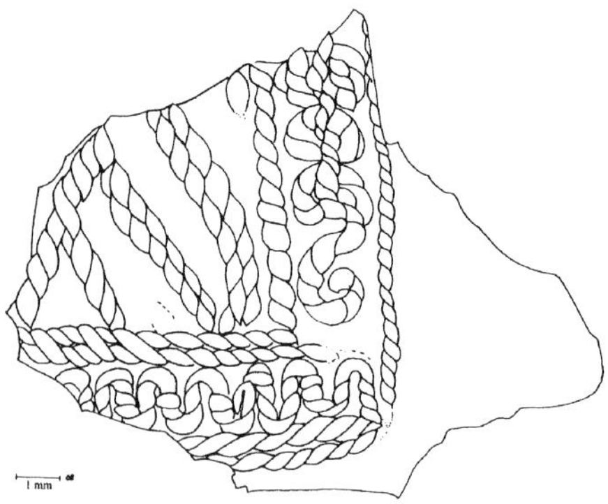

Figure 31 (above). Soumak textile from Tomba del Trono, Verucchio, 8th century BCE (© Stauffer 2004, 206 fig. 3).

another is looped (Stauffer 2002, 213; Stauffer 2003).

Central Italy

21. Pisa San Rossore, Etruscan-Roman shipwrecks: numerous cordage fragments and a fragment of z/z tabby made of wool (Lentini and Scala 2002; Lentini and Scala 2005; Camilli, De Laurenzi and Setari 2006, 52 no. 22). Cordage materials include hemp, *Hybiscus sp.*, *Calotropis sp.*, *Musa sp.*, *Asclepias sp.*, *Gossipium sp.*, *Spartium junceum*, *Lygeum spartum*, *Chamerops umilis*.
22. Casale Marittimo, tombs, 8th–6th century BCE: animal and bast fibre pieces of clothing and also utilitarian fabrics (Esposito 1999, 41 fig. 31, 68–69 fig. 65, 71 fig. 67, 93–94; Bottini *et al.* 2003, 58 fig. 5, 135 no. 80). In Tomb G, in which a child was buried, an iron dagger and an axe were wrapped in woollen fabric, while food offerings were covered with linen cloth (Figure 32). In Tomb A, a bronze cinerary urn contained textile remains that probably served as wrapping for cremated bones (Figure 33). Textiles from Tomb H2 were present in the form of pseudomorphs on iron spear counterweights. These textiles, conserved in the presence of copper and iron salts, present a range in fineness and complexity. The yarns vary in spin direction and are sometimes two-plied. While most of the examples are tabby, twill weaves are also present.
23. Volterra, Portone, 4th–1st century BCE: a large piece of linen (Fiumi 1976, 65; Stage 1985, 78–79).
24. Volterra: several fragments of light tabby with yellowish borders (pers. com. Larissa Bonfante, 2006).
25. Vetulonia, Tomba del Duce, 650–630 BCE: a linen cloth 10.6 × 9.4 cm, found inside a bronze and silver larnax, which contained cremated bones (Falchi 1887, 506; von Duhn 1924, 239; Camporeale 1967, 141; Banck-Burgess 1999, 232; Torelli 2000, 582 no. 130).
26. Sesto Fiorentino, Tomba della Montagnola, 7th century BCE (Figure

Figure 32 (left). Bronze patera with textile remains from Tomb G, Casale Marittimo, 7th century BCE (After Esposito 1999, 71 fig. 67).

Figure 33 (below). Linen fragments from Tomb A, Casale Marittimo, 7th century BCE (After Esposito 1999, 42 fig. 31).

34): several unbalanced tabby fragments were attached to the back portion of a piece of iron armour, while a twill fragment possibly was associated with the funerary bed (Rilli 1964, 35; Bonfante 1989, 1385).

27. Poggio Aguzzo (Murlo), tombs, early 7th century BCE: pseudomorphs on two spear counterweights from Tomb 1 and an iron knife blade from Tomb 4. The type of fibre has not been determined although "the weave appears very tight and the threads are fine" (Tuck 1996, 116–117).
28. Chianciano, Poggio alla Sala, 7th century BCE: a purple woollen mantle with adhering bits of gold was allegedly found wrapped around a bronze cinerary urn (von Duhn 1924, 321; Levi 1935, 94–95; Bonfante 1975, 11, 106 note 3; Stage 1985, 74; Barber 1991, 194; Banck-Burgess 1999, 229). Unfortunately, the fragments were destroyed by the flood of 1965 in the Florence Museum where they were kept.

Figure 34. Mineralised textile remains from the Tomba della Montagnola at Sesto Fiorentino, 7th century BCE: a–b) tabby from the bronze armour; c) twill from the funerary bed (Courtesy of Larissa Bonfante).

29. Chianciano, tumulus in loc. Morelli, 7th century BCE: textile remains covering bronze urn, were conserved by large iron nails; other fragments were found inside the urn and probably wrapped cremated bones (Paolucci 1997, 26; Haynes 2000, 106).
30. Chiusi, Tomb 878, 3rd–1st century BCE: two fragments of weft-faced tabby folded over on themselves (pers. com. Hero Granger-Taylor, 2003).
31. Chiusi region, 2nd–1st century BCE: textile fragment completely calcified inside a clay burial urn of a type peculiar to the Chiusi region; it is now in the Royal Ontario Museum (Hayes 1977, 144). The fragment is unusual in being of a very fine weave described by the analyst as 'twined'.
32. Orvieto, Cannicella Tomb A, *c.* 300 BCE: an iron ring which, the excavators suggested, functioned to hold a veil on the head of the deceased, since traces of both textile and hair remain on the object (Bonamici, Stopponi and Tamburini 1994, 222 no. 11, pl. XXXVIIa).
33. Orvieto?, early 6th century BCE: a pseudomorph of a piece of plain z/z tabby weave adhering to the bottom of an oinochoe of Italo-Corinthian style is attributed to Orvieto (Hayes 1977, 144). This completely calcified fragment is probably part of a cloth on which the jug was placed or in which it was wrapped.
34. Colfiorito di Foligno, Tomb 248, 6th century BCE: a loosely woven twill pseudomorph on an iron fibula (Bonomi Ponzi 1991, 150 no. 2.8, 151 fig. 2.8).
35. Maltignano di Cascia (Spoleto), 1st century BCE: a coarse fragment preserved on an iron strygil (pers. obs. Spoleto Museum 2004).
36. Perugia?,[5] 390 ± 45 years BCE (Figure 35): one of the largest and probably the most famous specimen of Etruscan cloth, the so-called Zagreb mummy wrapping, preserved by the dry climate of Egypt (Krall 1892; Roncalli 1980; *Liber Linteus Zagabriensis* 1986; Flury-Lemberg 1986;

Flury-Lemberg 1988, 344–357, 496; van der Meer 2007). The cloth survives in twelve strips, which originally comprised a part of a linen book, *liber linteus*. It bears the longest currently surviving Etruscan text, with letters painted on the cloth with black and red ink. It is believed that the book was taken to Egypt *c.* 150 BCE and after some time torn into strips to be used as mummy wrappings.

37. Perugia, various necropoleis: textile remains found on the bottom of travertine urns. Remains of burnt bones were wrapped in wool and placed on a linen cloth, which was folded several times (Cenciaioli 2004, 8).
38. Vulci, Osteria, Tomba del Carro di Bronzo, early 7th century BCE: textiles adhering to the inside of the bronze cinerary urn and the wrist of the bronze hand model (Cristofani 1985, 288 no. 108).
39. Vulci, Isis Tomb, 7th century BCE: thousands of small glass beads, some with threads still attached to them, indicating that at some point they composed a sumptuous garment, possibly a skirt for the bronze bust from the same tomb (Roncalli 1998, 20).
40. Vulci, Ponte Rotto, Tomb VI, 3rd–1st century BCE: several textile fragments as well as pseudomorphs (Falconi Amorelli 1987, 37 no. 29 and pl. III, 46 no. 27, 48 no. 30 and 31, 49 no. 34, 52, 55). Textile fragments were recovered from Camere B and D. Camera C yielded textile remains on the rim of bronze cist, 2 bronze fibulae and 2 iron strygils. In Camera I white fabric was visible on the skeletons at the time of excavation.
41. Vulci, Tomb 66, *c.* 680 BCE (Figure 36): short thread fragments preserved in the shanks of a set of hemispherical bronze buttons (Dohan 1942, 87 no. 35, Pl. 46; Turfa 2005, 68; Gleba and Turfa 2007, no. 1). The buttons are 5–7 mm in diameter and 2 mm in height. Of the 44 objects, 37 buttons have remains of white thread in the shanks; in other 7 the shanks are broken and the thread has been lost. All thread fragments are short, mostly less then 6 mm in length. The thread is about 1 mm thick, and is constructed of three loosely Z-plied s-twisted yarns.

Figure 35. Liber linteus zagrabiensis, *3rd–2nd century BCE (© Zagreb Archaeological Museum).*

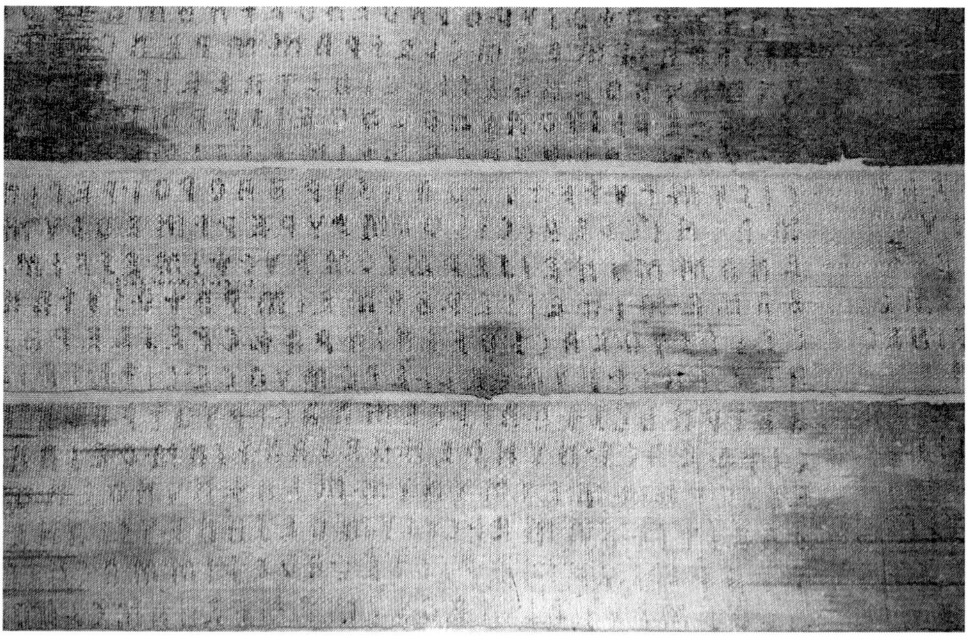

Microscopic examination indicates a fibre of plant origin.

42. Civita Castellana, Cappuccini, 7th century BCE: several mineralised textiles found in a multiple deposition tomb (De Lucia Brolli 1998, 186, 187, 189, 191, 193). An iron dagger and iron razor from loculus 1 both had traces of plant fibres; an iron sword from loculus 4 had traces of textiles of different fineness, possibly remains of different articles of clothing. In loculus 2, textile remains were noted on an iron belt. Finally, conspicuous traces of textiles were observed on an iron dagger sheath from the bench deposition.

Figure 36. Bronze button with thread, Vulci, Tomb 66, 680 BCE, under magnification. (© University of Pennsylvania Museum).

43. Civita Castellana, Valsiarossa Tomb 7 (CXXXIX), 3rd century BCE: gold thread that probably formed part of garment's decoration (pers. obs. Villa Giulia Museum, 2005).
44. Cogion – Coste di Manone (Civita Castellana), burial, *c.* 350 BCE (Figure 37): very fragile fragment is a fine s/s tabby and contains strands of human hair similar to a lock, which was also found to belong to the same burial, indicating that the textile is probably a piece of veil or sakkos that once covered the head of a dead woman (Turfa 2004; Gleba and Turfa 2007, no. 2).
45. Tuscania – Tarquinia, 2nd–1st century BCE (Figure 38): a clay urn with numerous calcified traces of a tabby textile that probably wrapped the urn (Gleba and Turfa 2007, no. 3).[6]
46. Sasso di Furbara (Cerveteri), Caolino, burial 42, 9th century BCE (Figure 39): large quantities of cloth fragments, most of which unfortunately no longer exist (Masurel 1982; Mames and Masurel 1992; Barber 1991, 194; Banck-Burgess 1999, 43, 231). The textiles are of high quality both technically and aesthetically, and illustrate a range in fineness and design. No complete piece has survived,

Figure 37. Textile fragment from Cogion-Coste di Manone, 4th century BCE. (© University of Pennsylvania Museum).

but both garments and utilitarian textiles seem to be present. Some of the textiles are of extremely fine quality; the patterns represented are often obtained by varying the spin direction of the thread, a technique similar to that observed in the Verucchio textiles. An important piece from Sasso di Furbara is a fragment with a herringbone pattern made in tablet weaving technique.

47. Cerveteri, Cava della Pozzolana, Tomb LXXII, middle 8th century BCE: pseudomorph on a bronze belt.
48. Tarquinia, Le Rose, Tombs XLIV and LIX, early 7th century BCE (Figure 40): pseudomorphs on fibulae, which often have traces of tabby on one side and traces of 2/2 twill on the other (Buranelli 1983, 129, figs. 102, 51.7, 52.6).
49. Tarquinia, Tomb M9, 8th–7th century BCE: bronze biconical cinerary urn with mineralised traces of diamond twill (pers. com. Christophe Moulherat, 2005).
50. Tarquinia, Monterozzi, Tomb 5740, 4th–3rd century BCE: a bronze mirror with attached textile fragment that probably constituted a garment (Cavagnaro Vanoni 1977, 162, 166 no. 10).
51. Tarquinia, Tomba del Guerriero, 7th century BCE (Figure 41): a linen armour lining, now lost (Montelius 1895–1910, pl. 287–283; Ducati 1927, pl. 40, 132; Hencken 1968a, 201, fig. 180; Helbig, 1974, 257–258; Kilian 1977, 26–27 no. 5; Barber 1991, 194; Banck-Burgess 1999, 231).
52. Veio, Tomb 1, 8th century BCE: fragments of iron spits and other objects with textile traces (De Santis 1997, 128 nos. 41 and 42, fig. 17). One of the fragments appears to be an open tabby; all fragments have z-twisted yarn.

Figure 38 (above). Calcified textile traces on a clay urn, Tuscania-Tarquinia, 2nd–1st century BCE. (© University of Pennsylvania Museum).

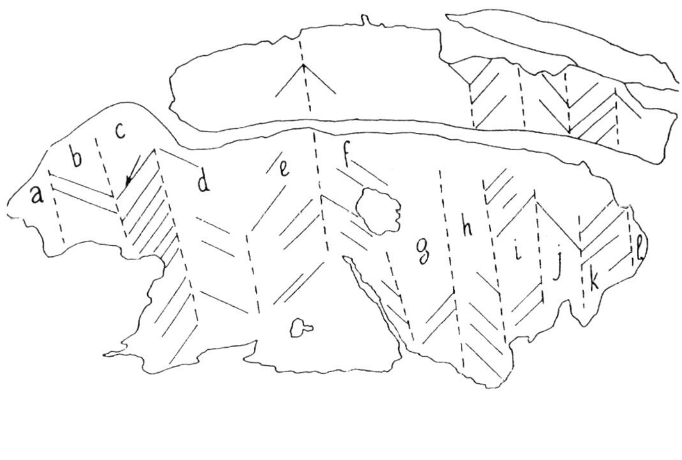

Figure 39 (left). Drawing of textile fragment 1 from the boat burial, Caolino, Sasso di Furbara, 9th century BCE (After Masurel 1982, fig. 2).

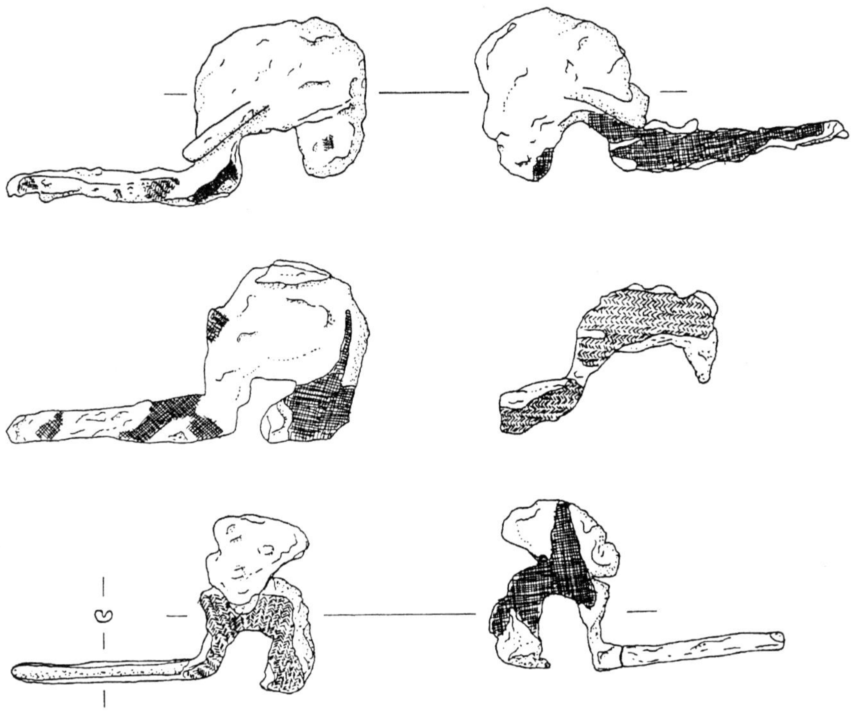

Figure 40. Textile traces on fibulae from Tombs LIX and XLIV, Tarquinia Le Rose, early 7th century BCE; top and bottom drawings show two sides of the same fibulae, preserving different textiles (After Buranelli 1983, fig. 102).

Figure 41 (below). Drawing of the linen armour lining from the Tomba del Guerriero in Tarquinia, 7th century BCE (After Strøm 1971, fig. 85).

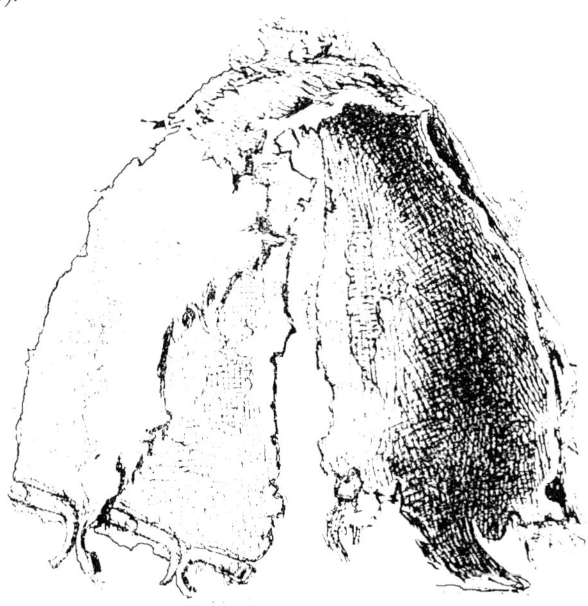

53. Veio?, 6th century BCE: a bronze bowl at the Newark Museum, USA, has a z/z tabby imprint on the bottom (Carroll 1973; Barber 1991, 194; Bender Jørgensen 1992, 106; Banck-Burgess 1999, 232).
54. Etruria or Agro Capenate, second half of 7th century BCE: a bronze *kardiophylax* preserving two linen balanced tabby fragments (Kilian 1977, 66; Jucker 1991, 90). One of the edges has well-preserved twisted fringes. Textile interpreted as a shroud.
55. Castel di Decima, 7th century BCE: dresses made of copper wire mesh decorated with amber, glass and faience beads (Bonfante 1975, 106 note 1; Bedini 1976, 287ff; Stage 1985, 74–75; Bonfante 1989, 1387; Barber 1991, 195; Banck-Burgess 1999, 230). Although none of the organic components of these dresses were preserved, the finds illustrate how sumptuous garments of the wealthy members of the society could be.
56. Tor de' Cenci, Tombs 3 and 8, late 8th century BCE: pseudomorphs on iron

fibulae (Bedini 1992, 231 fig. 13.2, 232 nos. 1, 2, 3 and fig. 14.2, 156 no. 3 and fig. 36.3).

57. Castel d'Asso, Tomb 30, 3rd–2nd century BCE: tabby pseudomorphs on at least three strygil fragments (Barbieri 1999, 171 no. 40, fig. 12).
58. Colle al Forno, tombs: pseudomorph on an iron *sauroter* from Tomb XXIII, dated to the middle 7th century BCE; pseudomorph on an iron fibula from Tomb XXII, dated to the 6th century BCE; pseudomorph on a small iron fibula from Tomb XX dated to the late 4th–early 3rd century BCE (Santoro 1983, 122 no. 66, 130 no. 11, 131 fig. 29.11, 139 no. 53). Objects were stolen and are now lost.
59. Borgo le Ferriere (ancient Satricum), Tomb 41, 5th century BCE: a pseudomorph on a metal object preserves two layers of fine z/z weft-faced tabby with 11 threads/cm in warp and 60–80 threads/cm in weft (Gnade 1992, 115 and fig. 35).
60. Palestrina, Bernardini Tomb, *c.* 650 BCE (Figure 42): a tabby pseudomorph on the iron part of the bronze and iron revetment (von Duhn 1924, 239, 502; Curtis 1919, 87 pl. 71.11; Steinberg 1966, 126–127 no. B20E; Cianciani and von Hase 1979, 58 no. 75 and pl. 51.5).
61. Palestrina, Barberini Tomb, *c.* 650 BCE: complex folded textile with two integrally-woven bands of tablet weaving, a wider incomplete band on the left (minimum 3 cm) and a narrower band (0.5 cm) in the centre (pers. com. Hero Granger-Taylor, 2003).
62. Alfedena, burials, 6th–5th century BCE (Figure 43): numerous pseudomorphs on iron fibulae and other objects, not always mentioned in object description but appearing in drawings (Bedini *et al.* 1975, 446 nos 1 and 2, 455 no. 3, 457 nos. 5 and 11, 465 no. 3, 470 no. 1; Parise Badoni and Ruggeri Giove 1980, *passim*).
63. Ascoli Piceno area, 7th century BCE (Figure 44): a twill pseudomorph, preserved on a fibula (Lucentini 1992, 485, 477 fig. 8).

South Italy

64. Cales, Tomb 1, 7th century BCE (Figure 45): two iron knives and a fibula with clear pseudomorphs of fine-woven textiles (Chiesa 1993, 73, 41 nos. 37, 39, 40; 43 no. 50; pls. X.37, XI.39, XI.40, XXXVIII).
65. San Marzano sul Sarno, Tomb 33, 7th–6th century BCE (Figure 46): a fibula with traces of a textile woven in twill (d'Agostino 1970, 609 fig. 10 no. 7).
66. Cuma, 8th century BCE: a bronze lebes from Tomb 2 preserving traces of fabric in which the bones probably had been wrapped (Pellegrini 1903, 226; d'Agostino 1977, 60; Stage 1985, 76). A similar small lebes found in Tomb 104 reportedly was covered with a fine coffee-coloured cloth (Gabrici 1913, 214–15; d'Agostino 1977, 69; Stage 1985, 76).
67. Capua, tombs, 7th century BCE: Tomb 25 contained an iron knife with visible folds of a tabby textile, while the slightly later tomb 1582, contained several iron fibulae with textile traces.
68. Pontecagnano, tombs, 8th–7th century BCE: several bronze objects, including two ax-heads, a lebes and a tripod, found in the 8th century BCE tomb 928 (Figure 47), have traces of textile adhering to

Figure 42. Textile remains from the Bernardini Tomb, Palestrina, c. *650 BCE (After Cianciani and von Hase 1979, pl. 51 no. 5).*

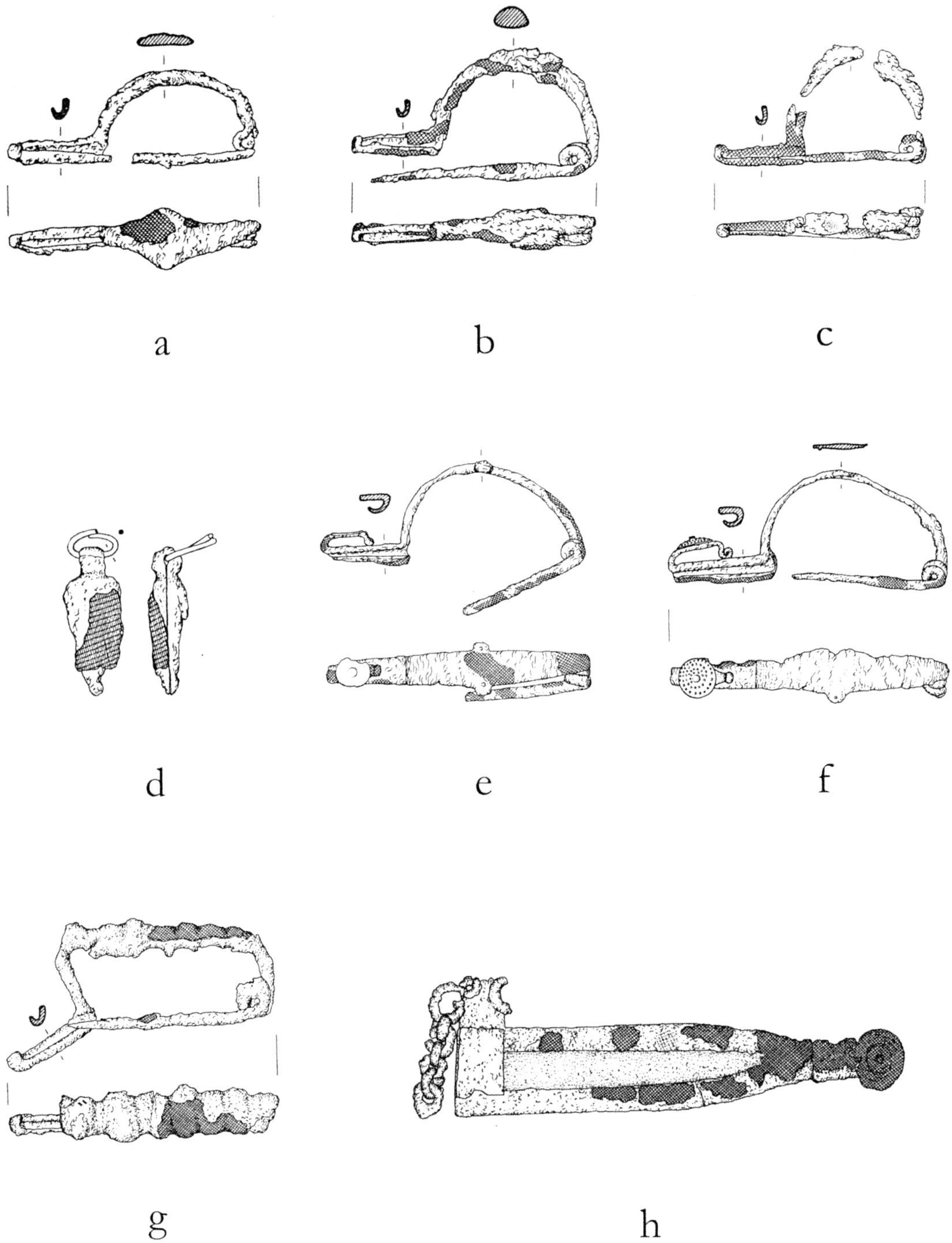

Figure 43. Iron objects with pseudomorphs from Alfedena, 7th–6th century BCE: a) Tomb 85; b–c) Tomba 78; d–f) Tomb 70 g) Tomb 104; h) Tomb 102 (After Parise Badoni and Ruggeri Giove 1980, pl. 29 no. 1; pl. 78 nos. 1, 7; pl. 20 nos. 1, 4, 6; pl. 39 no. 2; pl. 38 no. 2).

them (d'Agostino 1977, 14, 60; Stage 1985, 75). The lebes and the tripod most likely have different pieces of fabric since they were found in different sections of the tomb. Other pseudomorphs are preserved on three iron fibulae from Tomb 2145 (Figure 48a), and on a large iron fibula from Tomb 4870. Tomb 1697 had not only fine tabby traces on iron spits but also numerous small rings and bronze buttons that were probably attached to a sumptuous dress of the deceased. Tombs 661 (Figure 48b), 664, 677 yielded pseudomorphs on iron fibulae and weapons (Gastaldi 1998a, 74 no. 8 and pl. 94 no. 8, 75 no. 9, 84 no. 8 and pl. 99, 127 no. 6 and pl. 114). Another pseudomorph on iron fibula comes from

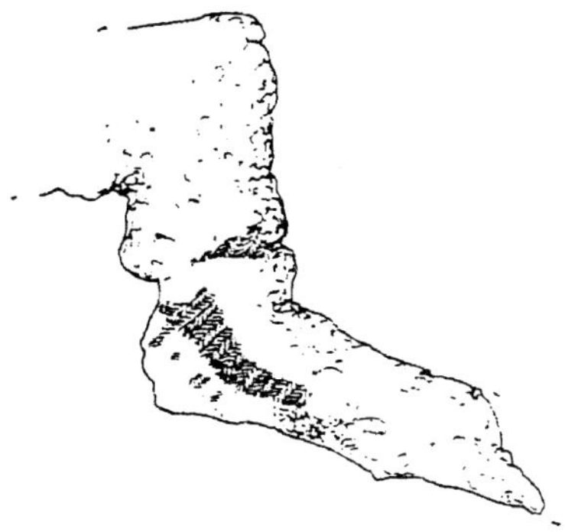

Figure 44. Minute textile remains on a fibula from Ascoli Piceno (After Lucentini 1992, 477 fig. 8).

Figure 45 (below). Textile traces on the knives from Tomb 1 at Cales, 7th century BCE (After Chiesa 1993, pl. X no. 37, pl. XI no. 39)

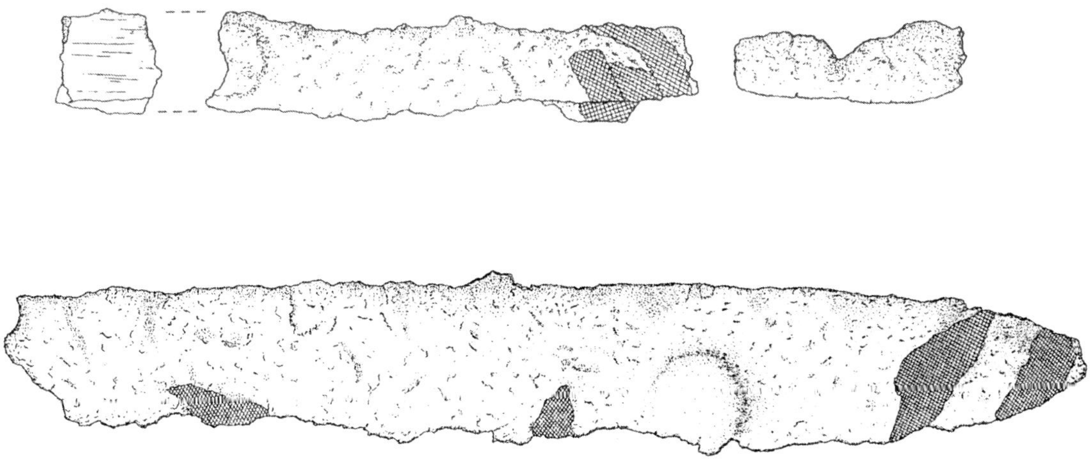

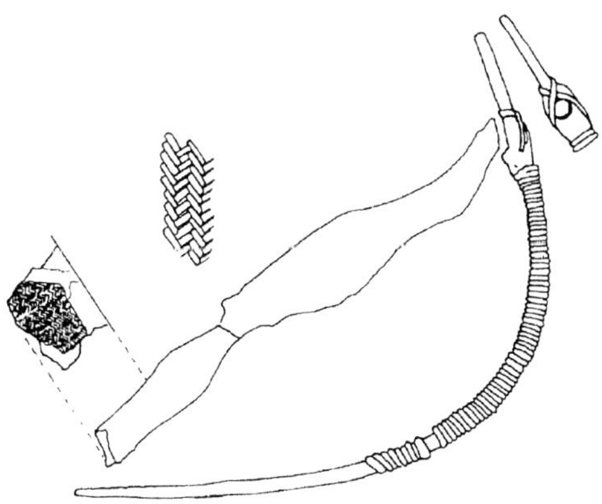

Figure 46. Textile remains on a fibula from Tomb 53, San Marzano sul Sarno, 7th–6th century BCE (After d'Agostino 1970, 609 fig. 10 no. 7).

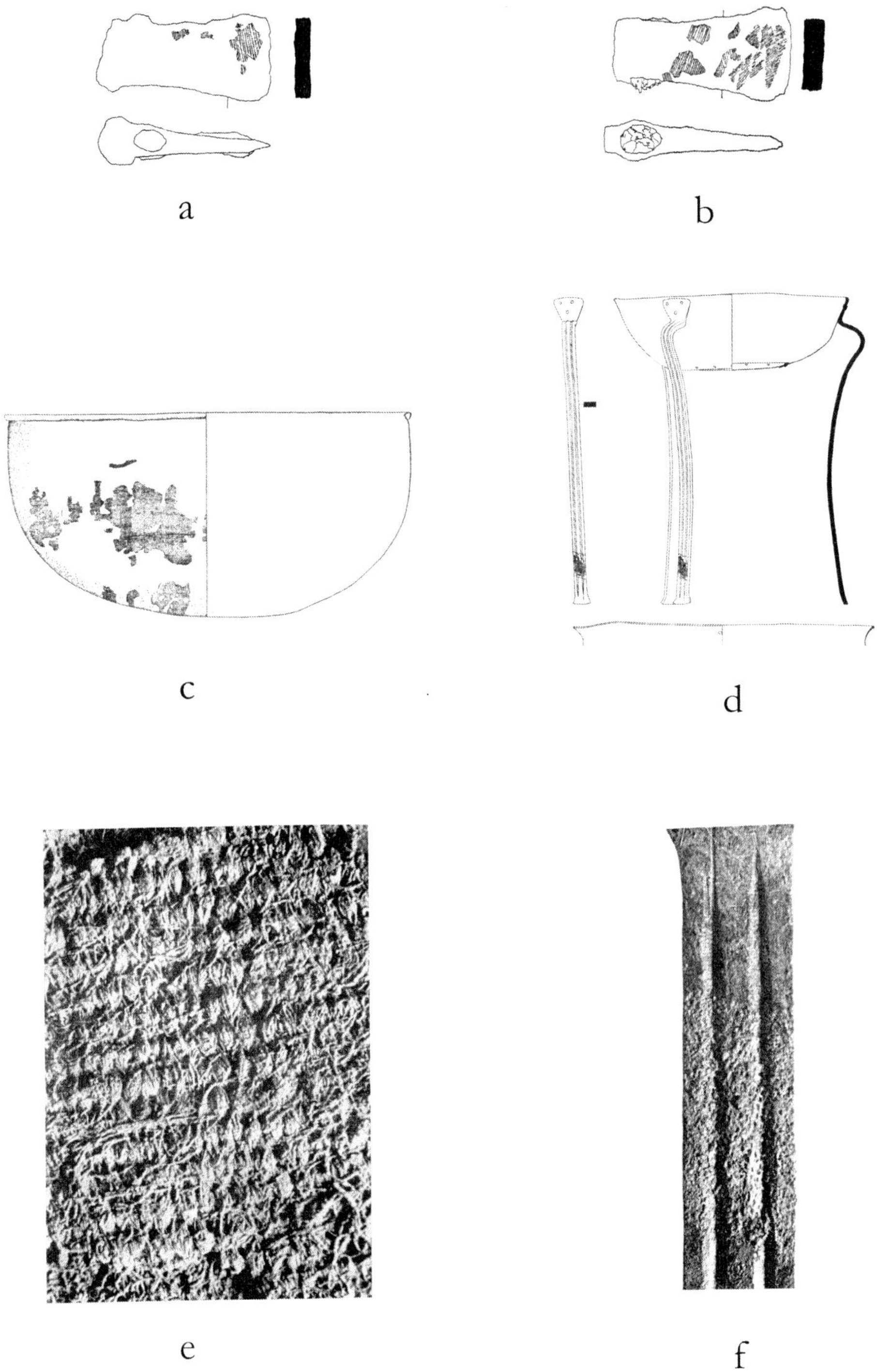

Figure 47. Textile remains on metal objects from Tomb 928, Pontecagnano, 8th century BCE: a–b) axe heads; c and e) lebes; d and f) tripod (After d'Agostino 1977, 92 fig. 14 R32–33, fig. 19 L62, R64).

Picentino Tomb 2145 (D'Agostino and Gastaldi 1988, 198 no. 12, pl. 162).

69. Eboli, Cataldo, Tomb 242, 6th century BCE: an iron spit with traces of tabby textile (pers. obs. Eboli Museum, 2003).

70. Sala Consilina, Iron Age: fibulae with textile traces, some of which are in a twill weave (pers. obs. Sala Consilina Museum, 2003).

71. Massa Lubrense, Tomb 36, 7th–6th century BCE: a plain-woven textile imprint on a fibula (pers. obs. Vico Equense Museum, 2002).

72. Vico Equense, Via Nicotera, 6th–5th century BCE: a bronze one-handled basin contained many textile fragments, all discoloured green from bronze corrosion products. Most pieces probably belong to a single textile of fine plain weave; however, several other fragments seem to have a more complex structure. An iron knife from Tomb 39 from the same necropolis had clear traces of a tightly wrapped textile; other sporadic material also had textile traces (Bonghi Jovino 1983, 18 no. 1, 41 no. 13, 36 nos. 1, 2, 5 and figs. 81.2.4, 82.2.4, 81.13, 38 nos. 27, 31, 36 and figs. 81.17 and 84.4, 39 nos. 40, 42, 43 and figs. 83.6, 83.8, 82.2).

73. Fratte, Tomb 134, 5th century BCE (Figure 49): a bronze olpe with a textile pseudomorph (Greco and Pontrandolfo 1990, 250 no. 8, fig. 421).

74. Paestum, Gando, burials, 4th century BCE: numerous pseudomorphs on iron objects (Pontrandolfo and Rouveret 1992, 369 no. 6).

75. San Salvatore, Tombs 8 and 12, 6th century BCE: pseudomorphs on iron fibulae (Lo Porto 1991, 18 no. 6).

76. Montescaglioso, Tombs 192 and 28, 4th century BCE: iron fibulae with pseudomorphs (*Siris – Polieion* 1986, pl. 57 no. 159058).

77. Difesa San Biaggio, Tomb 28, middle 8th century BCE: an iron fibula with pseudomorphs, one of which is tabby, *c.* 10 threads/cm; other piece looks finer with one much denser system.

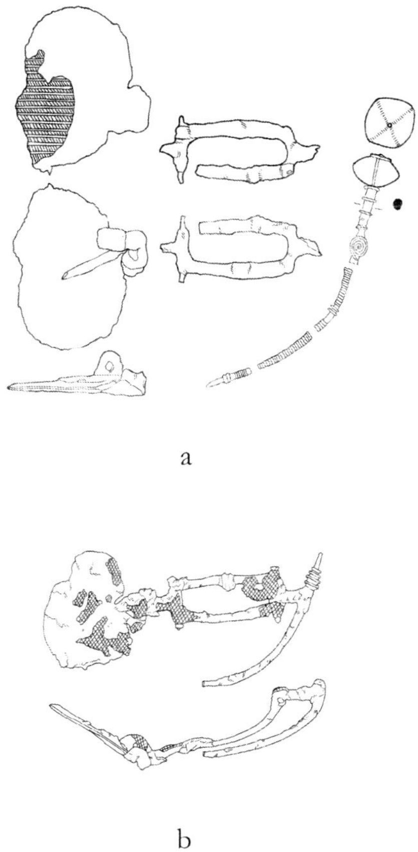

Figure 48. Textile remains on fibulae from Pontecagnano, 8th–7th century BCE: a) Tomb 2145; b) Tomb 661 (After Gastaldi 1998a, fig. 162 no. 12, pl. 94 no. 8).

78. Pisticci (Matera), Tomb 1, 8th century BCE: pseudomorph of fine textile on iron object (Lo Porto 1969, 144 no. 45).

79. Rutigliano, Purgatorio, Tomb 17, 5th–4th century BCE: textile traces found inside the sarcophagus (Lo Porto 1977, 739).

80. Bitonto, Tomb 12, 5th–4th century BCE: an iron fibula with well-visible traces of textile of plain weave in which one system is denser than the other.

81. Canosa, Tomba degli Ori, 3rd–2nd century BCE: numerous textile fragments in a very poor state of conservation but preserve traces of gold thread and are of extremely fine quality. At least three types of very fine tabby with one denser system. It has been suggested that the organic fibre was originally dyed purple (Duhn and Messerschmidt 1939, 315; De Juliis

Figure 49. Textile pseudomorph on a bronze olpe from Tomb 134, Fratte, 5th century BCE (After Greco and Pontrandolfo 1990, 250 no. 8, fig. 421).

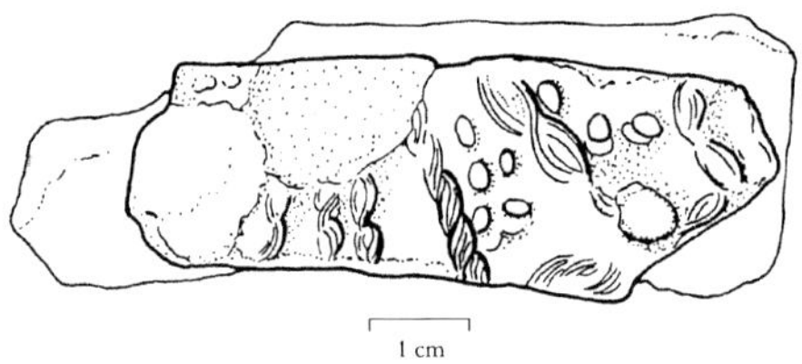

Figure 50. A rope pseudomorph on an iron object from Colle Madore, 6th–5th century BCE (After Vassallo 1999, 115 fig. 125).

1984, 329–330, 339 no. 277, 447–448 CXX nos. 24–26; Banck-Burgess 1999, 229).

82. Taranto, tombs, 2nd–1st century BCE: via Gorizia Tomb 14.V.1934 yielded gold thread without organic core, possibly border decoration; via Zara Tomb 16 had narrow bands with gold weft (De Juliis 1984, 330, 339–340 no. 278, 464 CXXX no. 5).
83. Torre Galli, Tombs 33, 34, 73, 114, 10th–9th century BCE: pseudomorphs on various iron objects, especially fibulae; tabbies and twills discernible (Pacciarelli 1999, 74, 151 no. 13, 151 nos. 7 and 10, 161 no. 5, 168 no. 6, 415 pl. 189.E–I).
84. Torano Castello (Cosenza), Tomb B1, 8th century BCE: a textile imprint on earth found next to an iron fibula, probably part of the funerary shroud that had bronze button attachments, 250 of which were recovered (De la Genière 1977, 406 fig. 20).

Sicily

85. Valle del Marcellino, Tomb 105, 9th century BCE: traces of a tabby textile have survived on a bronze fibula (pers. obs. Syracuse Museum, 2003).
86. Madonna del Piano, Tomb 26, 9th century BCE: bronze buttons for clothing decoration with Z-plied threads used for attachment still present; many 1–2 cm long fragments (pers. obs. Syracuse Museum, 2003).
87. Sabucina, Tombs 10 and 39, 7th–6th century BCE: iron knifes with tabby pseudomorphs (pers. obs. Caltanissetta Museum, 2003).
88. Colle Madore, 6th–5th century BCE (Figure 50): a small rope pseudomorph on an iron object is interesting in that its fibre has been identified as hemp (Vasallo 1999, 115 no. 33, figs. 125–126; Terranova and Lo Campo 1999).
89. Vassallaggi, tombs, 5th century BCE (Figure 51): South Necropolis Scavi 1961, pseudomorphs on iron knives from tombs: 10, 30, 31, 37, 52, 77, 81, 114, 144, 149, 150, 176, pseudomorphs on iron daggers from tombs 23, 45, 85, 167, 177. Objects were wrapped in bands of textile and usually deposited inside a crater or other vessel (Orlandini 1971, 28 no. 5, 52 no. 2, 59 no. 3, 63 no. 2, 72 no. 2, 89 no. 3, 118 no. 3, 127 no. 5, 147 no. 2, 176 no. 6, 180 no. 2 and no. 2, 201 no. 2, 82 no. 4, 135 no. 2, 197 no. 3, 203 no. 3, 217). Scavi 1956, Tombs 41A, 51A bis, 37A, 31A, 40A produced seudomorphs on iron knives (Pizzo 1999, 229 no. 14, 233 no. 6, 243 no. 9, 248 no. 5, 252 no. 3, 373).

Figure 51. Iron knife with textile traces, Tomb 41A, Vassallaggi, 5th century BCE (After Pizzo 1999, 230 fig. 15, 376 fig. 112 b).

90. Himera, burial, 5th century BCE: a fragment of a z/z tabby, found in an infant burial (Di Scalafani *et al.* 2005). Although quite small, the find is interesting for the fact that this simple balanced tabby is made of hemp.
91. Marsala, Tomb 40, late 4th century BCE: several plain-woven fragments were found in association with two mirrors, for which the textiles probably served as containers (Di Stefano 1993, pl. XVIII, XXVII).
92. Marsala, Punic shipwreck, 3rd century BCE: large quantities of string and rope made of esparto grass (Frost 1981). Strings of medium to thick plait; plaited cord; one rope piece is 30 m long and ending in an eye-splice.

Textiles without known or attributed provenance

93. A fragment of asbestos textile (Figure 52) supposedly from Etruscan area is currently in the collections of the British Museum (Granger-Taylor 1982, 23 note 2). The fragment is 13 by 14 cm, of relatively coarse tabby with possibly one selvedge preserved. The yarn is s-twisted with short fibres, about 1–2 mm in diameter. The thread count is about 5–6 for the warp and 4 for the weft.
94. Abruzzo?, 6th century BCE: a pseudomorph on a *kardiophylax* of a Picene type, now in the Louvre (Tomedi 2000, 46 no. 71).

The above list of Italian pre-Roman archaeological textiles, if tedious and certainly far from exhaustive, is intended to demonstrate the large number of textiles that have survived in Italy, whether in their original form or as pseudomorphs, despite the unfavourable archaeological conditions. It is now possible to examine the corpus of extant Italian archaeological textiles in relation to their raw material, structure and use.

Fibres

Fibre is a basic unit of raw material having suitable length, pliability and strength for conversion into yarns and fabrics. In the first half of the 1st millennium BCE, two basic fibre groups, divided on the basis of their origin and molecular composition into cellulose-based plant and protein-based animal fibres, were used to make textiles.

Vegetal fibres are derived from plants and include linen, hemp, nettle, esparto and bast fibres from trees such as linden,

Figure 52. A fragment of asbestos textile from Etruscan area (Author's photo, reproduced with permission of the British Museum).

Figure 53. Mineral amphibole with asbestos fibres, the Natural History Museum, London (http://en.wikipedia.org/wiki/Image:Asbestos_with_muscovite.jpg).

oak and willow. Tree-bast was more important in prehistoric times, but it occasionally turns up during the Iron Age as well (Rast-Eicher 2003, 48–50). Cotton arrived in Europe only during the Roman period (Barber 1991, 32).[7]

The major animal fibre of antiquity was sheep's wool, with occasional use of goat hair and other, more unusual fibres, such as horse (Wincott Heckett 1998) or Cashmere goat hair (Ryder 1993b; Good 1999, 64–65). The other important animal fibre, silk, did not come into common use in Italy until Roman times and hence, will not be considered here.[8] An unusual animal fibre, byssus or sea-silk, derived from the secretion of the mollusk *Pinna nobilis* L., was also spun in antiquity, although there is no direct evidence for its use before the late Roman period (Maeder 2002; Maeder, Hänggi and Wunderlin 2004).[9]

One last, unusual, textile fibre should be mentioned, asbestos (Pionati Shams 1987, 3–11). It is derived from a mineral amphibole (Figure 53) and has a unique quality of withstanding extremely high temperatures, a trait that was noticed and used in antiquity. Thus, Pliny the Elder (36.19–21) calls it live or incombustible linen and praises its usefulness for making funeral shrouds, napkins, lamp wicks, and fishing nets. The British Museum has an example of presumably Etruscan asbestos (Cat. No. 93).

The methods used in fibre identification vary depending on the state of textile's preservation and often a combination of methods is used. The biological source of fibres may be identified by chemical tests, such as zinc chloride iodide (cf. Stauffer 2005). Solubility measurements can distinguish cellulose fibres from protein-based ones but will not help to identify a specific species (Lambert 1997, 143). The new methods of amino acid composition (Good 1999, 87–88) and DNA (Hamlyn *et al.* 2000) analyses are being developed for protein-based fibres.

Optical and scanning microscopy allow identification of fibres by comparison with reference standards for the cross section and longitude of the fibre. Most frequently, a basic light microscope is used. Animal fibres, if well preserved, can be recognised by the scales or *cuticulae* on their surface; the shape, position and sequence of these scales present on the fibre surface can then be used to identify an animal species or its variety (Appleyard 1960). The identification of vegetal fibres is often more complicated since the distinction between flax, nettle, hemp and other plant species is difficult to make even in the laboratory, particularly when they are degraded. Cellular structure of fibres often needs to be ascertained, and this can only be accomplished with the light microscope. The similarity between plant materials was noted in ancient times as well: in the 5th century BCE, Herodotos (4.74) mentions how much hemp resembles linen and that garments made of it are so similar to linen that the two are difficult to distinguish one from another.

Flax

Origins

Since prehistoric times, linen has been one of the most widely used textile fibres (Helbæk 1959; Forbes 1956, 27–43; Renfrew 1973, 120ff; Alfaro Giner 1984, 49–53; Barber 1991, 11–15; Rottoli 2003). It is made of flax *(Linum usitatissimum L.)* (Figure 54). In fact, the earliest known textiles are made of flax, as illustrated by the 7th millennium BCE finds at Nahal Hemar (Bar-Yosef 1985; Schick 1988) and Jarmo in Israel (Good 1998, 657, but see note 2) and textiles from Çatal Hüyük in Anatolia, radiocarbon dated to *c.* 6000 BCE (Burnham 1965). There is also a pseudomorph from Çayönü composed of bast fibre and dated *c.* 7000 BCE (Good 1998, 657). The earliest evidence for cultivation of flax comes in the form of linseed from 8th millennium BCE Tell Aswad in Syria (Miller 1991, 142). In Central Europe, flax was cultivated by the second half of the 7th millennium BCE (Rast-Eicher 2005, 119).

Figure 54. Flax plant (Drawing by the author).

Due to the scarcity of archaeological evidence for linen use and production in the 1st millennium BCE Italy, for a long time the commonly held view was that linen had to be imported to the Apennine peninsula from other flax-producing regions of the Mediterranean (*e.g.* Bonfante 1975, 11). One of these was Asia Minor, especially Cilicia with Tarsus as main centre (Schmidt-Colinet, Stauffer and Al-As'ad 2000). Another was Egypt, where flax cultivation is recorded since the 5th millennium BCE (Barber 1991, 45ff). Egypt often has been assumed as the area of origin for linen found in Italy. The number and wide chronological and geographical distribution of surviving linen fragments, and especially Neolithic and Bronze Age evidence from North Italy, however, seem to indicate indigenous production. The suggestion was made already by Carroll (1973) based on the analysis of a mineralised linen fragment adhering to a 6th century BCE bronze bowl from Veio, now in the Newark Museum. She argued that the piece was an Etruscan product since both sets of thread were spun in a z direction, typical of European rather than Egyptian technology. The few other Italian textile fragments securely identified as linen are all z-twisted. In fact, archaeological, historical, and artistic evidence suggest that linen was used and produced in Italy long before the Roman period.

Flax appeared in Italy during the Neolithic period and, in some areas, may have been the dominant textile fibre (Rast-Eicher 2005). Yet, while there are many scholarly works on the economy of wool production in Italy, linen is barely mentioned in archaeological and historical literature, if at all.[10] The notable exception is a recent study of Roman textile production in Italy by Vicari (2001), who examines historic and epigraphic evidence for both wool and linen.

Archaeobotanical evidence

The cultivation of flax is best demonstrated by the archaeobotanical evidence, that is, the recovery of flax seeds. Unfortunately, one of the reasons for our lack of understanding of the role of flax in ancient Italian agriculture and economy has been the frequent lack of proper archaeological methodology and almost complete absence of any palaeobotanical data analysis from past excavations of Italian archaeological sites. Flax seeds are quite small (2.5–5 by 1.3 mm), and, even when preserved, they are almost impossible to isolate unless a technique like flotation is used (Renfrew 1973, 122–123). Poor preservation obviously aggravates the situation (McCorriston 1997, 519).[11] When linseed is found, it is most likely preserved as a by-product of oil or food processing, but in the archaeological literature, often only the presence of seeds is indicated with neither their quantity nor size given, making identification of intended cultivation product very difficult.[12] Still, there is sufficient evidence for early cultivation of flax in Italy.

The oldest find of flax in Italy, consisting of one seed only, comes from Sammardenchia (Udine), the site dated from the middle of the 7th to the middle of the 6th millennium BCE (Rottoli 1999; Rottoli 2003, 68). Seeds and a few seed-capsules have been found in the submerged 6th millennium BCE settlement of La Marmotta in Lake Bracciano (Rottoli 1993, 310). Numerous finds of linseed come from Late Neolithic sites Lagozza, Isolino di Varese and Settefonti (Renfrew 1973, 122; Rottoli 2003, 68). Late Neolithic levels at Palù di Livenza produced hundreds of flax seeds and capsule fragments, as well as seeds of the herbaceous plant *Silene linicola*, which is a characteristic weed species of flax crops (Rottoli 2003, 68). Later still, flax is attested among the crops of Bronze Age North Italy, the region that would be praised by Pliny for producing some of the best linens of the Roman world. The sites include Fiavé and Volano San Rocco in Trentino, Lazise and Valeggio sul Mincio in the lake Garda area, and Castellaro del Vhò in the province of Cremona (De Marinis 1988b, 22). In the 1st millennium BCE, *Linum usitatissimum L.* is attested in the 8th century BCE site Palse, the 5th century BCE settlements Casale di Rivalta and Montereale Valcellina, and at the slightly later site of Monte Bibele, all located in the Po valley – again the region famous for its linens in later times (*Protostoria tra Sile e Tagliamento* 1996, 465; Bertani

1995, 42). In the south of the peninsula the evidence is limited to only one site, Pratola Serra (Avellino), which yielded linseed (Rottoli 2003, 68).

Textile evidence

That linseed was not only related to food/oil production is substantiated by linen textile fragments recovered from various archaeological sites in Italy. The oldest linen textile in Italy, and indeed in Europe, comes from the submerged Neolithic settlement of La Marmotta in Lake Bracciano, and has been dated by radiocarbon to 5480–5260 BCE (Cat. No. 1). A large number of linen textiles have been found in the *terremare* in North Italy, dated to the 3rd and early 2nd millennia BCE (Rast-Eicher 1997). Well-known are the plain- and pattern-woven linen cloths from the lake-dwelling Molina di Ledro (Cat. No. 2), where balls of flax thread were also preserved (Figure 55), proving not just textile use but also textile production.

Unlike the prehistoric settlement material, the 1st millennium BCE finds come predominantly from funerary contexts. For example, fragments, identified as linen, were found in the 7th century BCE Tombs A and G at Etruscan Casale Marittimo (Cat. No. 22). Some of these were wrapped around weapons or contained cremated bones. In another case, the Tomba del Duce at Vetulonia, dated to the second half of the 8th century BCE, a plain-woven linen cloth was found inside a larnax that also contained cremated bones (Cat. No. 25). In fact, textile fragments described as linen are frequently mentioned in excavation reports as remains of the wrapping material for various objects. Unfortunately, very few of these have been analysed scientifically and the identification is often based on general observation or assumption.

Of undoubted importance is the famous *liber linteus Zagrabiensis*, dated to the 3rd–2nd century BCE (Cat. No. 36). It is the longest surviving Etruscan inscription and, as the only existing fragment of a linen book, it is one of the most important pieces for the textile history of Italy, reflecting the significance of linen as a ritual material. Production of textiles for the sacred books must have been strictly controlled and hence most likely local. Furthermore, given the vagaries of sea transport at the time, it would hardy have been prudent to rely on long-distance imports.

Iconographic and literary evidence

In addition to the archaeological evidence, a lot of information confirming the wide-spread use of linen in ancient Italy can be obtained from other sources. Indirect evidence for the existence of linen garments comes from the later

Figure 55. Ball of linen yarn from Molina di Ledro, Middle Bronze Age. (After Bazzanella et al. *2003, 173).*

artistic representations. The 'transparent' white dress with fine folds rendered in the 6th–5th century BCE Etruscan tomb paintings of Tarquinia is assumed to be an artistic convention for thin linen cloth (Bonfante 1975, 12). Wool, on the other hand, is usually shown with checked patterns.

Although later than the period of our interest, written sources describe events and practices that certainly harked back to earlier times. It is suggestive already that linen is mentioned seldom in texts, indicating that its use was taken for granted, except in unusual cases. Thus, Livy (4.20) recounts the story of the linen cuirass of Lars Tolumnius, king of Veio. Such linen corslets are mentioned by Homer (*Il.* 2.528, 830), Herodotos (2.182, 3.47), Xenophon (*Anab.* 4.7.15, *Cyrop.* 6.4.2), Aeneas Tacitus (*Tact.* 29.4), Livy (4.20) and Pausanias (1.3), and seem to have been used since proto-historic times.[13] Not only military attire was of linen; sacred garments required the 'purity and whiteness' of the material, as did the afore-mentioned Etruscan *libri lintei* containing lists of rituals (Apul. *Met.* 2.28, 11.10, 11.24; Livy 4.20). Philostratus (*Apol.* 8.7.5) in particular discusses the 'purity' of linen, which is due to its non-animal nature, making it an especially suitable material to wear when lecturing, praying, sacrificing, and to sleep under.

Properties and use

Several qualities made linen also a utilitarian fabric *par excellence*: it did not lint, absorbed water well and dried quickly and hence was especially suitable for tablecloths, towels and napkins (Vicari 2001, 2–3). Linen was also utilised as a filtering or packaging material for food preparation and storage. Columella recommends using linen bags for straining in the preparation of myrtle wine and other beverages (*Rust.* 12.38.6–7).

Linen certainly would have been a very suitable material for sail and tent making owing to its strength, low elasticity and durability. During the last centuries BCE, Tarquinia was especially renowned for raising flax and making sails. Thus, in 205 BCE, its citizens were required to manufacture such sails for the ships of Publius Scipio (Livy 28.45). Pliny the Elder (*NH* 19.22) implies that all sails in his time were linen. This is confirmed by Julius Caesar (*B. Gal.* 3.13) who notes that local tribes on the North Sea use hides and leather for their sails "either because they do not have linen and are ignorant of its use, or – and it seems more likely – they believe that such sails could not hold up against storm". In later periods wool was used for sails in Scandinavia (Möller-Wiering 2002).

The extravagant awnings for theatres and fora described by Pliny (*NH* 19.23–25) also seem to have been made of linen sailcloth,[14] as most likely were military tents,[15] cart covers, *vela* hung between temples' columns and *centones*, special textiles used for extinguishing fires. Linen was also suitable for paintings: the case in particular is the colossal portrait of Nero painted on a piece 40 meters high (Plin. *NH* 35.51). Finally, the medical usage of linen bandages is described by Galen in *De fasciis*, while Pliny the Elder (*NH* 19.21) states that the nap of linen cloths, especially that which comes from ships' sails, is used as a medicine. Most of these uses certainly predated the Roman period in Italy.

Pliny the Elder is the author who has left us by far the largest amount of written information about linen in antiquity. In the *Naturalis Historia*, he described in detail not only the various uses of linen but also how and where it was produced in his day. Thus, he claims that the best linen of his day, *i.e.* the 1st century CE, came from Saetabis in Spain (Plin. *NH*

19.9; *cf.* Sil. *Pun.* 3.377; Grat. Fal. *Cyn.* 41).[16] The second and third places in quality, however, are given to the linens of North Italy (Plin. *NH* 19.9).

Areas of cultivation

There appear to have been at least two main areas of flax cultivation in the Apennine peninsula. As mentioned before, Pliny the Elder (*NH* 19.9) singled out North Italy as the most important region: "the Alia district of Italy between the Po and the Ticino [produced linen which was] the third best of Europe, [...] as the second prize is won by the linens of Retovium near the Alia district and Faventia on the Aemilium road". Archaeobotanical and epigraphic evidence (Vicari 2001; Gleba 2004b), though scarce, also point to Lombardy and the Po valley as the main linen-producing regions not only in Roman times, but long before as well.

The second important region of linen production was located in Latium and Campania. As mentioned earlier, the south also seems to have had a flax-growing tradition going back to the prehistoric period, albeit the evidence is scarcer than in North Italy. By the middle of the 1st millennium BCE, the area of Tarquinia was famous for its flax, and linen cloth was considered characteristic dress for Faliscans during the 3rd century BCE (Sil. *Pun.* 4. 223 and Grat. Fal. *Cyn.* 40). Flax cultivated in the nearby Paelignian territory was praised for its whiteness (Plin. *NH* 19.13). In Campania, Cuma was renowned specifically for the production of nets for fishing and hunting (Plin. *NH* 19.10), while Capua made the best ropes (Cato *Agr.* 135). Also, according to Livy, in the war of 308 BCE, Samnite troops wore white linen tunics (9.40) and slept under linen tents (10.83), indicating that a large supply of the material was available. Much later, 10th century CE Arab accounts speak of Naples as the city of flax (Horden and Purcell 2000, 362). Thus, it seems that, in the south, the tradition of linen production continued well into the post-Roman period.

Cultivation techniques

Because of different climatic conditions, the northern and southern areas employed different cultivation technologies. When discussing flax growing in the north, Pliny (*NH* 19.6), states that it is sown in spring and plucked in summer. Other Roman authors, describing the cultivation techniques practiced in the south, around Rome, recommend sowing flax in late fall and winter (Virg. *G.* 1.208–14; Columella *Rust.* 2.17; Varro *Rust.* 28.XVI.16). The northern tradition of sowing flax in spring was similar to Central European practice and most likely was the original technology inherited when flax first was introduced into Italy during the Neolithic period.[17] In the south, where water is more abundant in winter and temperatures rise more quickly in spring, flax was planted in late fall, which was either an adaptation to climate or an influence from the east, brought by Greeks or Phoenicians during their colonisation of South Italy and Sicily.

Consumption

In the Roman period, the North Italian linens seem to have been destined for consumption by the wealthiest households. The thread strength, uniformity and luster, as well as cloth whiteness, fineness and nap presence are mentioned by Pliny as some of the qualities contributing to the price of North Italian linens (*NH* 19.9). He does not give the specific prices, but two centuries later, Diocletian's Price Edict (*Edict. Diocl.* 26–28) includes nine qualities of linens and the prices range from 70 to 1200 *denarii* per pound. It is the raw material itself that was the main factor in the price of the linen cloth as labour costs of the weavers were meagre: 20–40 *denarii per diem* (Jongman 2000, 191). It is more difficult to ascertain the value placed on linen in the Iron Age. The following considerations, however, would apply to earlier periods as well, and to some extent may explain why, already in the Bronze Age, linen was superseded by wool as the major textile fibre.

Several factors would have contributed to the cost of linen. First of all, flax requires relatively large areas of soil, and soil of good quality, as well as an ample supply of water. On the Italian peninsula, this certainly limits the cultivation to the Po valley, the rich soils of Campania and a few smaller areas on the coast of Tuscany and Latium. Flax also had to be rotated with other crops, as it was believed to deplete the soil.[18] For this reason Columella (*Rust.* 2.10.17) suggests even that flax should not be cultivated unless it yields a good crop and brings a good price in the area where it is grown.

Second, as with any crop, cultivation itself involves sowing the seeds, weeding and harvesting, all of which required time, labour and a certain amount of skill. In fact, Columella (*Rust.* 2.12.5) informs us that flax was one of the most labour-intensive crops to grow. Eleven work days were required for the cultivation of one *iugerum* (an area about 36 by 72 m) of flax: four days ploughing and planting, three days harrowing, one day hoeing and three days harvesting; which is less only than the amount of time needed for cultivation of sesame (fifteen days). In addition, a large amount of linseed was required to grow a good crop – according to Columella, 8–10 *modii* (about 80 liters) per *iugerum*.

Flax cultivation thus required considerable organisation and planning, especially for large quantities, as in the case of sail production or army supplies. Furthermore, preparation of the flax fibres for spinning is very time and labour consuming (McCorriston 1997). Only after a long and laborious process can the fibre be spun and the final product, the linen cloth, woven. Ultimately, the quality and cost of linen was affected by the climatic conditions in the region and the annual weather changes.

Hemp

Hemp, a fibre from *Cannabis sativa*, a tall annual plant with a coarse, strong stem (Figure 56), shares many of the characteristics of flax (Pionati Shams 1987, 25–37; Körber-Grohne 1991, 95–96; Barber 1991, 15ff). It is more robust, however, and was used extensively for ropes, nets, sails and other hardwearing uses (Plin. *NH* 19.54). The plant probably originated in Central Asia and the earliest finds of seeds are noted in Germany at a site dated to 5500–4500 BCE, and in Ukraine, where seed imprints were found in Tripolian contexts, dated to 5300–3500 BCE (Barber 1991, 16–17). The seeds however, do not prove that the plant was used for fibre, one of its early uses being associated with its hallucinogenic properties. While widely used for textiles by the inhabitants of Eurasian steppe zone since the Neolithic (Shishlina, Orfinskaya and Golikov 2002), the fibre did not become popular in the Mediterranean textile production until the Iron Age. The difficulty of distinguishing it from flax without special analytical techniques, however, makes the data on hemp scarce and unreliable. The situation is similar in the case of nettle (*Urtica diotica* and other species), another plant fibre, which however seems to have been more typical for northern Europe (Køie 1943; Barber 1991, 19); no sample has been identified in Italy so far.

The earliest evidence for hemp in Italy consists of a textile identified on a metal object found in an Early Bronze Age burial at Gricignano d'Aversa, Caserta (Cat. No. 9). The other securely identified Italian hemp finds include a 6th–5th century BCE rope pseudomorph from Colle Madore in Sicily (Cat. No. 88); various rope remains from the pre-Roman ships of Pisa (Cat. No. 21); and a small textile fragment from a 5th century BCE infant burial in Himera, Sicily (Cat. No. 90).

Figure 56. Hemp plant (Drawing by the author).

Tree bast and other plant fibres

Tree fibres were common in the prehistoric period (Körber-Grohne 1991, 98–101) and Antoinette Rast-Eicher (2005, 119) even coined the term 'bast-culture' for the Neolithic Europe. A bunch of carbonised linden fibres wound in a ring is documented at Early Bronze Age Molina di Ledro (Bazzanella *et al.* 2003, 174). Tree-bast was used in the textiles from the Early–Middle Bronze Age Valle delle Paiole (Bazzanella *et al.* 2003, 198). Various grasses were also in use. The most notorious find is the mantle of the Iceman Ötzi, made of twined grass (Winiger 1995).

In fact, any available material was used and it is not surprising that items made of Spanish broom (*Spartium junceum*), albardine (*Lygeum spartum*), various species of *Hybiscus*, *Calotropis*, *Musa*, *Asclepias* and other arboreal plants have been recovered from the ships of Pisa (Cat. No. 21) and in the Vesuvius area (D'Orazio 2000). A lot of these fibres were used for making a wide variety of ropes and cordage needed for transport, construction and sailing.

Esparto

Last but not least, esparto grass (*Stipa tenacissima* L.) should be mentioned (Figure 57) (Barber 1991, 33). It was primarily used for ropes in ancient times, especially those used for the rigging of ships. Pliny (*NH* 19.7–9) refers to the good quality of Spanish esparto (Alfaro Giner 1984, 59ff), but notes that the plant is used extensively in North Africa as well. Livy (22.60.6) mentions vast quantities of esparto grass collected by Hasdrubal, the founder of Carthage, to make ropes for his navy, while Athenaeus (*Deipn.* 206f) includes Spanish esparto among the materials used for the construction of the *Syracusia*, the great freighter built by Hieron II of Syracuse (Turfa and Steinmayer Jr. 1999a). So far, the only examples of esparto identified in Italy are known from shipwrecks. A variety of esparto ropes and cordage come from a 3rd century BCE Punic shipwreck excavated at Marsala (Lilibaeum) in western Sicily (Cat. No. 92). Other finds are known from the Roman ships found in the lake of Nemi (Ucelli 1950, 268), at Comacchio (Rottoli 2005, 70) and in river Stella (Rottoli 2005, 70). It is reasonable to assume that esparto ropes and cordage from these sites are of Spanish or African origin.

Figure 57. Esparto grass in Spain (Courtesy of Carmen Alfaro Giner).

Wool

Origins

The second major type of textile fibre in antiquity was wool (Forbes 1956, 2–27; Barber 1991, 20–30; Good 1999, 47ff). Wool is a hair fibre composed of the protein keratin, obtained predominantly from the coat of sheep, and contains three parts: kemp, coarser hair and the wool itself. Early varieties of sheep had coats that contained more hair and kemp than wool. Woolly sheep are believed to have developed by the middle of the 4th millennium BCE (Barber 1991, 22ff; Ryder 1992; Ryder 1993a; Good 1999, 42–47). The earliest textile remains made of sheep's wool from Shahr-I Sokhta, Eastern Iran, are dated to the middle of the 3rd millennium BCE (Good 1999).[19] Recently, however, wool was identified in textiles of Majkop Kulture in the North Caucasus, dated 3700–3200 BCE (Shishlina, Orfinskaya and Golikov 2003).

Properties and use

Wool has properties that differ markedly from plant materials (Wild 1970, 4–10; Barber 1991, 20ff). Its fibres have a scaly surface, which accounts for wool's ability to felt. They are also kinky, producing air pockets between the kinks and thus making wool an excellent insulating material. Wool fibres stretch much more easily than bast fibres and hence are very elastic. Wool is a much more suitable fibre for weaving in twill technique (Rast Eicher 2005, 128). Unlike plant fibres, which are generally light and difficult to colour, wool comes in a variety of natural shades and can be dyed a multitude of bright colours. All of these properties make wool one of the most useful types of fibre available and it is not surprising that in ancient times it was one of the main textile materials, particularly for clothing.

Archaeozoological evidence

As early as the beginning of the 5th millennium BCE, when the first evidence for farming in Italy appears, sheep are among the most important domestic animals (Barker 1975, 46). In the 3rd millennium BCE, the mortality data of faunal samples indicate the increasing importance of animal secondary products (Barker 1976, 301; Barker and Rasmussen 1998, 50; Rast-Eicher 2005, 127). The evidence consists of slaughter patterns, which indicate predominance in the flock of adult animals. Furthermore, according to this evidence, an important change in animal practices took place in Central Italy during the Middle and Late Bronze Age, with a clear increase in sheep and goat numbers (De Grossi Mazzorin 2004, 38). By the Early Iron Age, generalised stock-keeping had been replaced by a more intensive system, with stocks consisting mostly of sheep and goats, and with a distinct emphasis on wool production (De Grossi Mazzorin 2004, 39). A wider range of domestic resources and more specialised butchery systems were developed (Barker 1988, 782; Barker and Rasmussen 1998, 185).

Various sites began to specialize. Thus, in Central Italy, Cerveteri seems to have been producing wool, while in Populonia sheep were kept to supply the settlement with meat (Barker and Rasmussen 1998, 186). Smaller sites and farms had a less specialised sheep husbandry producing all needed subsistence products like meat, milk and wool.

Although the evidence for sheep husbandry and wool production is much more abundant than remains of flax cultivation, well-collected faunal assemblages are still rare on Italic sites. Not only identification of species but also slaughter patterns with age/sex distributions are needed in order to draw conclusions about wool flocks. This information, however, is often difficult to ascertain due to limitations of archaeological samples (cf. Barber 1991, 28). The few samples available come from major and minor centres rather than from farms and often reflect patterns of urban consumption (Barker 1988, 782).

Sheep husbandry

Traditional animal husbandry does not require complex technology and today sheep are still a common sight throughout Italy (Figure 58). Since grazing of the same small area is insufficient to support the flock all year round, the sheep are moved to different pastures. Animal transhumance has been a fundamental element of Italian economy since sheep husbandry began and the minimal change over time reflects the success of this original strategy (Barker 1976, 6, 304; Frayn 1984; MacKinnon 2004). Large scale transhumance was generally practiced over long distances, while small scale transhumance was restricted to smaller areas. It was the large scale transhumance that was "likely geared towards the production of wool with more animals reared to adulthood" (MacKinnon 2004, 56). Numerous transhumance routes existed from ancient times until quite recently,[20] especially in the south (MacKinnon 2004) and in the north of Italy (Bonetto 2004). Mediterranean islands may have been used for pastoral activities as well (Alfaro Giner 1998). Thus, textiles were not necessarily produced where the raw material was procured and wool and even sheep often travelled large distances (Bonetto and Ghiotto 2005, 53).

The advantage of sheep herding over flax cultivation is that wool, as a fibre, does not require prime agricultural land; in fact, non-agricultural areas can be utilised as pastures. Wool does not

Figure 58. Sheep on via Cassia at Casal Risieri, near Viterbo (Courtesy of David Morton).

require ploughing, sowing, weeding, or harvesting, and it is less affected by weather conditions. Fewer herders are needed to tend the flock in order to produce a greater volume of fibre than would be generated by the same people cultivating flax. The agricultural labour, which would have been necessary for flax growing, could therefore be diverted to other tasks. The preparation of wool fibres for spinning is a technology less time-consuming than the numerous stages of flax post-harvest processes (Barber 1991, 20–21; Good 1999, 79).

Sheep husbandry also has broader significance for the small household. A few animals can provide enough wool for household consumption (Barker 1976, 301). Although large-scale wool manufacture was probably controlled by the elite, small-scale production must often have been carried out for the needs of each individual household, at least in non-urban setting.

Literary evidence

Pliny the Elder (*NH* 8, 187–199) gives a very detailed description of sheep husbandry in his time. According to him (*NH* 8, 190), the most valued wool came from Apulia, although it was of short hair and was best suited for cloaks. Tarentine and Canosine wools in particular were regarded as some of the best and were exported raw (Plin. *NH* 8, 190–191; Morel 1978, 108). The second best was the wool of a Greek breed, also called Italian, and the third place was held by Milesian sheep. But the fleece most valued for whiteness and fineness came from the Po area. Pliny (*NH* 8, 190–191) informs us about this type of wool that: "none has hitherto gone beyond the price of 100 sesterces a pound". Confirmatory information comes from Columella (*Rust.* VII.2) who states that, Gallic races of sheep, especially that of Altino, were considered superior to those of Calabria, Apulia, Tarentum and Miletos. Martial (*Epigrams* XIV.155) and Tertullian (*De pallio* 3.6) also mention the high quality of wool from Altino (Bonetto e Ghiotto 2005, 52–53).

Sheep breeds

It is clear from Pliny's descriptions that specific breeds of sheep, differing in the colour, length and thickness of their fleece, were established in different areas by the 1st century CE (Plin. *NH* 8. 189–193). While little work has been done to date in differentiating ancient Italian sheep breeds archaeologically, some possibilities have been noted. Thus, analysis of bones from Picene sites Ancona, Osimo and Moie di Polenza has shown presence of one goat variety and at least two breeds of sheep, one of which is distinguished from the other by the absence of horns in female animals (Wilkens 1992, 42).

Textile evidence

The earliest actual find of wool documented in Italy comes from Molina di Ledro and is dated to the Early Bronze Age: it consists of some sewing threads used in a linen belt (Bazzanella *et al.* 2003, 162). Textiles from Valle delle Paiole (Cat. No. 4) constitute a mix of woollen and bast fibre, while a fragment of pure wool comes from a Middle Bronze Age *terramara* of Castione dei Marchesi (Cat. No. 5). There are, however, numerous woollens surviving from the Iron Age. The most complete items have been recovered at the 8th century BCE site of Verucchio (Cat. No. 20), where whole garments have been preserved. A large assemblage has also been found in the contemporary boat burial at Sasso di Furbara (Cat. No. 46). Unlike the 1st millennium BCE linens, which were preferably plain-woven, all of these textiles have complex weaves and patterns. Such a distribution may indicate that pseudomorphs with complex weaves are woollens as well, thus helping in fibre determination when more sophisticated analytical methods are unavailable.

Fibre diameter

Fibre diameter measurement is of great importance in fibre studies. In the case of wool, statistical analysis of many measurements provides information about fibre quality, selection and preparation. Based on fibre diameter measurements, Michael Ryder has established an evolutionary scheme for wool development and demonstrated that, in the course of history, selective breeding has produced increasingly finer and more uniform wool (Ryder 1964; 1974, 109; 1992, 133 fig. 3). Recent work in fibre analysis shows that situation is even more complex as ancient varieties of sheep had very heterogeneous wool (Good 1999, 139–142; Christiansen 2004; Rast-Eicher forthcoming). Fibre in textile is a product of numerous processes: breeding, selection, processing and finishing. Ryder's scheme however, does not take into account what happens between raw wool and finished textile.

In Italy, only wool from Castione dei Machesi has been analysed thus far (Figure 59). It has fine underwool and very thick hairs, a distribution typical of Bronze Age wool in other areas of Europe (Rast-Eicher 2005, 127). Further work is much needed in this direction to see possible chronological developments of wool fibre in Italy, which may help differentiating sheep breeds.

Goat wool

Sheep were not the only animals that could provide fibre for textiles. Well preserved Early Iron Age goat-wool garments have been found in the Alpine glacier of Vedretta di Ries (Cat. No. 10). Furthermore, presumed Cashmere goat wool was recently identified to be the main fibre of the six textile fragments excavated at the Etruscan site of Lattes, France, in a context dated 470–460 BCE (Moulherat and Vial 2000; Landes 2003, 137–138 no. 10–6.2). While both derived from goat, the quality of these two finds differs significantly, as regular goat hair is quite coarse whereas cashmere wool is among the finest known.

Dyes

While plant fibres resulted in uniformly coloured shades of grey-white, wool came in a variety of natural colours, which could be used for decorative purposes. Pliny (*NH* 8, 191) describes black wool of Spain and red wool from Asia and Baetica, tawny from Canosa and dark from Taranto. Silius Italicus (*Pun.* 8.436–437) praises Ancona for its coloured wools. The Vedretta di Ries textiles, in fact, have patterns created by the naturally pigmented wools of light and dark colour (Bazzanella *et al.* 2003, 181). Although the natural wool colours are limited in range, white fibres can be dyed easily. Virgil even dreamed of sheep with naturally purple, saffron and scarlet fleece (*Ecl.* 4.42–45).

Figure 59. Fibre from woollen textile found at Castione dei Marchesi, Middle–Late Bronze Age, under magnification (Courtesy of Antoinette Rast-Eicher).

The use of colour is probably no less ancient than the weaving of textiles themselves, as humans strove to improve their aesthetic quality. Dyeing can be imparted in several different ways (Barber 1991, 225–226):

A. Throwing finished cloth into a dye pot: this is the simplest way of dyeing and will result in the entire fabric being coloured;
B. Dyeing raw fibres or yarn for polychrome weaves or embroidery;
C. Resist-dyeing: placing cloth into a dye vat after parts of it have been treated with a dye-resistant substance;[21]
D. Painting or printing patterns.

Mineral pigments, such as ochre, were probably the first materials used to transform the colour of textiles. Pigments, however, are insoluble in water and do not chemically bind to the textile. As such, they cannot produce a long-lasting colour. Dyes, in contrast, are substances that can be chemically bound to a substrate – in our case, a fibre. Almost all dyes are organic materials. Direct dyes are water-soluble and can bind chemically to the substrate itself. These dyes are usually fugitive since the binding forces are relatively weak, deteriorating with time. Vat dyes, on the other hand, are generally water-insoluble and have to be chemically altered to a soluble and often colourless form (Lambert 1997, 73). Once the soluble form is bound to the fabric, treatment with air regenerates the original dye yielding a very stable or fast colour.

The earliest evidence for true organic dyes comes from late 3rd millennium BCE Mesopotamia (Forbes 1956, 98–148; Barber 1991, 223–235; Walton Rogers 2003). In Europe, they were certainly used by the 1st millennium BCE (Walton Rogers 2003, 25). Thus, numerous dyes have been identified in Hallstatt textiles, proving that complex dyeing techniques were known and used in the Iron Age (Walton Rogers 1999; Hofmann-de Keijzer, van Bommel and Joosten 2005).[22]

Levelers and mordants

Most organic dyes require certain additional ingredients to make them colourfast. Dyes can be classified as acidic or basic. Acids were obtained by using vinegar (acetic acid) or by fermenting the dye plant. Basic dyes, on the other hand, were treated with ammonia or salt. Woad and indigo, for example, required reduction in an alkaline vat. Human or animal urine was often used for the purpose. Lime and potash, the latter made by leaching wood ashes, were other substances commonly utilised.

Mordants are also classified as acidic or basic, and acidic mordants are used to bind with basic dyes and *vice versa*. Acid mordants are generally derived from tannins available from tree bark (*e.g.* oak, alder, chestnut, *etc.*) and gall. Basic mordants come from salts of metals such as iron, aluminium and copper. These salts are often ingredients of certain soils, such as alum and natron. Scraps of metal could also be used in a vat. The presence of the mordant may change the colour of the dye and different mordants used with the same dye may produce different colours. The mordant may be bound to the fabric before dyeing or it may be added to the dye bath. Mordant dyes are usually very colourfast.

Washing

Usually wool was coloured, since linen is difficult to dye and would have been used in its natural colour. Unlike proteinatious wool, cellulose, the main chemical constituent of bast fibres, does not have acid or basic radicals required for the binding of dyes and mordants (Barber 1991, 236 note 17). Often, before yarns or textiles were treated with dyes and mordants, they were washed in order to remove impurities and facilitate the dyeing process. This could be carried out with the help of soapwort *Saponaria officinalis* L. (Figure 60). Pliny (*NH* 19.18) mentions that juice obtained from its root was used for washing wool and made it whiter and softer. There were also various clays used for cleaning, such as Fuller's earth that has been found in Pompeii. Pliny the Elder (*NH* 35.196–198) describes several types of clays used for cleaning and whitening textiles. Robertson (1949, 52) identified them as varieties of clay minerals bentonite and montmorillonite. It has been suggested that blocks of clay deposited in one of the Orientalising tombs at Terni may have been Pliny's *creta Umbrica* or Umbrian earth and, thus, connected with the washing of textiles (Broncoli 2001, 351).

Dye plants and animals

A number of plants could have been used for dyeing (Leadbeater 1976, 84–88; Firmin 1989; Cardon 2007). Theophrastus, Vitruvius (7.13) and, particularly, Pliny the Elder (*NH* 21.45–46) left detailed descriptions of the sources of materials and the procedures for the preparation of numerous dyes.

Blue

Blue colour could be obtained from woad (*Isatis tinctoria* L.) and Asian indigo (*Indigofera tinctoria* L.), both of which contain the active ingredient indigotin. Indigo was probably not used in Europe until Roman times. Woad, however, was one of the most common dye plants in Europe, in use until modern times (Figure 61). The dye was reduced in a woad vat, an anaerobic fermentation of processed woad leaves assisted by the microorganism *Clostridium isatidis* and others (Padden and John 2000).

Red

Reds and oranges were some of the most sought after and expensive colours. The root of dyer's madder (*Rubia tinctorum* L. and other similar species like *Rubia peregrina* L.), native to southern Europe, was probably the earliest and most commonly used (Figure 62). Its main colorants are alizarin and purpurin. It was used and even cultivated already during the Bronze Age in some areas of eastern Mediterranean (Nosch 2004). Pliny the Elder (*NH* 19.17) mentions that in his time the most esteemed kind of madder was Italian, especially that grown for the purpose near Rome. Kermes (*Kermes vermilio*) and lac (*Kerria lacca*), small insects growing respectively on the kermes oak and various Asian trees and shrubs,

Figure 60. Soapwart plant (Drawing by the author).

produced the most brilliant reds but were considerably less common and very expensive. While kermes was collected in the Mediterranean area, lac was imported from the east and most likely did not reach Italy until the Late Antique period. Kermes was, however, already used in some of the 6th century BCE textiles of Hochdorf (Walton Rogers 1999, 241–243).

Yellow

Yellows could be obtained more easily and from a larger variety of plants, such as weld or dyer's weed (*Reseda luteola L.*), saffron (*Crocus sativus L.*), dyer's greenweed (*Genista tinctoria L.*), safflower (*Carthamus tinctorius L.*), and others. On the other hand, active ingredients of yellow dyes are often less stable and are more difficult to identify in archaeological textiles.

Other colours

In addition to the bright colours, often requiring expensive and rare dyes, a variety of shades could be obtained from plants common throughout the Italian peninsula (Bazzanella and Mayr 1995b; Cardon 2007). Many trees produce tannin-bearing dyes, which are also used for the curing of hides. Thus, oak and sumac can be used to dye textiles yellow, brown, or black, while pine produces brown and yellow dyes. Galls and nuts were commonly used for brown. The berries of *Sambucus nigra L.*, an arboreal plant common throughout the Italian

Figure 61 (below left). Woad plant (Drawing by the author).

Figure 62 (below right). Madder plant (Drawing by the author).

peninsula, produce green or black colour. Blueberries, on the other hand, create a violet, pink, or blue-grey effect. River cane can be used to dye fibre green or yellow and nettle yellow-green or gray-green. Lichens produce brown, yellow and purple shades. Investigation of pollen and seeds often helps in identification of particular plants used (Bazzanella and Mayr 1995b, 127).

Archaeological evidence for dyes in Italy

The studies of dyes in ancient Italy have been limited by rare survival of actual textiles and the often badly degraded state of their colour. Dye and mordant identification usually requires sophisticated chemical analyses (Firmin 1989; Raheel 1994, 148ff). Furthermore, positive identification of a particular dye present on a fibre is extremely difficult since the results of these analyses show only the more stable chemical components of the dyes and not their sources, so it is often impossible to tell what specific source or combination of thereof was used in the dyeing process. Most dyes, however, were obtained from locally available plants, animals and minerals.

Information about ancient mordants is even scarcer than our knowledge of ancient dyes. Pliny the Elder mentions that alum, for example, was mined in Mesopotamia, Asia Minor, Egypt, Melos, Cyprus, Sardinia and Aeolian islands (*NH* 35.52). At the present time, however, no study of mordants in ancient Italy exists. Provenance studies may hold great potential, as with all metals.

The most extensive dye analyses of Italian pre-Roman material have been performed for the Villanovan Verucchio textiles (Vanden Berghe 2002, 220; Stauffer 2004; pers. com. Annemarie Stauffer, 2007). These results are of great importance, since they indicate the use of complex dyeing technologies and of possibly imported dyes in the Early Iron Age. Thus, Verucchio Mantle 1 was dyed with madder, while the fibres of its blue border were most likely treated with madder and woad. Mantle 2, on the other hand, was found to contain purpurin and luteolin in the base textile, indicating the use of a variety of madder, probably *Rubia peregrina*. Its border was also dyed with woad, creating a purple-red effect. Other textiles contained purpurin, suggesting the use of madder, and indigotin, indicating woad. This preliminary analysis demonstrates that several different dyes were used to add colour to the Verucchio textiles, while their combination in some textile parts shows an understating of a complex, multiple-stage dyeing process.

Indigotin has also been identified in the sewing thread of the shoe from Vedretta di Ries (Dal Rì 1996, 375; Bazzanella *et al.* 2005, 158). Blue colour was observed on some threads of a textile from Chiusi (Cat. No. 30).

Iconographic evidence

While little information is available at the moment regarding original dyes and colours of ancient Italian textiles, some indication is provided by artistic representations. Tomb paintings of Etruria and Campania in particular are useful: garments and utilitarian textiles are resplendent in reds, blues, greens, yellows and purples. Care must be taken in order not to over-interpret this information since, of course, the colour schemes were highly dependent on the availability of the pigments to be used by the artists. Still, certain tendencies are evident: tunics are generally shown light, while cloaks are depicted in darker colours. The decorative elements such as borders are often red.

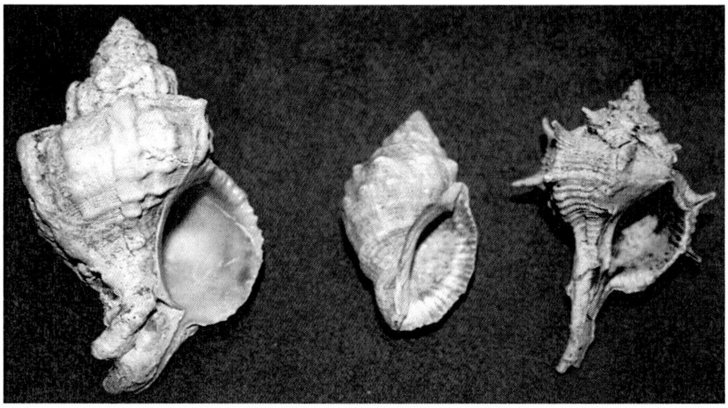

Figure 63. Purple shells, from left to right: Murex *or* Hexaplex trunculus, Stramonita haemastoma, Bolinus brandaris *(Courtesy Carmen Alfaro Giner).*

Purple

Among ancient dyes, in a category of its own stands Royal or Tyrian purple, the most famous dye of antiquity, obtained from a variety of marine molluscs of *Muricidae* family, such as *Murex* or *Hexaplex trunculus*, *Murex* or *Bolinus brandaris*, *Purpura* or *Stramonita haemastoma*, and other species (Figure 63). Its main colorants and chemical tags are brominated indigoids (Cardon 2007, 554). Various molluscs produce different amounts of these together with the blue indigotin; hence, depending on the percentage of each species in the dye vat, the final shade could range from redish purple to blue, purple being the most sought colour.

Because of the value of true purple, it was frequently imitated by combining red and blue dyes or by using certain lichens. Pliny the Elder (*NH* 9.124–141) left a detailed and generally accurate description of purple production (Steigerwald 1986; Doumet 1999; Macheboeuf 2004; Haubrichs 2005).

Origins

Purple production is usually associated with the Phoenician cities of the Levant, especially Tyre and Sidon, although the earliest archaeological evidence is found on Crete, dated to the 1800–1600 BCE. Substantial quantities of shells have been found at Palaikastro, Kommos, and Kouphonisi (Cardon 2007, 571). In the Near East, the production of shellfish purple was established by 1500 BCE, exemplified by crushed shell heaps and written documents at Ugarit (Lambert 1997, 87–88). The heaps of crushed mollusc shells in Syria, Lebanon, Israel, Crete, Spain, North Africa and elsewhere in the Mediterranean, testify to the scale of the industry in antiquity (Jensen 1961, 105–106; Barber 1991, 228–229; Sagona 1999; Alfaro 2002).

Italian evidence

The earliest evidence in Italy comes from Coppa Nevigata in South Italy, where an accumulation of *Hexaplex trunculus* was found already in the Early Proto-Apennine layers (Figure 64), reaching highest levels in the Apennine and Late Apennine periods, indicated by tens of thousands of mollusc remains (Cazella *et al.* 2005; Reese 2005, 111). Later evidence comes from two sites situated at the opposite ends of the Italian peninsula: Broglio di Trebisacce in northern Calabria and Frattesina in Veneto (pers. com., Renato Peroni 2003). All three sites had connections with the Aegean in the Bronze Age and it is possible that the technology was brought to Italy by the Mycenaeans.

More definitive evidence indicating purple production in Italy is recorded a few centuries later. Several centres of purple production have been identified in Italy and Sicily on the basis of archaeological and literary evidence: Ancona, Monte Circeo, Pozzuolli, Canosa, Taranto, Siracusa, Marsala, Segesta and Mozia (Forbes 1956, 136, 162; Morel 1978, 104ff; De Juliis 2000, 81; Macheboeuf 2004; Reese 2005, 111). In Mozia, significant accumulations of crushed muricid shells were found in 6th–5th century BCE layers, in association with stone hammers and whale vertebrae; the latter presumably were used as crushing platforms for the shells (*Mozia* 1989, 40; Reese 2005). In Taranto and Marsala, purple production can be dated back to at least 5th–4th century BCE, attested by coins with shell depictions from both sites (Macheboeuf 2004, 28–29). Textual sources and mounds of crushed shells documented in the 19th century indicate that Taranto was probably the most important purple production centre in Italy.

Concentration of purple production sites in South Italy and Sicily suggests that the technology may have been introduced there by Phoenicians. In fact, Mozia and Marsala are Punic sites, with Segesta located nearby, while in Siracusa, the presence of Phoenicians is recorded before the arrival of the Greeks.

Other interesting finds are known from North Italy. Three shells of *Bolinus brandaris* came from the 8th century BCE Tomb 139 at Casa di Ricovero in Este (Chieco Bianchi and Calzavara Capuis 1985, 56 no. 11, 76), while another specimen was found in a slightly later Tomb 28 of Casa Alfonsi necropolis at the same site (Chieco Bianchi and Calzavara Capuis 1985, 414 no. 8). Shells of *Bolinus brandaris* and *Hexaplex trunculus* were also found in Tomb 37 at Ognissanti in Padova dated to the first half of the 5th century BCE.[23] During the Roman period, several purple merchants (*purpurarii*) were established in North Italy, attested by the inscriptions from Parma, Faenza (Faventia), Mevaniola, Piacenza (Placentia), and Aquileia (*CIL* XI 1069a, 645, 6604; Macheboeuf 2004, 29–33).[24] The latter site also produced thousands of crushed muricid shells.

None of the surviving pre-Roman textile fragments found in Italy have tested positive for purple so far. However, the precious purple and golden textiles were famous at least since the Hellenistic period.[25] It may thus be possible to trace no longer existing purple through its golden counterpart.

Gold thread

Gold is one of the most difficult colours to reproduce. In the case of textiles, the ancient solution was to use flat gold strips or to wind very thin strips of gold around the yarn.[26] Because of the material, the resulting thread was especially costly. It is unclear how early the practice began in Italy, but Pliny (*NH* 19.57) quotes Verrius Flaccus who recorded that Roman king Tarquinius Priscus (traditional dates 616–578 BCE) celebrated a triumph wearing a golden tunic. The earliest archaeological finds, however, date to the 4th century BCE. Recent conservation of the Tomba François in Vulci yielded the remains of thin golden strips, which originally were integrated into a textile, possibly a cloak not unlike the one worn by Vel Sathies in the fresco in the same tomb (Moretti Sgubini 2004b, 23, 28 fig. 12). At Canosa, in the contemporary Tomba degli Ori, fine golden strips were found

Figure 64. Hexaplex trunculus *remains from Coppa Nevigata (Courtesy of Alberto Cazzella).*

inserted into linen textile (Cat. No. 81). In several 2nd–1st century BCE tombs at Taranto (Cat. No. 82), strips of golden weft have been recovered that may have been garment fringes. Apart from another find from a 3rd century tomb in Civita Castellana (Cat. No. 43), the rest of the gold thread-containing textiles in Italy date to the Roman period.[27]

The production of gold thread must have been specialised and limited to few places. The possibility has been suggested that ancient Taras/Tarentum may have been an Italian production centre for gold thread and textiles in Hellenistic times (De Juliis 1984, 331). This is not unlikely, given the early date for many of the finds and the fame of Tarentum as a centre of gold crafts. Furthermore, the Tarentine purple industry would have provided the resource needed for production of gold and purple textiles.

Textile technology

As the material presented in the catalogue illustrates, by the Bronze Age, sophisticated technologies were being utilised by ancient inhabitants of the Italian peninsula for textile production and, by the Early Iron Age, Italic populations were familiar with diverse fibres, dyes and weaving techniques. Despite the fact that the majority of textiles listed in the catalogue have not been analysed scientifically, some conclusions can be drawn on the basis of the corpus reviewed above. It should be stressed that the following considerations are based on the current state of publication and in no way are intended as definitive.

Variety of fibres

A variety of fibres were utilised in textile production in ancient Italy. These include: linen, hemp, esparto, various tree basts, sheep wool, goat hair, and possibly asbestos. This variety reflects not only availability of raw materials – whether locally or through exchange – but also the knowledge of technologies to convert them from their raw state to usable fibre.

Yarn

While plied yarn is characteristic for the early periods, single yarns are most common during the 1st millennium BCE, as can be observed from the available data, collected in Table 2.

Z-twist seems to be prevalent throughout all periods but it is not exclusive. The woollen fragments from the Bronze Age Castione Marchesi (Cat. No. 5) and Valle delle Paiole (Cat. No. 4), are Z2s twisted, in contrast to the S2z twist of the contemporary linen textiles. Bender Jørgensen (1992, 118) sees in this contrast an indication of a new technology developed for the woollen fibre. The evidence is, however, too limited at present. Plying is a feature observed in the Iron Age Hallstatt (Grömer 2005, 22) and Hochdorf (Rast-Eicher forthcoming) textiles.

During Early Iron Age, there is a clear tendency to combine yarns of opposite twists to create spin-patterned textiles, such as those of Verucchio (Cat. No. 20) and Sasso di Furbara (Cat. No. 46). Close parallels can be found among Hallstatt textiles (Grömer 2005, 23). While the reason for this tendency may be aesthetic, it reflects the knowledge of the technique and appreciation of the subtlety of spin pattern.

Another observation, albeit based on very few examples, is that Bronze Age woollen yarn has a significantly larger diameter than any of the early linen textiles. This tendency can also be observed in contemporary Scandinavia, where the more abundant woollen textiles also have larger yarn diameters than the few plan fibre finds (Mannering and Gleba forthcoming). The question must remain whether this phenomenon reflects different techniques developed for different fibres or preference and tastes of the Bronze Age populations.

Thread count

The thread counts in textiles from pre-Roman Italian sites are relatively high even during early periods (Table 2). The majority of the Bronze Age textiles have a thread count above 10, with the exception of woollen textiles from Valle delle Paiole (Cat. No. 4) and Castione dei Marchesi (Cat. No. 5). During the Early Iron Age, the thread counts go up to 30–40 with even higher numbers in later periods. For comparison, among the Danish Pre-Roman Iron Age textiles, none have thread counts above 20 (Mannering and Gleba forthcoming). The thread counts are more uniform among the Bronze Age textiles, ranging from 10 to 20 and the textiles are relatively balanced. Later periods show more variation in count and balance, with many weft-faced tabbies.

Weaves

A variety of techniques were used to create textiles, including basic weaving, tablet weaving, soumak and some type of twining.

The basic weaves include tabby and a variety of twills. Tabbies, characterised by plied yarn in one or both systems and defined by Bender-Jørgensen (1992, 122) as 'Döhren type', are characteristic for the Neolithic/Bronze Age Italian finds. They are also common in Central Europe, where they are present from the Neolithic and the Bronze Age through the Iron Age (Bender Jørgensen 1992, 117). Plain linen z/z tabby seems to be prevalent in Italy during the Iron Age, just as in the eastern area of Central Europe (Bender Jørgensen 1992, 125). Many of the tabbies are not balanced and especially weft-faced tabbies are common.

Although regarded as an Iron Age feature of textile technology, twill developed during the Bronze Age (Rast-Eicher 2005, 128). By the Early Iron Age, complex twills are ubiquitous throughout Europe. The sophistication of twills from Verucchio (Cat. No. 20), Sasso di Furbara (Cat. No. 46), and Tarquinia (Cat. No. 49), point to a well-established and settled technology. Among the twills, prominent is Bender-Jørgensen's (1992, 122) 'Vače type', a spin-patterned twill in single yarns. This technique, encountered in the Hallstatt textiles (Grömer 2005, 23), is probably illustrated in the numerous Etruscan artistic representations showing a variety of plaids, diagonals, chevrons, diamonds, and elaborate borders, and, to a certain extent, is one of the defining features of Etruscan textiles.

The diagonal twill, observed among the Vedretta di Ries (Cat. No. 10) and Sasso di Furbara (Cat. No. 46) textiles, is common among the textiles found in the salt mines of Hallstatt and Dürnberg in Austria (Hundt 1967, 58–60; Barber 1991, 186–188; Bichler *et al.* 2005). Diamond twill is also present in the Early Iron Age finds from Verucchio (Cat. No. 20) and Tarquinia (Cat. No. 49).

Tablet borders

Borders woven in complex tablet technique are found on textiles from the princely burials in Italy (Verucchio, Sasso di Furbara) as well as in Central Europe (Hallstatt, Austria; Hochdorf and Hohmichele, Germany). The textiles of Verucchio have been demonstrated to be ceremonial garments and their tablet-woven borders appear to be status markers with not only their presence but also their width bearing significance (Stauffer 2002). The specifics of spool presence in Early Iron Age burials (see p.175) of Italy and ubiquitous presence of borders in Etruscan garment

Cat. No.	*Site* object	Fibre	Weave	Warp twist	Warp D	Warp count	Weft twist	Weft D	Weft count
2	*Ledro*								
	N.5	linen	tabby	S2z	0.5–0.6	16	S2z	0.5–0.7	12
	n. 6	linen	tabby	S2z	0.4–0.5	18	S2z	0.4–0.6	12
	n. 7	linen	tabby	S2z	0.4–0.5	12	S2z	0.3–0.4	10–14
	n. 8	linen	tabby	S2z	0.4	16	S2z	0.5	18
	n. 9	linen	tabby	S2z	0.6–0.7	14–16	S2z	0.6	10
	n. 10	linen	tabby	S2z	0.4	14–16	S2z	0.6	12
	n. 11	linen	tabby	S2z	0.6	14	S2z	0.5	16
	n. 12	linen	tabby	S2z	0.3–0.4	18	S2z	0.4–0.5	14
3	*Lucone di P.*								
	n. 20	linen	tabby	S2z	0.4–0.5	18	S2z	0.6	16
	n. 21	linen	tabby	S2z	0.3	18	S2z	0.6	12–13
	n. 22	linen	tabby	S2z	0.4	12	S2z	0.4	16
	n. 23	linen	tabby	S2z	0.4	12	S2z	16	16
	n. 24	linen	tabby	S2z	0.3	16–17	S2z	0.4	12
	n. 25	linen	tabby	S2z	0.5	12–13	S2z	0.4–0.5	12–13
4	*Valle delle Paiole*	bast wool	tabby	Z2s	0.5–1	7–10	Z2s	0.5–1	5–7
5	*Castione Marchesi*	wool	tabby	Z2s	1.3	8	Z2s	1.3	6
9	*Gricignano*	hemp	tabby	S2z	0.2–0.4	–	S2z	0.2–0.3	–

Table 2a. Technical data for Bronze Age textiles from Italy.

representations further argue that that these borders were not purely decorative but communicated a very clear and important message of status not only to the Etruscans but among other European Early Iron Age cultures.

Dyeing

Dyeing techniques were known in Italy at least since the Early Iron Age. Among the identified dye sources are woad and madder. Indirect evidence, however, points towards existence and use of numerous other dye sources of plant and animal origin. In the later periods, gold thread was also used to add colour to textiles.

Sewing

Although ancient Italy is commonly regarded as the area of the 'off-the-loom' or 'woven to shape' textiles (Granger-Taylor 1982), many finds indicate that needle was in use for both structural and decorative purposes. Some of the finds have seams and hems, while others have mends and patches sewn onto them (Vedretta di Ries, Cat. No. 10). Sewing is also prominent among the textiles from Hallstatt (Mautendorfer 2005). Applied decoration was used in Italian textiles at least since the Bronze Age and could consist of seeds, glass beads, amber or metal attachments.

Textile function and context

Last but certainly not least important aspect of textile investigation is determination of their function, which is often provided by their context.

Garments

The vast majority of surviving textiles come from burial contexts, thus demonstrating that they constituted a very important part of broader mortuary practices. Vedretta di Ries (Cat. No. 10) finds constitute at present the only

Cat. No.	*Site*/object	Fibre	Weave	Warp twist	Warp D (mm)	Warp count	Weft twist	Weft D (mm)	Weft count
20	*Verucchio*								
	Mantle 1	wool	2/2 twill	z, s	–	12–14	z, s	–	24
	Mantle 2	wool	2/2 twill	z, s		12–16	z, s		22–26
	Garment	wool	2/2 twill	z, s		14–16	z, s		19–20
	T1	wool	tabby	z		4–6	z		4–6
	T2	wool	2/1 twill	z, s		10–12	z, s		18
	T3	wool	2/2 twill	z, s		20	z, s		20
	T4	wool	tabby	z		18–20	z		30–32
	T5	wool	2/2 twill	z, s		22–24	z, s		20–24
	T6	wool	2/2 twill	z, s		12–18	z, s		12–18
	T7	wool	2/2 twill	z, s		16–20	z, s		16–20
	T8	wool	2/2 twill	z, s		20–26	z, s		20–26
	T9	wool	tabby	z		6	z		6
	T10	wool	2/2 twill	z, s		16	z, s		26
45	*Sasso di Fur.*								
	Group 1	wool	shevron twill	z, s	0.2	35–45	z, s	0.1–0.2	35–40
	Group 2	wool	tabby	S2z	0.3–0.4	7–8	z	0.2	35–40
	Group 3	wool	tabby	S2z	0.6–0.8	6–7	z	0.6–0.7	17–20
	Group 4	wool	tabby	z	1.5	3	s	0.8–0.9	18
	Group 5	wool	tabby	z	0.4–0.6	8	z	0.5–0.6	25–30
	Group 6	wool	tabby	Z2s	0.7	7–8	z	0.3	32
	Group 7	wool	twill	z, s	–	–	z, s	–	–
10	*Vedretta*								
	Under-leggings	wool	tabby	s	–	18	s	–	8
	Over-leggings	wool	shevron twill	Z2s		8	s		14
	Patch	wool	tabby	z		9	z		6
	Inner shoes	wool	2/2 twill	z		12	s		8
	Patch	wool	2/2 twill	z		10	z		8
	patches	wool	2/2 twill	z		6	z		6
36	Perugia?	linen	warp-faced tabby	z	–	23	z	–	12
43	Cogion-Coste di Manone	linen?	weft-faced tabby	s	–	10	s	–	60
58	Borgo le Ferriere	–	weft-faced tabby	z	–	11	z	–	60–80
33	Orvieto?	–	tabby	z	–	15–16	z	–	15–16
90	Himera	hemp	tabby	z	–		z	–	
52	Newark	linen?	weft-faced tabby?	z	–	6–8	z	–	16
30	Chiusi	wool	weft-faced tabby	–	–	35	z	–	60
13	Este	–							
	n. 105		tabby	z	0.5	20–25	z	0.5	20–25
			tabby	z	0.3	15–20	z	0.3	15–20
48	Tarquinia	–	diamond twill	S2z	0.5	18	s	0.3	14
14	San Basilio	–	2/2 twill	z	0.4	16	s	0.3	20
93	Unknown	asbest.	tabby	s	1–2	5–6	s	1–2	4

Table 2b. Technical data for Iron Age textiles from Italy.

exception. Their owner died, probably in an avalanche, while attempting to cross the Alps (Dal Rì 1995–1996, 385). Of the surviving textiles, most are garments. The luxurious mantles and other clothing articles of Verucchio (Cat. No. 20) are the only almost complete garments and provide a glimpse of what some of the ancient costumes looked like. Apart from them, a set of under- and over-leggings and shoes from Vedretta di Ries (Cat. No. 10) and a sock from Padova (Cat. No. 12) are evidence for the lower body attire, probably during cold periods. Otherwise, costume reconstruction on the basis of textile finds is difficult, since most of them are small fragments.

Figure 65. Reconstruction of female funerary dresses from Alianello, 7th century BCE (After De Juliis 1996, 115 fig. 109).

Still, pseudomorphs on metal articles of personal decoration such as bracelets, belts, and, most commonly, fibulae also provide an idea of clothing. Thus, when traces of different weaves are found on the same fibula, as for example in some of the Alfedena finds (Cat. No. 62), a conjecture can be made that the deceased was provided with several layers of garments or a garment and a shroud. If a number of pseudomorphs are present in a burial, it is possible to reconstruct how the garment was located based on their distribution, position and direction of the weave (Rast-Eicher 2000b).

Even if textiles do not survive, their presence is indicated through other evidence, such as fibulae and other decorative ornaments. In some cases, reconstructions of garments are possible, based on the position of surviving decorative elements in relation to the skeleton, sarcophagus, or trench, as in the case of female costumes from Allianello (Figure 65) (De Juliis 1996, fig. 108–109).

Occasionally textile remains are found on the inside of the armour, suggesting that they may have belonged to a lining or a garment worn under the metal, as in the case of the Vetulonia (Cat. No. 25) and Sesto Fiorentino (Cat. No. 26) finds.

In some cases, the dead were provided with very sumptuous dresses, as attested by the finds from Verucchio, Castel di Decima (Cat No. 55) and the Isis Tomb at Vulci (Cat. No. 39). Whether such garments would have been used in life is a debatable issue. Still, the majority of clothing articles found in graves could have been, and most likely were, used in life. Confirmation comes from the Verucchio garments, which have wear marks and, hence, must have been used by the deceased in life.

Shrouds

The shrouds, on the other hand, were most probably made specifically for the funerary purposes. Traces of one such cover have been noted in Tomb 3 at Osteria dell'Osa (Figure 66), thanks to the rectangle of small bronze rings and buttons that probably were sewn onto the edges of the cloth (Bietti Sestieri 1992b, fig. 3a.391). The dark trace of the shroud is still visible in the soil of the burial, which is on display in the Museo Pigorini in Rome. A simple textile identified as a shroud was preserved on a metal breastplate from Etruria or Capena area (Cat. No. 94).

Wrappings

Another use of textiles, particularly linen, in burial ritual was to wrap cremated bones. Many scholars connect this practice to a ritual described by Homer for the burials of Hector and Patroklos (*Il* 34.796 and 23.254), which is assumed to have been adopted by the elites throughout the Mediterranean during the Iron Age (Bérard 1970, 28; d'Agostino 1977, 59–60). It is probable that the rite was adopted by the native populations of Italy from Greek, specifically Euboean, colonists. In fact, in Greece, textile fragments have been found in similar circumstances in Eretria.[28] The earliest evidence of this practice in Italy comes from Cuma (Cat. No. 66) and Pontecagnano (Cat. No. 68), sites that were first to come into contact with Euboeans. This 'Homeric' ritual then spread quickly among the Etruscan elites, as attested by the textiles from Casale Marittimo (Cat. No. 22), Vetulonia (Cat. No. 25) and Veio (Cat. No. 53).

A different and possibly native Italian tradition seems to be attested by the finds from Central and North Italy. The cremated remains or the cinerary urns, like the ones from Chiusi and Este, as well as Verucchio and Bologna (Eles 2006b, 73) were wrapped in textiles that were then fastened with fibulae, as if wearing a garment (Figure 67). These textiles have been interpreted as clothing for the urns, and thus as representing the deceased (Bonfante 1975, 106 note 3). Anthony Tuck (1994, 626) argued that if Villanovan biconical urns represent the deceased, the incised geometric patterns similar to Hallstatt woven motifs are intended to represent a garment. Chiusine 'canopic' urns may strongly support this theory. The tradition was probably much

Figure 66. Traces of a rectangular shroud in Tomb 3 at Osteria dell'Osa, 9th century BCE (After Bietti Sestieri 1992b, fig. 3a.391).

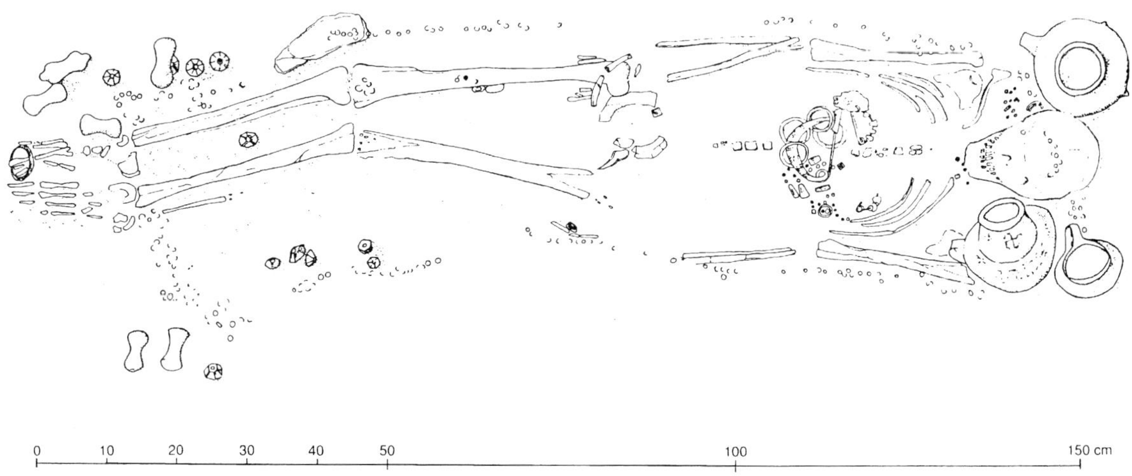

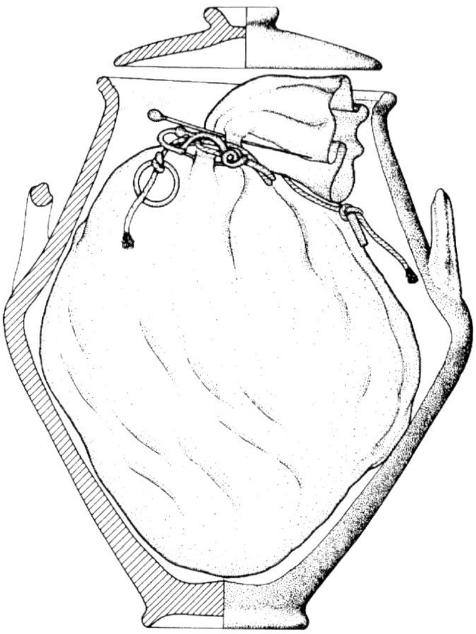

Figure 67. Reconstruction of the deposition of cremated remains wrapped in cloth (After Bianchin Citton, Gambacurta and Ruta Serafini 1998, 92 fig. 39).

more widespread. While textiles do not usually survive, the position of various small decorative objects, such as fibulae, around the urn (Figure 68), may indicate presence of a cloth, which was fastened with them. Such, for example, is the case of several cinerary urns at Tarquinia (Trucco 2006, 98–99).

Many of the pseudomorphs indicate that, apart from cinerary containers, certain metal objects came into close contact with textiles because they were intentionally wrapped or enclosed in fabric. Already Lenticchia (1921, 34) noted the use of textiles for protecting metal objects. Knives, weapons, strygils, spits and mirrors are among the most common objects to bear textile traces. At Cales (Cat. No. 63) and Vassallaggi (Cat. No. 89), for example, most of the knives were subjected to this procedure (Figures 45 and 51). In some cases, the traces are clear enough to reconstruct the direction of a cloth band wrapping a particular object. The deposition of thus 'enclothed' objects in urns excludes the possibility of accidental contact with textiles. It is unclear whether this phenomenon has a ritual significance in funerary context or represents a regular practice of safekeeping of precious metal objects. The pseudomorph on a bronze dagger blade from the Montemerano hoard (Cat. No. 8) suggests not only that the practice of wrapping weapons may go beyond the burial context but also that such a practice existed already in the Bronze Age. The finds indicate that wrapping was common throughout Italy, and the evidence from the princely burial at Hochdorf, in Germany, where all objects were carefully wrapped in cloth (Banck-Burgess 1999), may even suggest that it had a wider, pan-European significance.

Utilitarian textiles and furnishings

Colourful bed covers, cushions and other utilitarian textiles are frequently represented in Etruscan and Campanian tomb paintings (Steingräber 1985; Pontrandolfo and Rouveret 1992). Unfortunately, they survive even less frequently than garments and even when they do, are rarely recognised as such. So far, only the pseudomorph from La Montagnola tomb at Sesto Fiorentino (Cat. No. 26) has been interpreted as a fragment of a bed cover.

Although no direct evidence for wall-hangings in Italy comes in the form of textile remains, a clear indication of their existence is given by the ceiling and wall paintings in the tombs of Tarquinia. Checks and small flowers on the ceilings are unquestionably textile patterns, the entire tombs representing tents or pavilions with roofs made of cloth (Holloway 1965; Stopponi 1968; Tonini 1970; Stopponi 1983, 39, 41; Naso 1996, 349–352). Hangings on the walls and furniture of the Hochdorf burial suggest yet further possible uses of textiles within a funerary context (Banck-Burgess 1999).

Ship rigging and sails

Another important use of textile fibres – especially linen – was for sails and ship rigging. Etruscans in particular were notorious throughout the Mediterranean as sailors and – according to their enemies – as pirates. Their ship building technology was among the most sophisticated for the period, including the use of the earliest foresail (Hagy 1986; Bonino 1989; Turfa and Steinmayer 1999). While no sail fragments survive, pieces of rigging have turned up in shipwrecks. Etruscan-Roman shipwrecks from Pisa San Rosore yielded many cordage elements (Cat. No. 21), while the Punic shipwreck of Marsala contained numerous rope elements (Cat. No. 92).

Books

A rather unique use of textiles known specifically in the Etruscan area is for books. Called *libri lintei* by the Romans, they were made of linen and used for recording religious rituals (Roncalli 1980; van der Meer 2007). Fragments of one such book, the so called Zagreb mummy wrappings (Cat. No. 36), were preserved in Egypt. Evidence for their use also exists in Etruscan art and Latin literature.

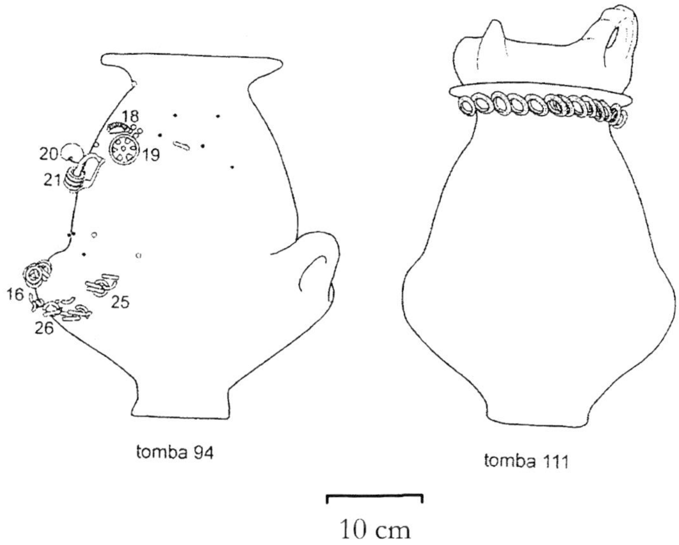

Figure 68. Urns from Tombs 94 and 111 at Tarquinia, Villa Bruschi Falgari, with fibulae positioned around, indicating original presence of textiles (After Trucco 2005, 97 fig. 1B).

Conclusions

Surviving evidence demonstrates the abundance and variety of textiles and their uses in pre-Roman Italy. Garments, utilitarian fabrics, sails, even books were made of textiles. Different functions required different properties, which in turn demanded different choices and careful planning at different stages of textile production.

PART 4

Techniques and Tools

This chapter addresses the process of textile production and the implements used in its various stages, specifically, their function, types, geographical variations and chronological development. Production may be defined as the totality of operations necessary to transform one object into another, different from the first (Mannoni and Giannichedda 1996, 3). Archaeological indicators of productive activities include installations, tools, residues or by-products, unfinished or defective products and raw materials (Mannoni and Giannichedda 1996, 169ff). The vast majority of surviving evidence for textile production consists of tools. Occasionally, installations, such as dyeing vats can be recognised. The heaps of *Murex* shells remaining from the purple production can be regarded as residues or by-products of the later process. Raw materials and unfinished or defective products are almost non-existent. It is thus textile tools that provide the main data for this study.

The sequence of production begins with the choice of raw material and continues with its transformation using various tools and 'recipes for actions' until a final product is obtained. In the case of textile production, raw fibres have to be prepared, spun, woven and finished. Each of these processes requires a particular set of tools.

Fibre preparation

Before fibres can be spun and woven into cloth, the raw material has to be subjected to certain preparatory treatments. Fibre preparation is essential for the good quality of finished yarn and, ultimately, the cloth itself. As we have seen in the preceding chapter, the most common raw materials used in the 1st millennium BCE Italy were flax and wool. Because of their different origin and properties, flax and wool require different procedures and tools to make them usable in textile production.

Flax

Flax preparation takes time and a variety of tools, which did not change significantly until industrialised processing was developed (Moore 1922, 82–83; Barber 1991, 13–14). Before the plant becomes over-ripe, flax plants are pulled up by the roots and gathered into bundles. The later the flax is pulled, the tougher the fibres will be. Yellow, or ripe, flax is occasionally used for utilitarian fabrics and ropes, which require strength.

Once dry, the stalks are rippled to remove seed capsules. Then, they must be retted in standing or running water or dew in order to rot away pectin that holds the fibres together. The retting can take two to three weeks and has to be closely controlled in order for the fibres not to be weakened or destroyed. Using shallow water pools where water is heated by the sun can speed up the process, as higher temperature increases the rate of decomposition. We know that, at least during the Roman period, the advantages of warm water were known (Plin. *NH* 19.17). At present, there is little or no archaeological or literary evidence to show that any sort of installations were built and used for the purpose in ancient Italy. However, at the site of Statonia recent excavations uncovered an installation with basins and channels, for which the use in wine or olive oil production can be excluded, and which might have been utilised for flax retting (pers. com. Emanuele Papi, 2005). For the Roman period, processing of linen and possibly esparto has been identified at a villa near ancient Saetabis, an area famed for its linen in antiquity (Albiach Descals, Gallego Úbeda and García Prósper 2006). The production area consisted of a long channel connected with two rectangular pools.

After the stalks have dried they must be broken with a wooden club to break the pith around which the fibres cling. In recent times, breaking of flax by foot or by horses has been used in some areas of northern Europe (Višniauskaitė and Laniauskaitė 1977, 52). Such a procedure would not require any tools and would have been performed before retting. After that, the flax is scutched with a broad wooden knife to remove the broken pith from the stalk. Finally, the fibres are hackled with a tool that has long teeth to remove the pith.

Unfortunately, the tools used for flax preparation were all wooden and none survive from the 1st millennium BCE contexts in Italy. The only tools believed to have been used for flax preparation are scutching knives and a flax comb (Figure 69) recovered from the Neolithic – Bronze Age lake-dwellings in Italy and Switzerland (Bazzanella *et al.* 2003, 138 and 236).[1] The process of flax preparation, thus, is known mainly from ethnographic examples and experimental archaeology (Višniauskaitė and Laniauskaitė 1977; Rast-Eicher and Thijsse 2001; Baines 2003).[2]

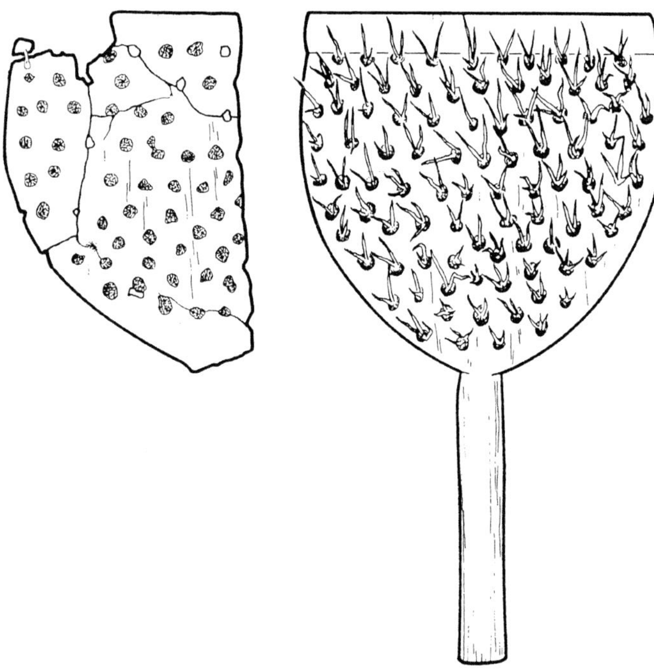

Figure 69. Fragment of wooden flax hackle from Sutz-Lattrigen, Switzerland, Late Neolithic period, and a possible reconstruction (After Vogt 1937, fig. 72).

Wool

Wool had to be removed from the animal, which could be accomplished by plucking or shearing. Plucking was an older method, used on primitive breeds of sheep which would moult their fleece. In some areas, the practice was known until the Roman period (Plin. *NH* 8.191), although it probably disappeared almost everywhere in Italy by the 1st century CE (Wild 1970, 23). The practice survives to this day in the Shetland Islands, where it is known as rooing (Christiansen 2004, 12–13).

More developed sheep breeds do not moult and their fleece has to be cut off, a process accomplished with the help of shears or a knife. Shears appeared during the Iron Age and all of the examples known are iron. In fact, their invention is tied to the use of iron, which is more springy than bronze (Forbes 1956, 8; Barber 1991, 29). It is possible that, before the appearance of shears, knives were used for the purpose but since they could be utilised for a variety of other functions, it is impossible to prove that certain knives were used specifically to shear wool. Unlike plucking, shearing allows the entire fleece to be removed in one piece. Shearing is done in late spring or early summer, when the new spring wool starts growing and can easily be distinguished from the winter wool.

Before the fleece is removed from the animal, the sheep is usually washed in fresh water to remove dirt and grease. Natural bodies of moving water were most likely used for the purpose, such as rivers and streams, although they could also be "manipulated by dams and channels… creating artificial pools to facilitate the procedure" (Santillo Frizell 2004b, 83).

Shears

All ancient shears found in Italy are of the same design, with two triangular blades (knives) on a simple U-shaped spring (Figure 70) (White 1967, 119–120). Their size ranges from 10 to 25 cm in length. The optimal size for the blade is about 15 cm (Wild 1970, 22; Alfaro Giner 1984, 41). The same type of wool shears has been used until recently in some areas of Europe. In Sardinia, for example, I observed such shears in a variety of sizes in the Ethnographic Museum of Sant Antioco.

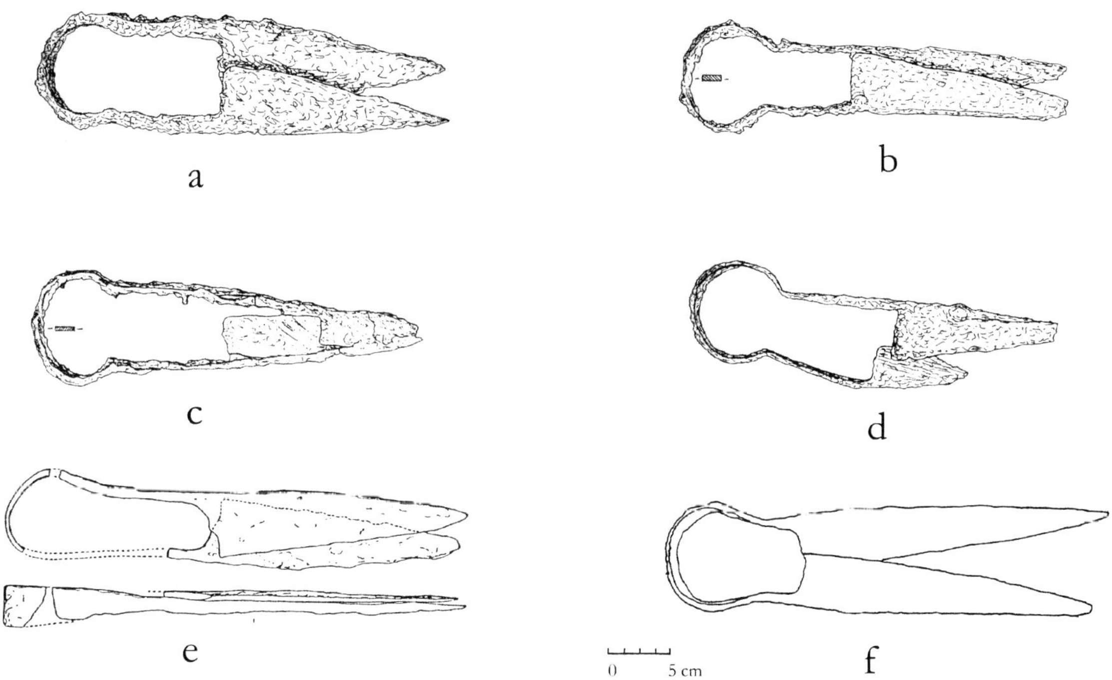

Figure 70. Iron shears from burials, 4th–1st century BCE: a) Tomb 2, Persona, Ornavasso (After Graue 1974, pl. 52); b) Tomb 52, Persona, Ornavasso (After Graue 1974, pl. 61); c) Tomb 11, S. Barnardo, Ornavasso (After Graue 1974, pl. 14); d)) Tomb 32, S. Barnardo, Ornavasso (After Graue 1974, pl. 32); e) Tomb 4, Arquà Petrarca (After Salzani 1987, fig. 5.1); f) Tomb 8, Cardano al Campo (After Dejana 1980, pl. I).

Function

Shears with blades much longer or much shorter than 15 cm are unlikely to have been used for clipping sheep. The smaller examples could have been used for toiletry purposes, while the larger ones were utilised, for example, for cutting large pieces of cloth, as depicted on the Roman relief from the Museum of Sens, France (Roche-Bernard 1993, 119).

Chronology

The shears start appearing during the 8th century BCE and become common in Celtic contexts in Central Europe from the late 4th century BCE (Callegari 1940, 163; Carter 1998, 818). Stylistically, the object is almost undatable although later examples seem to be more elongated in shape. The use of shears requires skill and it is not surprising that often this tool accompanied its owner to the afterlife, possibly indicating his or her ability to use it and the status of the occupation. The vast majority of surviving examples, in fact, have been found in burials.

Distribution

The distribution of these finds merits attention (Map 3), as they are concentrated in two areas of the Apennine peninsula and Sicily. The first area is North Italy, the regions of Piemonte, Lombardy, Veneto and parts of Romagna and Marche. Many of the published finds are collected in Table 3a.[3]

The burials are dated to the 4th–1st centuries BCE and some are even later. It is significant that all the shears for which the length is published are over 20 cm long. All examples were associated with weapons in burials and are assumed to belong to men.[4] It has been suggested that the practice of deposition of shears in male burials in North Italy, populated at the time by Celtic tribes, may express the wool-based wealth of the Celts, who appeared in the area in the 4th century BCE (Vicari 1994, 241). This hypothesis is strengthened by the finds of shears in earlier and contemporary sites in Europe. The sites where they have been found include La Tène (Neuchâtel) in Switzerland (Munro 1890, 288; Egloff 1991, 369), Numancia in Spain (Arslan and Vitali 1991, 736 no. 454 a), Manching in Germany (Arslan and Vitali 1991, 743 no. 547 f), Halish Lovachka in Ukraine (Maier 1991, 416), and Lochlet and Lagore in Britain (Munro 1890, 356, 416). Shears are also common finds in later Longobard sites and cemeteries in the same geographic area (5th–7th centuries CE).

The second area of shear deposition in burials is Sicily (Table 3b). There, too, they are found in relatively late contexts, dated to the 4th–1st centuries BCE. The largest sample comes from Marsala, ancient Lilybaeum (Bisi 1970, 550 no. 79; Tusa Cutrini 1971; Bechtold 1999), where shears were found in 23 out of 450 burials. The objects are relatively few, suggesting that they were regarded as a prestige item (Bechtold 1999, 232). Shears are deposited in both male and female burials, although their frequent association with mirrors indicates women as the more usual recipients.[5] In contrast to the shears found in North Italy, the southern examples rarely exceed 10 cm in length.

Similar items have been found in contemporary Sicilian necropoleis at Palermo, ancient Panormos (Gàbrici 1941, 270 fig. 13), and Olbia on Sardinia (Levi 1950, 32, 62 no. 6, pl. XI a). On the mainland South Italy, shears were found in a female Tomb 6 in Taranto, dated to the last quarter of the 4th century BCE (De Juliis 1984, 407 XXXIX no.4). Since ancient Tarentum had a flourishing wool industry in antiquity (Morel 1978; De Juliis 2000, 81), it would not be surprising to find shears there. Another example

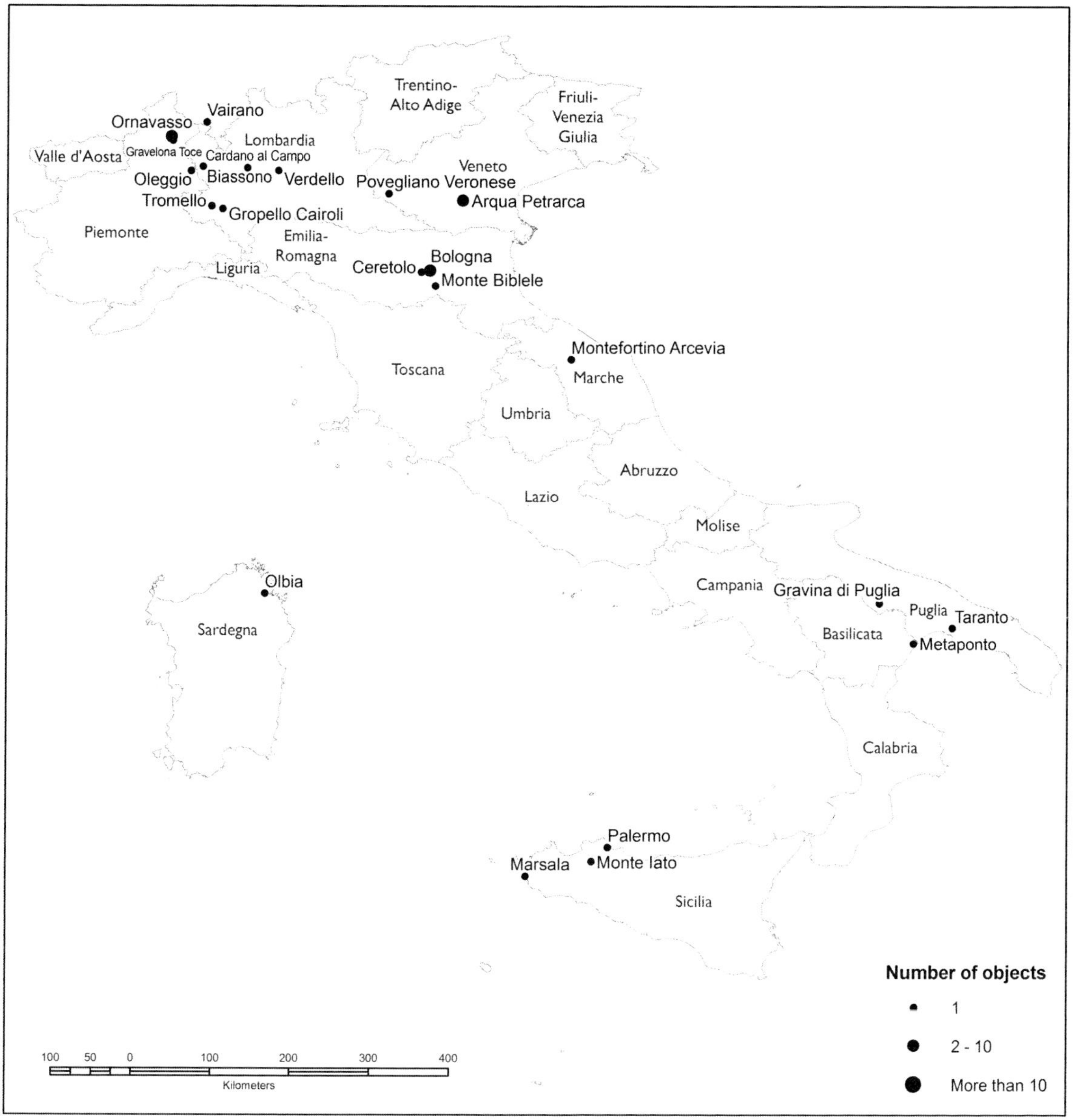

Map 3. Distribution map of shears.

comes from a contemporary female Tomb T 76 at Metaponto (Carter 1998, 266 no. 51, 818, 819 H8, fig. 20.19).

While the northern examples range in length between 18 and 30 cm (Table 3a), most of the southern items are smaller (Table 3b) and it is likely that they were used for purposes other than wool clipping. Unfortunately, no attempt has been made to summarise evidence for shears in Italy until now. The frequent lack of information on an instrument's size in publications makes it difficult to draw conclusions about its function. Future work will help in differentiating the sheep shears from other types of shears.

Site	Date	L (cm)	Bibliography
Montefortino Arcevia (Ancona), Tombs 26, 35, 46	4th c. BCE	-	Brizio 1899, 746–747, pl. X.3, XI.7; Montelius 1895–1905, pl. 153.2.
Monte Bibele, Tomb 132	4th c. BCE	30	Arslan and Vitali 1991, 725 no. 281 g.
Bologna, Benacci, Tomb 185	3rd c. BCE	26	Brizio 1899, 746 pl. VI.28; Grassi 1991, 85; Vitali 1992, 171 no. 2, pl. 19.
Bologna, Benacci, Tomb 934	3rd c. BCE	27	Vitali 1992, 274 no. 3, Pl. 29.
Bologna, Benacci, Tomb 953	3rd c. BCE	26	Vitali 1992, 292 no. 19, pl. 38.
Bologna, Benacci, Tomb 968	3rd c. BCE	23	Vitali 1992, 331 no. 5, pl. 45.
Bologna, De Luca, Tomb 83	3th c. BCE	-	Vitali 1992, 347 no. 1, pl. 51.
Bologna, De Luca, Tomb 85	3th c. BCE	-	Vitali 1992, 348 no. 1, pl. 51.
Bologna, De Luca, Tomb 105	3th c. BCE	23.4	Vitali 1992, 350 no. 1, pl. 53.
Ceretolo	3th c. BCE	24	Brizio 1899, 746; Santoro 1979, 132 no. 392; Grassi 1991, 96; Vitali 1992, 388 no. 9, pl. 60.
Cardano al Campo, Tomb 8	3rd–2nd c. BCE	27	Dejana 1980, 129, 135, pl. I.
Verdello, Tomb in Via Galilei	3rd–2nd c. BCE	11.8	Tizzoni 1981, 25, pl. 16 a.
Biassono	3rd–2nd c. BCE	-	Negroni Catacchio 1975, pl. 171 top.
Povegliano Veronese, Madonna di Via Secca	3rd–2nd c. BCE	-	Cippola 1880, 240, pl. VIII no.4; Aspes, Rizzetto and Salzani 1976, 181 no. 4, fig. 43.
Oleggio, Tombs 53, 106	2nd–1st c. BCE	-	Grassi 1991, Pl. 59.
Gropello Cairoli, Tomb IX	2nd c. BCE	18	Fortunati Zuccàla 1979, 23 no. 5, fig. 12,5.
Ornavasso, S. Bernardo Tomb 11	2nd–1st c. BCE	30	Bianchetti 1895, 105; Graue 1974, 216, pl. 14 no. 9.
Ornavasso, S. Bernardo Tomb 126	2nd–1st c. BCE.	27.5	Bianchetti 1895, 164; Graue 1974, 234, pl. 32 no. 6.
Ornavasso, S. Bernardo Tomb 147	2nd–1st c. BCE	25	Bianchetti 1895, 176; Graue 1974, 216, pl. 35 no. 3.
Ornavasso, Persona, Tomb 2	2nd–1st c. BCE	24	Bianchetti 1895, 198; Graue 1974, 246, pl. 52 no. 1.
Ornavasso, Persona, Tomb 52	2nd–1st c. BCE	24	Bianchetti 1895, 224; Graue 1974, 254, pl. 61 no. 6.
Ornavasso, Persona, Tomb 59	2nd–1st c. BCE	26.5	Bianchetti 1895, 230.
Ornavasso, Persona, Tomb 80	2nd–1st c. BCE	26.5	Bianchetti 1895, 239.
Ornavasso, S. Bernardo Tomb 6	1st c. BCE	-	Bianchetti 1895, 96; Santoro 1979, 89 no. 10.
Valeggio Lamelia, Tessera, Tomb 189	2nd–1st c. BCE	21.2	Santoro 1979, 108 no. 323.
Arquà Petrarca, Monte Ricco, Tomb L	Late 2nd c. BCE	24.5	Callegari 1940, 150, fig. 5; Gamba 1987, 258–259 no. 30, fig. 16.5.
Arquà Petrarca, Tomb 4	1st c. BCE	23.5	Salzani 1987, 278 no. 38,fig. 5.1.
Gravelona Toce, Tomb 96	1st c. BCE	-	Personal observation.
Lonato	1st c. BCE	-	Grassi 1991, 106.
Tromello, Cascina Negrina	1st c. CE	-	Invernizzi 1998, 123.

Table 3a. Shears in North Italy.

Apart from the numerous shears found in burials, a few examples are also known from habitation sites.[6] In North Italy, a pair was found in Vairano in the 3rd–2nd century BCE settlement context (De Benedittis 1980, 337 no. 34, fig. 101.34; Capini and Di Niro 1991, 174 d87, pl. 16d). In South Italy, three pairs of shears were recorded in Gravina di Puglia, although only one has a secure provenance (Macnamara 1992, 239 no. 1916, fig. 109). One pair of shears was

Site	Date	L (cm)	Bibliography
Metaponto, Tomb T 76	5th–4th BCE	11	Carter 1998, 266 no. 51, 818, 819 H8, fig. 20.19.
Palermo, 1883 scavi Tomb 5	4th c. BCE	-	Gàbrici 1941, 270, fig. 13.
Olbia, Joanne Canu (Fonteleoni), Tomb 36	4th c. BCE	10	Levi 1950, 32, 62 no. 6, pl. XI a.
Taranto, Tomb 6	4th c. BCE	11.5 and 8	De Juliis 1984, 407 XXXIX no.4.
Marsala, Tomb VIII	4th–1st c. BCE	12, 16, 12, 9.5	Bisi 1970, 550 no. 79.
Marsala, Tomb II	4th–1st c. BCE	16	Tusa Cutrini 1971, 679, 691, 699, 719, 743, 759, fig. 23.
Marsala, Tomb 27	4th–1st c. BCE	12	Bechtold 1999, 172, tav. XXXVI, 297.
Marsala, Corso Gramsci, hypogeum 40, room E	4th c. BCE	8.2 fr.	Bechtold 1999, 335 no. IP 40/E-53.
Marsala, Via Cattaneo, Hypogeum 2	4th–3rd c. BCE	8.5 fr.	Bechtold 1999, 303 no. IP. 2–41.
Marsala, Via De Gasperi, Tomb 97	3rd c. BCE	10.5 fr.	Bechtold 1999, 317 no. T. 97–11, pl. LVIII.2.
Marsala, Via De Gasperi, Tomb 102	3rd c. BCE	6.2 fr.	Bechtold 1999, 317 no. T. 102–7.
Marsala, Via De Gasperi, Tomb 124	3rd c. BCE	11.9	Bechtold 1999, 320 no. T. 124–3.
Marsala, Corso Gramsci, hypogeum, burial 2	3rd c. BCE	-	Bechtold 1999, 328, no. 1029–21p.
Marsala, Corso Gramsci, hypogeum 38, burial 1	3rd c. BCE	6.8, 11 frr.	Bechtold 1999, 331 no. IP. 38/1–26, pl. LX.2.
Marsala, Via Berta, Tomb 180	3rd c. BCE	8 fr.	Bechtold 1999, 367 no. T. 180–14.
Marsala, Via Berta, Tomb 194	3rd c. BCE	fr.	Bechtold 1999, 373, T. 194–4.
Marsala, Via Berta, Tomb 197B	3rd c. BCE	8.4 fr.	Bechtold 1999, 376 no. 197B-12.
Marsala, Via Berta, Tomb 190/N/2	3rd c. BCE	fr.	Bechtold 1999, 370 no. T. 190/N/2–1.
Marsala, Via Berta, Tomb 201A	3rd c. BCE	11 fr.	Bechtold 1999, 376 no. 201A-17.
Marsala, Via Berta, Tomb 74	3rd c. BCE	6.5 fr.	Bechtold 1999, 355 no. T. 74–6.
Marsala, Via Berta, Tomb 93b	3rd c. BCE	16.5	Bechtold 1999, 357 no. T. 93B-8.
Marsala, Via Berta, Tomb 95	3rd c. BCE	3.5 fr.	Bechtold 1999, 357 no. T. 95–5.
Marsala, Via Berta, Tomb 117	3rd c. BCE	7.6 fr.	Bechtold 1999, 359 no. T. 117–1.
Marsala, Via Berta, Tomb 61	2nd c. BCE	16.5 fr.	Bechtold 1999, 353 no. T. 61–15, pl. LXVI.1.
Marsala, Via Berta, Tomb 153	2nd c. BCE	fr.	Bechtold 1999, 363 no. T. 153–11.
Marsala, Via Berta, Tomb 11	1st c. BCE	3.1 fr.	Bechtold 1999, 350 no. T. 11–7.
Marsala, Via Berta, Tomb 60	1st c. BCE	Frr.	Bechtold 1999, 353 no. T. 60–10.
Marsala, Via Berta, Tomb 145	1st c. BCE	10.4	Bechtold 1999, 362 no. T. 145–4.
Marsala, Via Berta, Tomb 155	1st c. BCE	6 fr.	Bechtold 1999, 364 no. T. 155–5.

Table 3b. Shears in South Italy.

found in room 17 of a villa at Monte Iato, in Sicily (Isler 1986, 39, 41 fig. 26). The later find is considered a toiletry object by the excavator due to its small size (16.5 cm long).

It should be noted that shears have been found on various sites in Spain (Alfaro Giner 1984), and in Carthage, Gunugus, Tipasa and Leptis Magna in North Africa (Levi 1950, 32 note 96; Bechtold 1999, 185). Most of these sites are associated with Punic culture and probably so is the practice of shear deposition in graves in these areas. Finds of shears in contemporary burials in Greece suggest that the practice of using them as burial goods is not uncommon in Hellenic contexts.[7]

Sorting and washing

The quality of a finished textile greatly depends on how well the fibre was prepared. The wool, hence, has to be sorted, which is done by hand and takes a considerable amount of time. The fineness of some of the ancient textiles demonstrates that the procedure was a regular practice in ancient times. The criteria for sorting may be colour, fineness, crimp, length, strength, or texture (Leadbeater 1976, 21–26). The quality of wool greatly depends on which part of the animal it comes from, the sides and the shoulder area being the best. Furthermore, there are differences in quality between different animals and between male and female animals. The quality is also affected by the age and physiological state of the animal, such as whether a ewe has lambs or if the animal is sick, as well as by the food they consume (Andersson 2003, 17–18).

Although the wool can be spun immediately after it has been plucked or shorn from the animal, it is usually first teased by hand or combed to remove the impurities. Combing also straightens and arranges the fibres and separates the long hairs from the wool.[8] Hackles and combs with long teeth were used for the purpose. Another process, carding, probably was not used until Roman times. Carding introduces air in the wool and removes dirt and tangles, helping to produce consistent yarn. The commonly held view that early carders were probably made with dried teasels (*Dipsacus fullonum*), whose prickly heads have flexible bristles with hooks (Figure 71), is not likely since the hooks break off easily. Rather, they may have been used for raising nap on the surface of a textile (Goldmann 2007).

Figure 71. Teasel (Author's photo).

The prepared wool was often in the form of roving, which is rope-like in appearance. Numerous depictions on Attic vases, such as the Amasis lekythos in the Metropolitan Museum of Art, show the preparation and use of roving during spinning. There existed even a special tool, *epinetron* (Figure 72), a clay thigh-protector used to prevent the grease spoiling the clothes. *Epinetra* are often decorated with wool-working activities. In Italy examples have been found in Sicily in Hellenic contexts (Mercati 2003).

In most cases, wool was spun unwashed, since the lanolin, naturally present in the coat, helps in the process. Occasionally, however, it was scoured to remove the impurities, which could constitute up to forty percent of the wool's weight (Wild 1970, 23; Morel 1978, 108). Washing is, for example, recommended by Varro (*Rust.* 2.2.18) and Columella (*Rust.* XI.11.35). Ancient scouring agents were urine or soapwort. The wool had to be washed/scoured if it was to be dyed before spinning. As noted before, sometimes this was done 'on the hoof', but wool could also be washed after it had been removed from the animal. Natural water sources were probably used for the procedure but occasionally special

Figure 72. Epinetron with spinning scenes, early 5th century BCE (After Forbes 1964, vol. 4, 164 fig. 4).

installations were built. Thus, Barker (1988, 782) reports on a 4th century BCE farmhouse in Central Italy: "In the central room at Podere Tartuchino were a hearth, a sunken *dolium*, and an associated system of channels for liquids, at first thought to be for olive oil production but now identified more satisfactorily as a facility for the large scale processing of wool". Numerous installations that might have been used for the same purpose have been identified at Roman Pompeii (D'Orazio and Martuscelli 1999, 92; Borgard and Puybaret 2004). Identification of such installations is problematic, however, since similar constructions were used for other purposes.

Unlike in the case of flax, some of the tools used for wool preparation do survive.

Comb

Wool combs in ancient times were made of wood, bone or horn. Horn items from the Early–Middle Bronze Age settlement of Fiavè-Carera (Figure 73) may have been used for wool combing (Bazzanella *et al.* 2003, 141–142). Three 3rd century BCE tombs at Este contained pairs of combs, each made of bronze sheet and attached to a rounded handle (Figure 74); they have been identified as wool combs (Chieco Bianchi 1987, 213 nos. 98–99 and 215 Fig. 33 nos. 98–99, 227 nos. 36–37 and 229 Fig. 55 nos. 36–37; Capuis and Chieco Bianchi 2006, 282 nos. 55–56). More work needs to be done on various types of excavated combs in order to assess if they were used for the purpose of wool preparation.

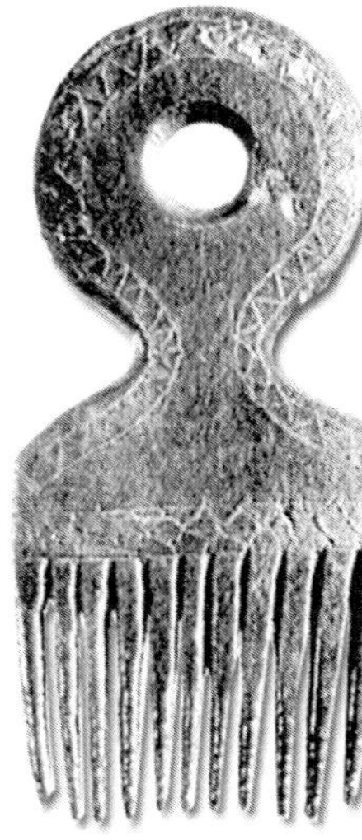

Figure 73. Horn comb from Fiavè-Carera, Middle Bronze Age (After Bazzanella et al. *2003, 142).*

Hook

The use of this implement, a large curved metal strip about 30 cm long with a handle on one hand and a hook on the other is not certain and the only interpretation offered so far has been that it was used for carding (Pacciarelli 1999, 99). All examples have been found in burial contexts and always come in pairs. Hooks (*uncini*) were found in Tombs 16 and 63 of Torre Galli (Figure 75), in both cases associated with spools and spindle whorls (Pacciarelli 1999, 99, 147 nos. 16–17, pl. 18, 159 nos. 17–18, pl. 54A), Tombs 123 and 178 in S. Valentino Torio (Gastaldi 1979, 23, fig. 30 nos. 4 and 6) and Tomb 5946 at Pontecagnano (pers. obs. Pontecagnano Museum, 2002), the latter associated with nine spindle whorls. Apart from the Early Iron Age Torre Galli examples, all other implements have been dated to the Orientalising period.

Situla

The prepared wool had to be stored and

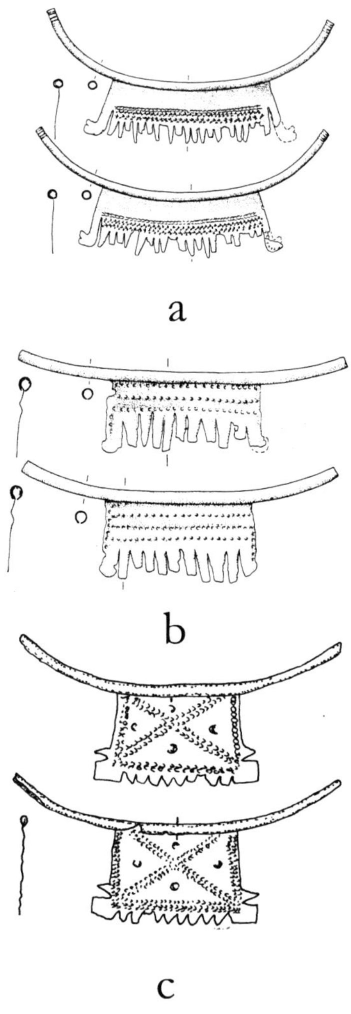

Figure 74. Bronze combs from Este, 3rd century BCE: a) Tomb 23 Casa di Ricovero; b) Tomb 36 Casa di Ricovero; c) Tomb 123 Villa Benvenuti (After Chieco Bianchi 1987, 215 fig. 33 nos. 98–99, 229 fig. 55 nos. 36–37; Capuis and Chieco Bianchi 2006, pl. 155 no. 56).

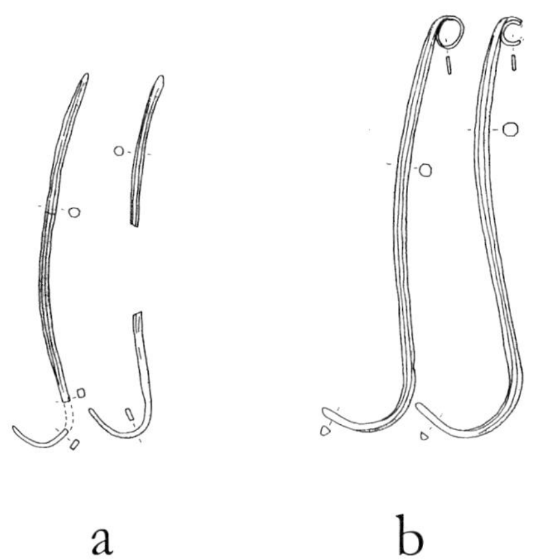

Figure 75. Hooks from Torre Galli, 9th century BCE: a) Tomb 16; b) Tomb 63 (After Pacciarelli 1999, 244 fig. 18 nos. 16–17, 280 fig. 54A nos. 17–18).

rest before and during its use. While various kinds of containers may have been utilised for the purpose, situla in particular was the vessel of choice in North Italy. The bottom scene of side A of the bronze *tintinnabulum* from Bologna shows two seated women dressing the distaffs from a large situla placed on the ground between them. Like its Greek equivalent *kalathos* or wool basket, often depicted in wool-working scenes on Attic vessels, situla could be used for several diverse functions and may have had a wider ceremonial significance. It is therefore difficult to associate it definitively with wool-working unless found together with other textile tools. Such is the case of the bronze situla containing bronze spindle, distaff and two combs, which was found in a 4th century BCE Tomb 23 in Casa di Ricovero necropolis of Este (Figure 76) (Chieco Bianchi 1987, 213 no. 97, 215 fig. 33 no. 97).

Spinning

Once the fibre mass has been prepared, it can be converted into a yarn through twisting and drawing out, or drafting, the fibres – the process known as spinning (Forbes 1956; Ryder 1968; Leadbeater 1976; Crockett 1977; Barber 1991, 39–78; Wild 2003, 11–13). During spinning, yarn acquires tensile strength, which is a prerequisite for a woven textile. In order to achieve control over the fineness and evenness of the thread, the fibres must be drawn and twisted simultaneously and with the same speed. The suspended or drop spindle, which consists of a small rod with a whorl that sustains rotary movement, accommodates all these requirements very efficiently. The type of spindle used in ancient Italy, and the north of the Mediterranean in general, was the low-whorl spindle, in which the whorl was attached to the lower end of the spindle rod. The alternative, high-whorl spindle was used in Egypt and parts of the Near East. The two techniques may have overlapped in Anatolia, where 'middle-whorl' spindles have been found (Barber 1991, 60–61). During spinning, raw fibre could be held in the hand or fastened to a distaff. A spindle and a distaff, thus, are two spinning tools and we have archaeological evidence for both.

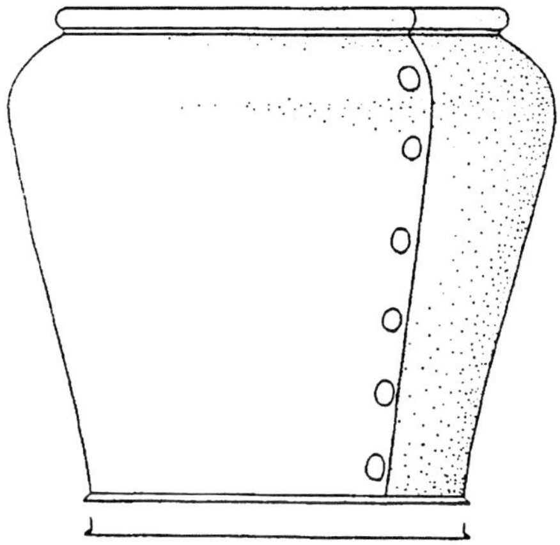

Figure 76. Bronze situla from Tomb 23, Este Casa di Ricovero, 3rd century BCE (After Chieco Bianchi 1987, 212 fig. 32 no. 90).

Spindle

A spindle is a simple rod with a hook or dent on one end to attach the thread. Spindles were usually made of wood or bone and rarely survive in archaeological contexts. The earliest finds come from the Bronze Age Terremare settlements of North Italy (Gastaldi 1865, 45), and numerous examples were recovered from Fiavè-Carera (Figure 77) (Perini 1987; Bazzanella *et al.* 2003, 137–138), Ledro (Rageth 1974), and Castione dei Marchesi (Mutti *et al.* 1988, 243–246 nos. 98–107). Several wooden spindles are also known for the Iron Age. A long spindle, with its whorl still on the elaborately decorated wooden shaft, was recovered from the underwater village of Gran Carro, dated to the early 9th century BCE (Figure 78a) (Tamburini 1995, 169 no. 2081 Fig. 51 no. 2081). Wooden spindle shafts have been identified at the waterlogged Iron Age site of Longola di Poggiomarino (pers. com. Claude Albore Livadie, 2007). At least four other items were found in a 6th century BCE context in the Sant Omobono sanctuary area of Rome, their length ranging from 16 to 30 cm (Pisani Santorio 1977, 57, fig. 18; *Enea nel Lazio* 1981, 147 no. C64 a–b). Incomplete examples were found in burials such as Moroni Tomb 24 at Verucchio (Gentili 1985, 80 no. 40, pl. XXXVII), or Tomb 7 at S. Ilario, Bettolino, dated 550–525 BCE where some spindle whorls have parts of the spindle shaft preserved (Malnati and Manfredi 1991, pl. 30). These later examples are a good reminder that all spindle whorls deposited in burials were probably originally attached to wooden spindles, which do not survive.

Figure 77. Wooden spindle fragments with wooden spindle whorls from Fiavè Carera, Middle Bronze Age (After Bazzanella et al. *2003, 137).*

In addition to the rare surviving wooden and bone items, there are examples made of metal, all of which were found in burial contexts. While their social significance will be discussed in the next chapter, these luxury spindles are instructive for the reconstruction of the length and shape of ancient spindles.

Figure 78. Spindles: a) Gran Carro, 9th century BCE (After Tamburini 1995, fig. 51 no. 2081), b) Tomb 56 Benacci-Caprara, Bologna, 7th century BCE (© Bologna, Museo Civico Archeologico); c) Tomb B, Casale Marittimo, 7th century BCE (After Esposito 1999, 62 fig. 55); d) Tomb 23, Este Casa di Ricovero (After Chieco Bianchi 1987, 215 fig. 33 no. 97).

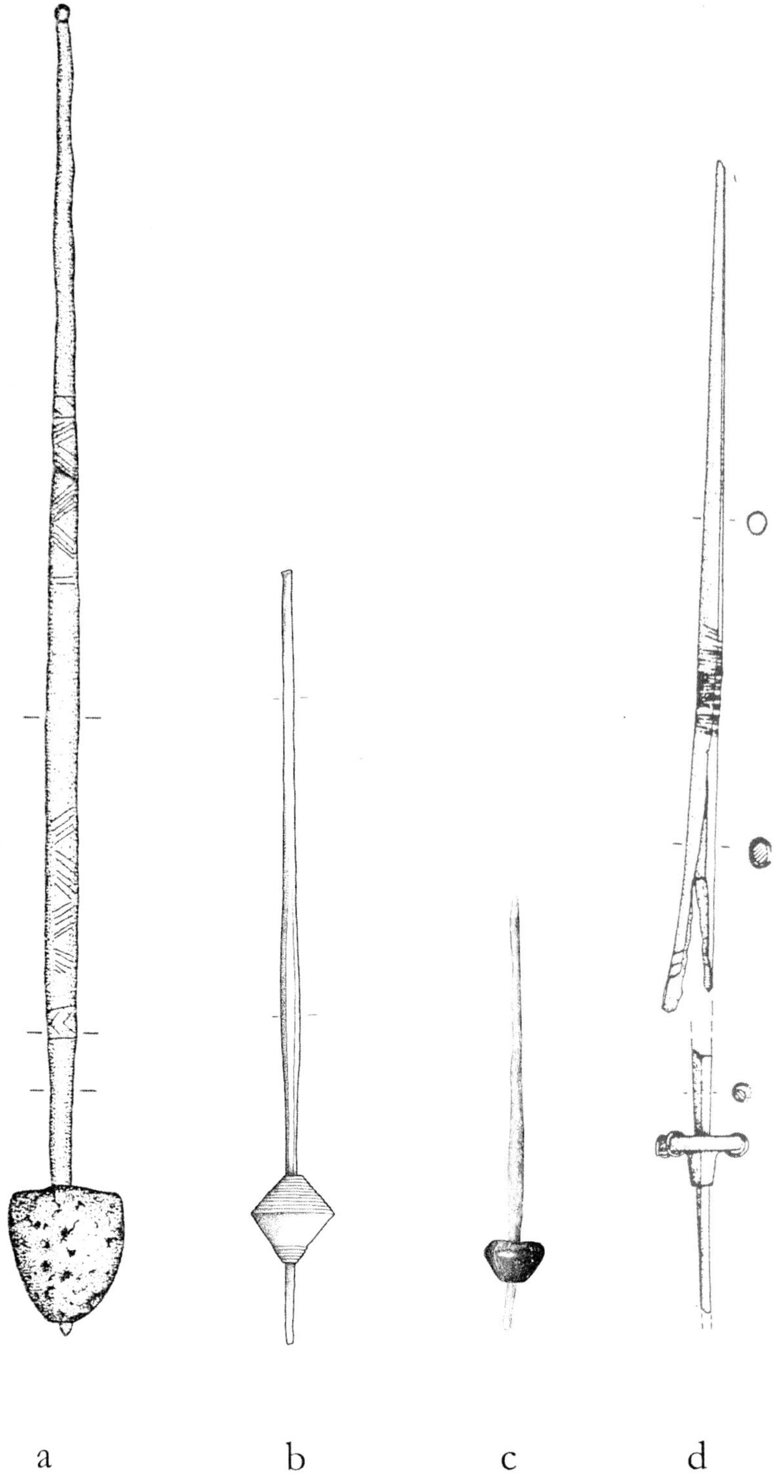

A bronze spindle with bronze biconical whorl from the 8th century BCE Benacci-Caprara Tomb 56 in Bologna still had some thread wound around it at the time of discovery (Figure 78b) (Brizio 1889, 329 no. 50; Morigi Govi 1982, 234; Rallo 1989b, pl. XIII–4; Forte and von Eles 1994, 55 no. 32). It is 25 cm long and was found in association with a bronze distaff and five terracotta spindle whorls. Another such spindle was found in Cuma (Kilian 1970, pl. 267 no.3). In the 7th century BCE Tomb B at Casale Marittimo, a bronze spindle with a conical bucchero whorl was found (Figure 78c) (Esposito 1999, 62 Fig. 55; Bottini *et al.* 2003, 133). This spindle is 14 cm long and is most likely a display object. A later, 4th century BCE burial, Tomb 23 in Casa di Ricovero necropolis of Este, yielded a bronze spindle with a bronze whorl in the shape of a wheel and remains of thread wound around its central part (Figure 78d) (Chieco Bianchi 1987, 213 no. 97, 215 fig. 33 no. 97). This spindle is 30 cm long and was found in a bronze situla together with a bronze distaff and two combs.

Length is an important parameter in a spindle since it determines the ease of its use. Giovanangelo Camporeale estimates the average length of a spindle during the 1st millennium BCE to be about 30 cm (Camporeale 1997, 197). Bronze Age and Roman spindles ranged from 12 to 27 cm (Ryder 1968, 79). The Gran Carro spindle however, is 40 cm long. The length most likely depended on the weight of the spindle whorl. Thus, very light spindle whorls require short spindle shafts in order to rotate evenly (Mårtensson 2007). A model spindle from Casale Marittimo may reflect the actual length of a spindle needed for such a spindle whorl.

Whether made of wood or metal, spindles often bear incised decoration, which may have a magical significance but could also be functional in keeping the wound thread in place.

Spinning hook

As mentioned before, a spindle needs a hook on top to keep the yarn from slipping off the spindle shaft. In most cases it was a simple indentation in wood, but occasionally a special element was added for the purpose, a spinning hook. It is a small object made of thin sheet of metal shaped into a cone and ending in a solid hook about 5 cm long, which can be mounted on a wooden spindle shaft. Several iron spinning hooks were found in Gravina di Puglia (Figure 79) (Macnamara 1992, 238 nos. 1905–1906). One such bronze implement was found in the 3rd century BCE Tomb T 258 at Metaponto (Carter 1998, 328 no. 179, 818, 819 H10).[9]

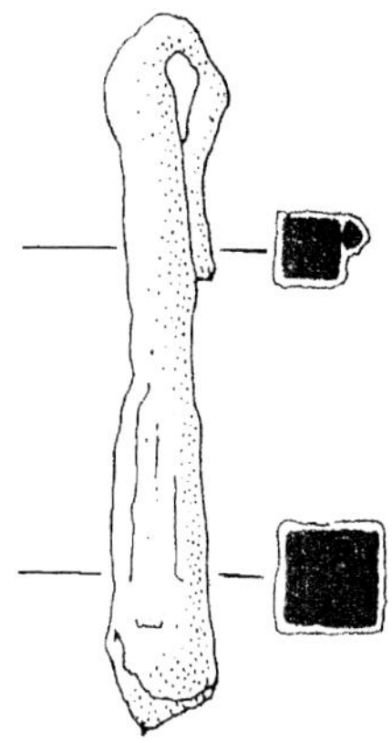

Figure 79. Spinning hook from Gravina di Puglia, 4th century BCE (After Macnamara 1992, 238 no. 1906).

Spindle whorl

Function

Since most spindles in antiquity were made of wood, often, the only evidence for their use consists of the less perishable spindle whorls. A spindle whorl is a symmetrical, centrally pierced object, often made of fired clay, stone, bone or wood. The vast majority of spindle whorls in Italy are made of fired clay. Wooden and stone whorls are known from Bronze Age sites (Bazzanella *et al.* 2003, 137–138). Bone items are common in the Roman period. Whorls made of luxury material such as glass and amber have been found in burial contexts throughout the 1st millennium BCE.

A spindle whorl is positioned on a spindle, and provides weight and tension for spinning fibres into thread (Ryder, 1968; Crewe 1998, 5–8). In a spindle, the spindle whorl has the function of a fly-wheel. According to Barber (1991, 303), "there are few practical requirements for their shape: they need to be broad enough to help maintain the momentum; they may be heavy enough to help with drafting but must not be so heavy that they break the thread; and it helps if they are symmetrical about a central axis so as not to introduce an inefficient and irritating wobble into the rotation".

Spindle whorls, often in large numbers, have been found almost at every settlement site in Italy dating from the Neolithic period onwards, so only a selection of examples will be noted below. A variety of whorl shapes are known; this may be due to the fact that different whorls were required for different types of yarn. Overall, since these shapes do not change dramatically over time, whorls are nearly impossible to date unless they come from a datable context or have a specific decorative scheme.

Typology [10]

The following typology is intended as a general guide since every site has its own whorl classification. Nevertheless, it should be pointed out that, frequently, site typologies are constructed on the basis of non-functional parameters, like decoration, as, for example, in the case of Tarquinia (Bonghi Jovino 2001, 64–65) and Osteria delll'Osa (Bietti Sestieri 1992b, 310–314), rather than functional traits: basic shape, weight, diameter and its ratio to height. Using functionality as the main criterion, the whorl shapes can be subdivided into the following basic types (Figure 80):

A. Discoid

The whorls of discoid shape have a high diameter to height ratio and rotate slowly but for a relatively long time. Often, discoid whorls have a slightly lenticular section. Chronologically, this is probably the earliest type, and it almost disappears by the beginning of the 1st millennium BCE. The early discoid whorls are especially wide and flat, probably because they were often made of pottery sherds. Over time, the diameter seems to decrease.

B. Cylindrical

The cylindrical type possibly was developed from the discoid shape. This type is common in the Bronze Age but becomes scarce by the 1st millennium BCE.

C. Globular

1. spherical
2. lenticular or depressed globular
3. ogival

The globular type is present from the early periods on, but never in significant numbers. The three subtypes indicate that the diameter to height-ratio and, hence, the rotation speed, could be varied by changing the height of the whorl.

D. Conical

1. conical
2. truncated conical

This type is asymmetrical, with the top of the whorl considerably wider than the bottom. This asymmetry makes the whorl rotate faster due to the action of centripetal force. An additional advantage is that the large top surface helps to keep the spun fibre on the spindle. These factors may account for the popularity of this shape, especially during the 1st millennium BCE. The maximum diameter of conical whorls is the same as or only slightly larger than their height, a proportion that appears to be optimal for spinning yarn of medium twist.

It should be noted that, frequently, this type of whorl is represented incorrectly in publication drawings and photographs, that is, with its top inverted.

E. Biconical

The maximum diameter is overall slightly greater than the height. As in the case of conical whorls, this proportion appears to be optimal for spinning yarn of medium twist.

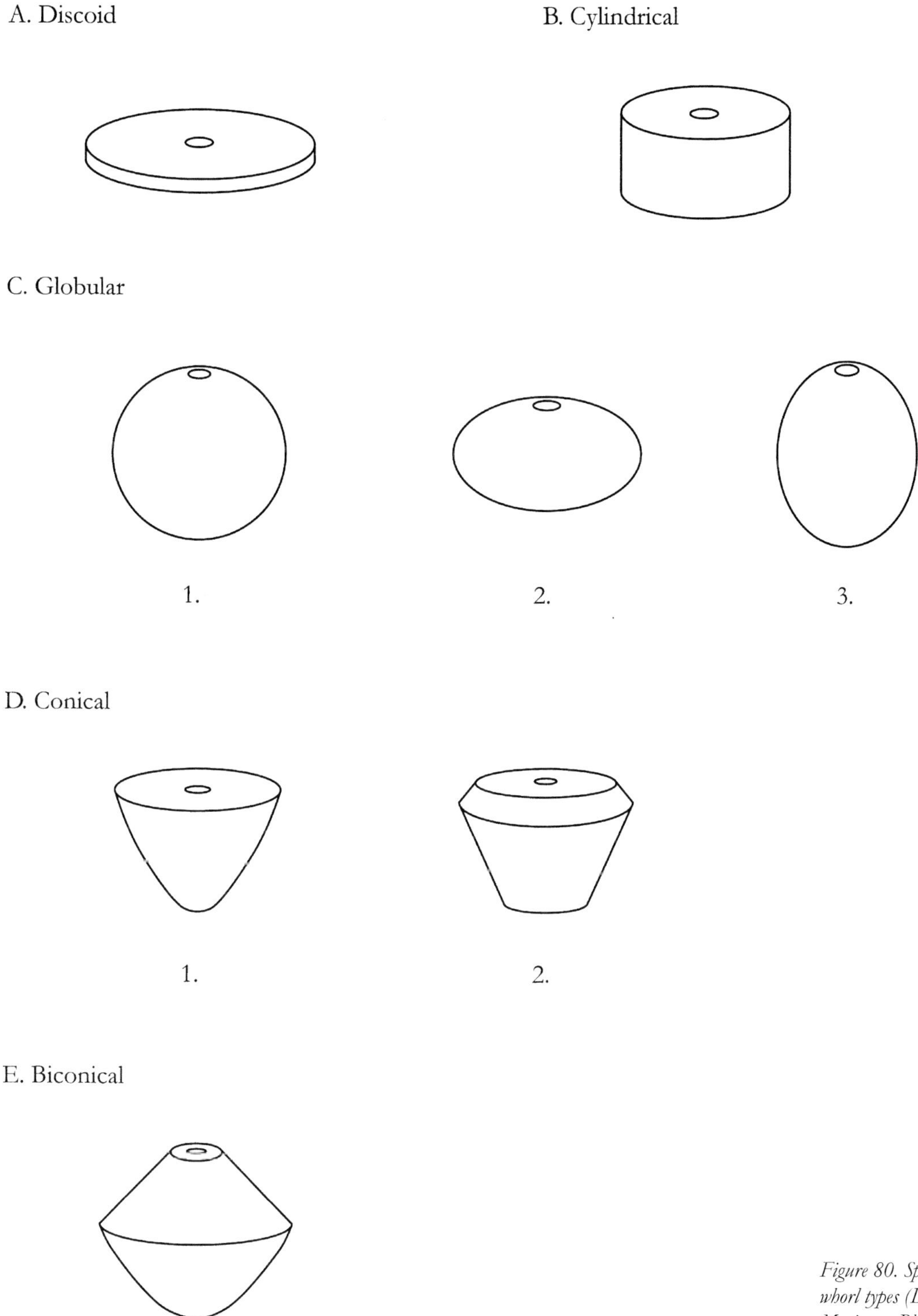

Figure 80. Spindle whorl types (Drawing by Marianne Bloch Hansen).

F. Cuboid
This relatively rare type of spindle whorl occurs in early contexts (Macnamara 1976, 169, fig. 54).

Functional features
Shape
The shape of the whorl does not influence the yarn significantly. More important is the ratio of diameter to height, which may affect the speed with which the spindle rotates.

The majority of the pre-Roman sites yield a variety of whorl shapes (Table 4). As we shall see in the next chapter, the situation is similar in individual burials where a number of whorls have been deposited: they range in shape and size, showcasing the skill of their owner in spinning different yarns. This variety of shapes illustrates variety of yarns and, consequently, textiles produced.

Although the data are very general and few, Table 4 also shows some tendencies in the change of spindle whorl types over time. Type A (discoid), for example is more typical for the Bronze Age and disappears after the Early Iron Age. This type might have been preferable for spinning bast fibres. Type E (biconical), on the other hand is more prevalent (and, in fact, it is often predominant) in later periods. The largest variety in types of spindle whorls on individual sites seems to be present during Bronze Age and Early Iron Age. Towards the later periods, most sites have only two or three types, suggesting that these types were quite universal and could be used to make a variety of yarns.

Predominance of a certain type may also indicate that it was traditional at the site or that the site was specialising in a certain yarn and, consequently, a certain type of textiles.[11] Thus, at Acquarossa, the biconical form predominates (Wikander 1986, 112), while at Murlo, more than 90% of the whorls are truncated conical (Gleba 1999, 62).

The depression on the top, especially in asymmetrical types, is a common feature of the whorls from many Italic sites and is already present in the whorls from the 13th century BCE site of Crostoletto di Lamone (Negroni Catacchio 1982, 86, fig. 49 nos. 19 and 20). Because so many whorls do have this depression, it is likely that it provided some sort of advantage during spinning. Possibly it secured the spun yarn on a spindle.

Weight
The weight of the whorl is a significant feature and can indicate the quality and, sometimes, type of fibre being spun and the quality of the thread produced, as demonstrated by experimental studies (Ryder 1968, 81–82; Parsons 1975; Barber 1991, 52; Andersson 2003; Grömer 2005, 109). For fine, delicate threads, small, light whorls are used during spinning (Liu 1978, 99). On the other hand, for stronger, coarser fibres or for plying, heavier whorls are needed. The recorded upper limit for the weight of the whorl is 140–150 g (Liu 1978, 90; Barber 1991, 52). The smallest spindle whorls known come from the Middle East during the Islamic period, when implements as small as 0.8 cm in diameter and under 1 gram of weight were used for spinning cotton (Liu 1978, 90–91). Thus, for particular types of yarn, spindle whorls fall within a certain range of weights. The following numbers are given by Barber (1991, 52). For spinning long-staple wool, heavy spindles are used, ranging 100–150 g. Even heavier implements are utilised for spinning full-length flax or for plying wool yarns. On the other hand, short wool, flax tow or cotton, are spun with a light spindle. The weight of the spindle used for short fine wool is about 8 g, while for the medium to heavy wool it is about 33 g (Ryder 1968, 81; Barber 1991, 52). By knowing the range of whorl weights at a particular site, one could infer what

Site	Date	Types	Source
Molina di Ledro	EBA–MBA	A, C, D	Bazzanella *et al.* 2003, 152–155.
Castione dei Marchesi	MBA	A, B, C, D, E	Mutti *et al.* 1988, 120–125.
S. Rosa di Poviglio	MBA	A, B, C, D	Bernabó Brea, Bianchi and Lincetto 2003, 113.
Cisano	BA	A, B, D	Fasani 1980, 154–161.
Castellaro del Vhò	BA	A, B, C, D	Frontini 1997.
Gorzano	BA	A, D, E	Säflund 1939, 146.
Lipari	BA	A, B, C, D	Bernabó Brea.
Sorgenti della Nova	FBA	A, B, C, D, E	Negroni Catacchio 1995, 369–372.
Narce	8th–4th c.	A, C, E, F	Macnamara 1976, 166.
Tarquinia	BA–EIA	A, C, D, E	Mandolesi 1999.
Sala Consilina	EIA	A, C, D, E	Kilian 1970, 198–202.
Gran Carro	FBA–EIA	A, B, C, D, E	Tamburini 1995, 246–248.
Torre Castelluccia (TA)	FBA–EIA	B, C, D, E	Gorgoglione *et al.* 1993, 43.
Frattesina di Frata Polesine	FBA	B, D, E	Bellintani 1992.
Ficana	950–500 BCE	B, C, D2, E	Brandt 1996, 269–271.
Fonte Tasca	FBA	C, D, E	Di Fraia 1995, 458, fig. 8 nos. 1–9 and 14–15.
Osteria dell'Osa	EIA	C, D, E	Bietti Sestieri 1992, 309–314.
San Giovenale	EIA	C, D, E	Berggren and Berggren 1980, 18; 1981, 42.
Luni sul Mignone	EIA	C, D, E	Östenberg 1967, 127; Hellström 1975, 63–64.
Torre del Mordillo	EIA	C, D, E	Colburn 1977, 515–516.
Pontecagnano	EIA	C, D, E	d'Agostino and Gastaldi 1988, 37–38k.
Borgo le Ferriere	EIA–6th c. BCE	C, D, E	Maaskant-Kleibrink 1992, *passim*.
Accesa	7th–6th c. BCE	C, D, E	Camporeale 1997, 57.
Poggio Civitate di Murlo	7th–6th c. BCE	C, D, E	Gleba 1999, 92.
Acquarossa	7th–6th c. BCE	C, D, E	Wendt 1982, 46.
Roselle	6th–5th c. BCE	C, D, E	Donati 1994, 139–140.
Vulci	6th–5th c. BCE	C, D, E	Falconi Amorelli 1983, 151–158.
Cetamura	5th–2nd c. BCE	C, D, E	De Grummond 2000, 29.
La Piana	4th–3rd c. BCE	C, D, E	Pers. obs. La Piana excavation, 2003.
Satrianum	5th–4th c. BCE	C, E	Holloway 1970, 113.
Colle Madore	6th c. BCE	C, E	Vassallo 1999, 143.
Incoronata	EIA	D	Pizzo 1986, 179–180.
Padova	FBA–EIA	D, E	*Padova Preromana* 1976, 118.
Cavallino	6th–4th c. BCE	E	Pancrazzi 1979, 190.
Gravina di Puglia	4th–1st c. BCE	E	Tatton-Brown 1992, 220.

Table 4. Types of spindle whorls at Pre-Roman sites in Italy.

types of yarn that were being spun there. Unfortunately, excavators often fail to publish the weight measurements making such estimates impossible.[12]

In Italy, the data have been analysed only for a handful of sites that are far removed chronologically from each other (Table 5). The numbers indicate, however, that relatively fine and possibly woollen yarn was spun at these sites.

Hole diameter and centring

Another important feature of a whorl is the diameter of the hole for the spindle rod. The hole must be large enough to accommodate the shaft. The expected range of the hole diameter for spindle whorls is 0.3–1 cm (Liu 1978, 97). Here, again, the published data are relatively scarce. The range of the maximum hole diameter in the Murlo whorls is 0.2–0.8 cm, with half falling between 0.4 and 0.5 cm, that is, in the lower end of the expected range (Gleba 1999, 62). This indicates that the majority of the spindles were quite delicate, confirming

Site	Date	Number	Weight (g)	Mean	Bibliography
Lagozza	Neolithic	-	30–40	-	Baioni *et al.* 2003, 101.
S. Rosa di Poviglio	MBA	100+	3-64	-	Bernabó Brea, Bianchi and Lincetto 2003, 111.
Ficana	950–500 BCE	23	6–53	23.9	Lipponen 2007, 2.
Poggio Civitate di Murlo	7–6th c. BCE	441	2–48 (most 4–12)	9.6	Gleba 1999, 60; Gleba 2000, 78.
Cetamura	4th–2nd c. BCE	18	4.5–36	-	Hackworth 1993, 25–32.
La Piana	4th–2nd c. BCE	40	6–56.6	-	Pers. obs. La Piana excavation, 2003.

Table 5. Weight range of spindle whorls.

that the yarn produced at Murlo was of fine quality.

The centring of the hole is extremely important. Even if the whorl is not perfectly symmetrical, the hole must be centred to avoid wobble during spinning (Barber 1994, 38; Crewe 1998, 13). Another important point is that the perforation in whorls is often slightly tapering, to help wedge the spindle shaft in (Liu 1978, 97; Barber 1991; Crewe 1998, 12). This feature often helps to distinguish small whorls from beads, the latter having no need for a tapering hole.[13] The taper is often very slight, but it is usually visible with the unaided eye and can be measured. For example, the mean minimum diameter of the hole of Murlo whorls is at least 0.1 cm less that the mean maximum diameter (Gleba 1999, 62).

Decoration

While their basic type remained the same, spindle whorls could be fashioned into certain forms for decorative purposes, as in the case of polygonal and unusually shaped whorls (Figure 81). Wheel-shaped whorls are common during Bronze Age. Fluted or ribbed whorls appear in the later part of Iron Age and become prevalent in some areas of Central Italy, particularly in burial contexts. Alternatively, or in addition, the whorl surface may be decorated with incised, impressed, or stamped patterns. Already during the Neolithic period, most whorls were incised with simple geometric patterns (Baioni *et al.* 2003, 100 fig. 3). In the case of asymmetrical whorls, the decoration most frequently occurs on the wider top of the whorl, which may be visible during spinning. Incised decoration is the most common. Impressed designs were made with a cord, or some small tool. Stamps are comparatively rare and simple. In the south, the whorls were occasionally painted, although this practice seems to have been typical for the Hellenic populations. In contemporary Greece, spindle whorls with painted decoration are common (Davidson and Burr Thompson 1943, 95).

Occasionally, graffiti (both alphabetic and non alphabetic) appear on spindle whorls, a practice documented as early as the 7th century BCE. These have been mainly recovered from burial contexts, suggesting a special significance for the objects (Govi 1994, 222).

In addition to the more common terracotta examples, glass spindle whorls were deposited in rich burials during the Orientalising period. They have been found predominantly in North Italy, for example, at Verucchio (Gentili 2003, 178 no. 41, pl. 87, pl. CLXVII, 217 no. 18, pl. 103), Este (Chieco Bianchi and Calzavara Capuis 1985, 291 no. 69, pl. 195; 325 no. 42, pl. 220), Montescudaio (Camporeale 1985, 62 no. 120) and Bologna (Bartoloni

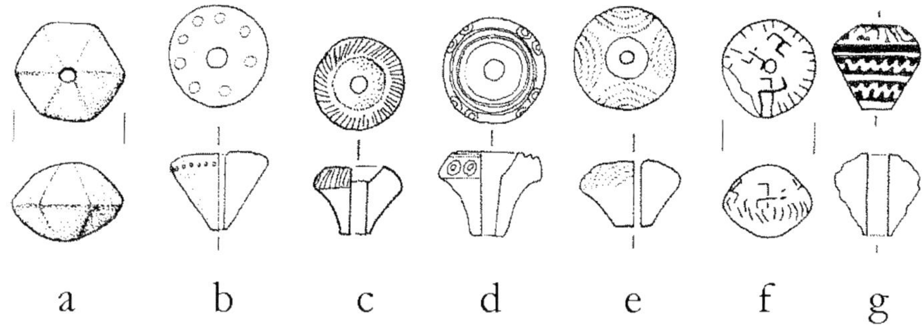

Figure 81. Decoration of spindle whorls: a) faceted spindle whorl, Tarquinia (After Bonghi Jovino 2001, pl. 53 no. 221/2); b) spindle whorl with circular impressions, Tomb 3, Villanova-Caselle San Lazzaro (After 245, pl. VI no. 31); c) spindle whorl with incised decoration, Tomb 13, Lippi, Verucchio (After Forte and von Eles 1988, 112 no. 222); d) spindle whorl with impressed concentric circles, Tomb 13, Lippi, Verucchio (After Forte and von Eles 1988, 112 no. 220) e) spindle whorl with concentric semicircles in false cord, Tomb 3, Villanova-Caselle San Lazzaro (After 245, pl. VI no. 37); f) spindle whorl with graffiti, Tarquinia (After Bonghi Jovino 2001, pl. 53 no. 56/3); g) glass spindle whorl, Tomb 13, Lippi, Verucchio (After Forte and von Eles 1988, 112 no. 238).

2000b, 279–280 nos. 351–355). The majority of extant examples are in blue glass paste with decorative patterns in white/yellow infill. These spindle whorls are often misidentified as beads, especially in the case of globular types (cf. Gentili 2003, 47 nos. 14–14b, 59 no. 16). The features, which may help in distinguishing glass whorls from beads, are their larger size, presence of remains of a shaft in the hole, number and position in the burial (whorls are rarely found in multiples and on the body). Being made of luxury material, glass whorls probably served as status indicators, rather than as functional implements.

Distaff

Function

A distaff is designed to hold the prepared fibre during spinning. Short hand-held distaffs were used for spinning short-stapled fibres, while longer ones, held under the arm or in the belt, were used for longer fibres. Distaffs were usually made of wood; even a simple forked stick would have been sufficient for the task. Considerably less frequent were items made of metals, including bronze, iron, silver and precious materials like glass and amber. Spinning tools made of gold, bronze and ivory are also known in the Near East and Greece (Barber 1991, 60–65). Yet, while wooden tools have not survived in archaeological contexts, numerous examples of objects identified as distaffs made of luxury materials have been found in tombs of notable wealth (Bartoloni 1989, 42–44). The majority of these distaffs are dated to the Early Iron Age and Orientalising period. Although such objects are not as frequent in later times (cf. Gottschalk 1996), Roman sources suggest that distaff had important social connotations, as will be discussed in the next chapter. Distaffs found in early Italian tombs, thus, most likely were a status symbol. The following typology of distaffs is thus created not so much from a functional point of view but rather as a guide to the cultural and geographical diversity of these objects.

Typology

I. Long distaff with forked top.

1. Length 100–150 cm (Figure 82).
   Very long staff forked on one end, made of wood and covered with bronze; the number of appendages in the bifurcation varies from two to eight.

This type is much longer than other types and was not held in the hand but rather fixed in the belt and supported by left shoulder, or even stuck into the ground next to the spinner (Bartoloni 1989, 43). These distaffs were constructed by winding bronze wire (two examples) or narrow bronze strip (three examples) around a wooden core. One example from Civita Castellana (ancient Falerii) was in addition decorated with amber

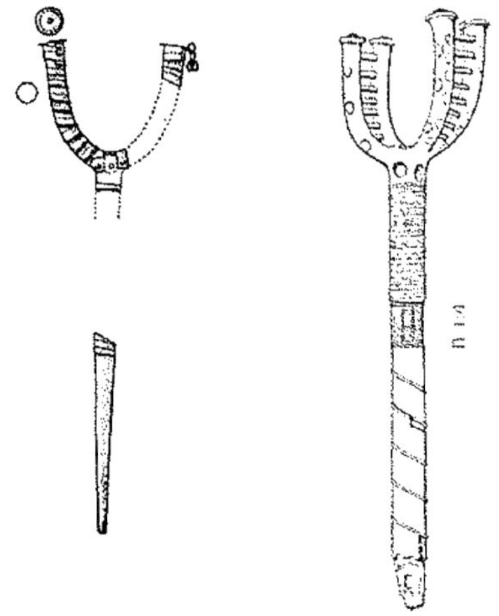

a b

Figure 82. Distaffs of type I: a) Tomb JJ 17–18, Quattro Fontanili, Veio, 8th century BCE (From Close-Brooks 1963, fig. 93); b) Tomb 15, Civita Castellana, 8th century BCE (After Barnabei 1894, 222 fig. 99 n).

studs (Barnabei 1894, 222; Randall-MacIver 1924, 180). Apart from the unprovenanced item in the Fiesole Museum, all extant examples come from a relatively small geographical area in South Etruria and *Agro Falisco-Capenate* (Map 4): six from Veio (Close-Brooks 1963, 220; Franco *et al.* 1970, 223–225 no. 7; Fabricotti *et al.* 1972, 384 no. 10; Bartoloni 1989, 43 note 65; Berardinetti and Drago 1997, 60), four from Civita Castellana (Barnabei and Pasqui 1894, 388), three from Narce (Barnabei and Pasqui 1894, 389; Pasqui 1894, 409 no. 11, 422 no. 8; Dohan 1942, 22 no. 18), and single examples from Capena (Stefani 1958, 133), Nazzano (Stefani 1911, 440) and Osteria dell'Osa (Bietti Sestieri 1979, 84). The distribution suggests that this type of distaff may have been typical for the Faliscan area.

Such long distaff may have been utilised for spinning a particular type of fibre, for example long linen. Written sources indicate that Faliscan territory had fame for producing linen textiles and other products.

II. Short hand-held distaff.

A. Made of hammered sheet metal, with wooden core.

1. Conical top, convex shaft. Length 20–25 cm (Figure 83 A1).

This is by far the largest group of distaffs known. The greatest number of type A1 distaffs (more than 40) was found in the necropoleis of Bologna, including the Villanova area (Brizio 1889, 330; Müller-Karpe 1959, pl. 79C no. 11; Pugliese Carratelli 1986b, fig. no.201; Tovoli 1989; Forte and von Eles 1994, 55 no. 32, 123; Bartoloni 2000, 279 no. 350).

Outside the sphere of Bologna, numerous examples have been found in Veio (Stefani 1928, 103; Palm 1952, 69 no. 17 pl. XXV, 70 no. 7 pl. XXVI; Close-Brooks 1963, 231 i, 237 fig. 195, 238 fig. 106; Cavallotti Batchvarova 1965, 132 aa fig. 54; Falconi Amorelli 1967, 247 no. 10, fig. 98; Bartoloni *et al.* 1994, fig. 6; Buranelli, Drago, and Paolini 1997, 69, 73 fig. 22; Bartoloni 2001, 101 no. I.G.5.28).

Others have been found in the burials of Verucchio (Gentili, 1985, 80 no. 41, 104 no. 17), Volterra (Cateni 1998, 43, 47), Cerveteri (Vighi 1955, 132 no. 2; Pohl, 1972, 105, fig. 88), Tarquinia (Pasqui 1885, 512; Hencken 1968a, 188, 264; Randall-MacIver 1924, 55), Vulci (Falconi Amorelli 1983, no. 357, fig. 86; Fugazzola Delpino 1984, 88), Bisenzio (Paribeni 1928, 347, fig. 2; Bottini *et al.* 2003, 134 no. 77), Sasso di Furbara (Brusadin Laplace and Patrizi Montoro 1992, 279 fig. 28, 288 no. 5), Narce (Pasqui 1894, 441 no. 16, 445 no. 16; Dohan 1942, 17 no. 22, Pl. VI; Turfa 2005, 82–83 no. 3), Capena (Stefani 1958, 92, fig. 23) and Rome (Pinza 1905, 158 b; Colonna *et al.* 1976, 137 no. 7). Two examples are known from Este, in the Paleo-Venetic area (Soranzo 1885, 71; Tosi 1992, 70, fig. 53; Chieco Bianchi and Tombolani 1988, 79 fig. 99), and another

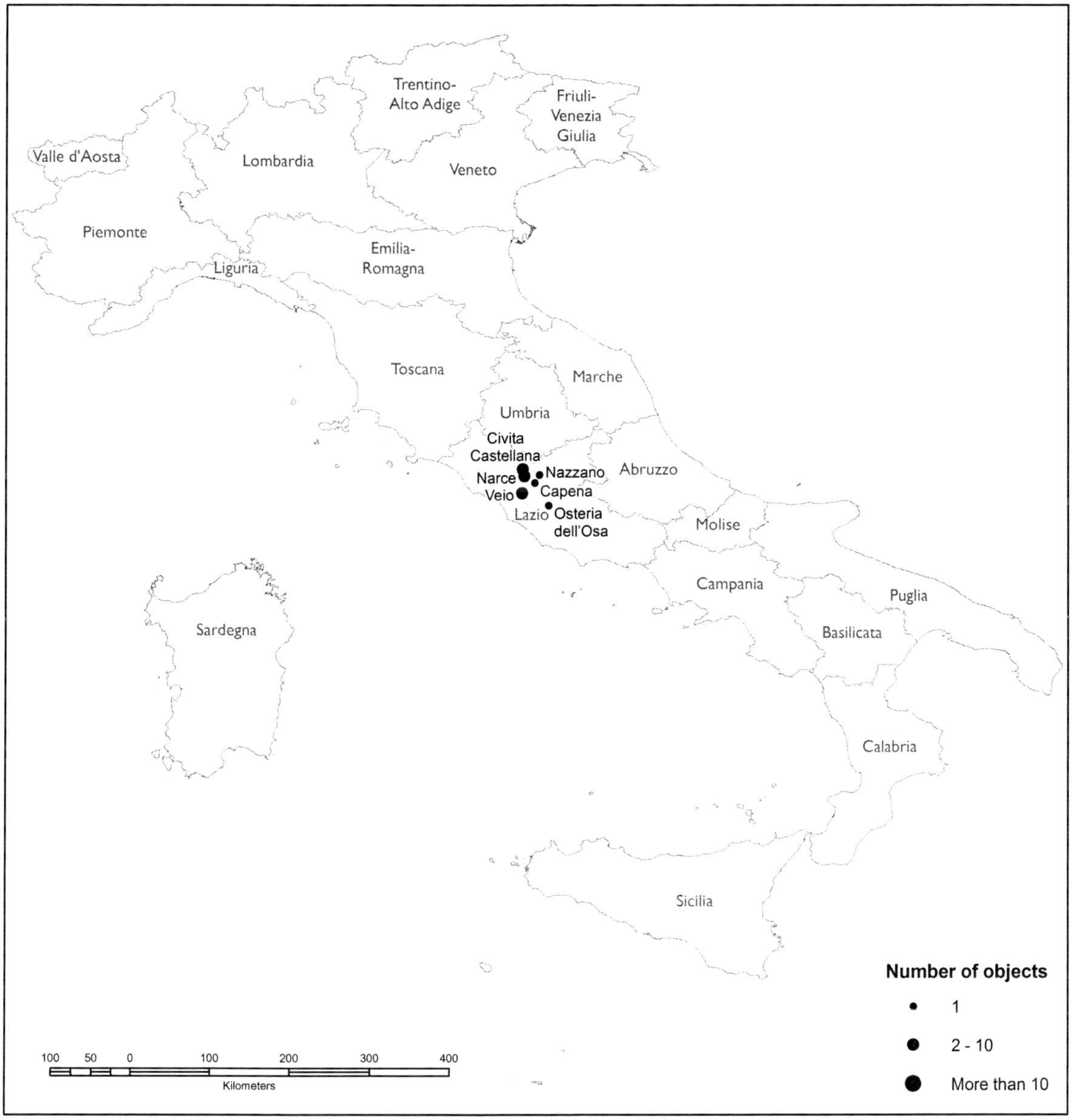

Map 4. Distribution map of distaff type I.

two from the Picene Novilara (Brizio 1895, 352 no. 9, 254 no. 11).

The distribution suggests that type A1 distaffs were produced in Bologna, ancient Felsina (Map 5). Identification of Bologna as the place of manufacture of these distaffs is confirmed by other types of objects from contemporary contexts, such as fibulae, bracelets, pins *etc.*, which bear conical appendages similar to the ends of the type A1 distaff. Such, for example, are bracelet of type 123, pin of type 138, and pin of type 111 from the Benacci Caprara necropolis (Tovoli 1989, 274, 278, 285, pls. 119 and 121). The distribution of this type of distaff further

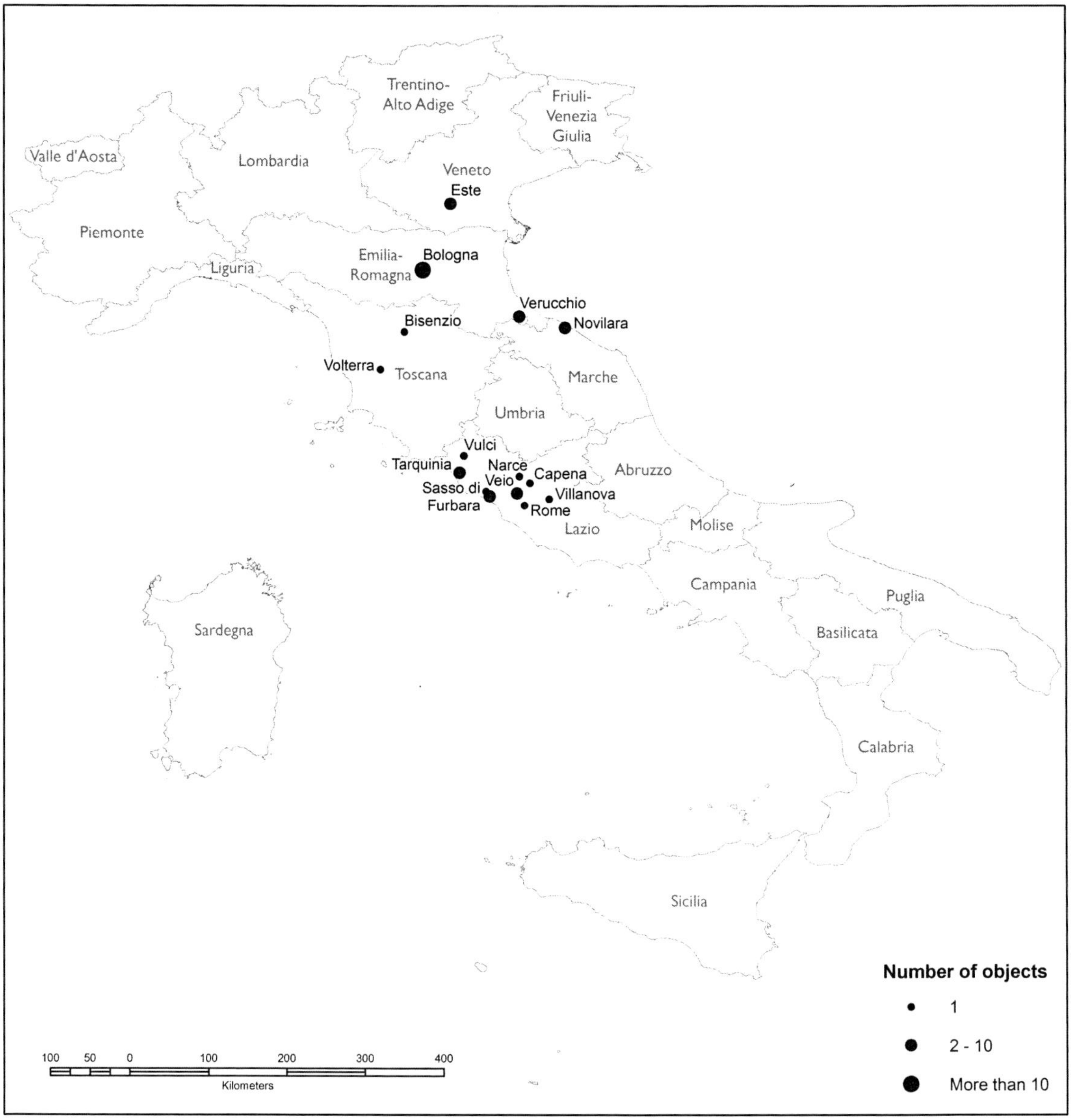

Map 5. Distribution map of distaff type II A1.

demonstrates that it was 'exported' to other sites throughout Central Italy, probably through gift exchange and/or intermarriage.[14]

2. Spherical top, convex shaft and disc at the bottom. Length *c.* 20 cm (Figure 83 A2).

Only a few examples of this subtype are known and they come from Bologna (pers. obs. Bologna Museum, 2002), Verucchio (Gentili, 1985, 70 no. 30) and the Siena area (Cimino 1986, 122 no. 352). This type probably can be regarded as a variation of A1, also produced in the Bologna area (Map 6). To the same type probably belongs a silver sheet distaff from the Regolini-Galassi Tomb

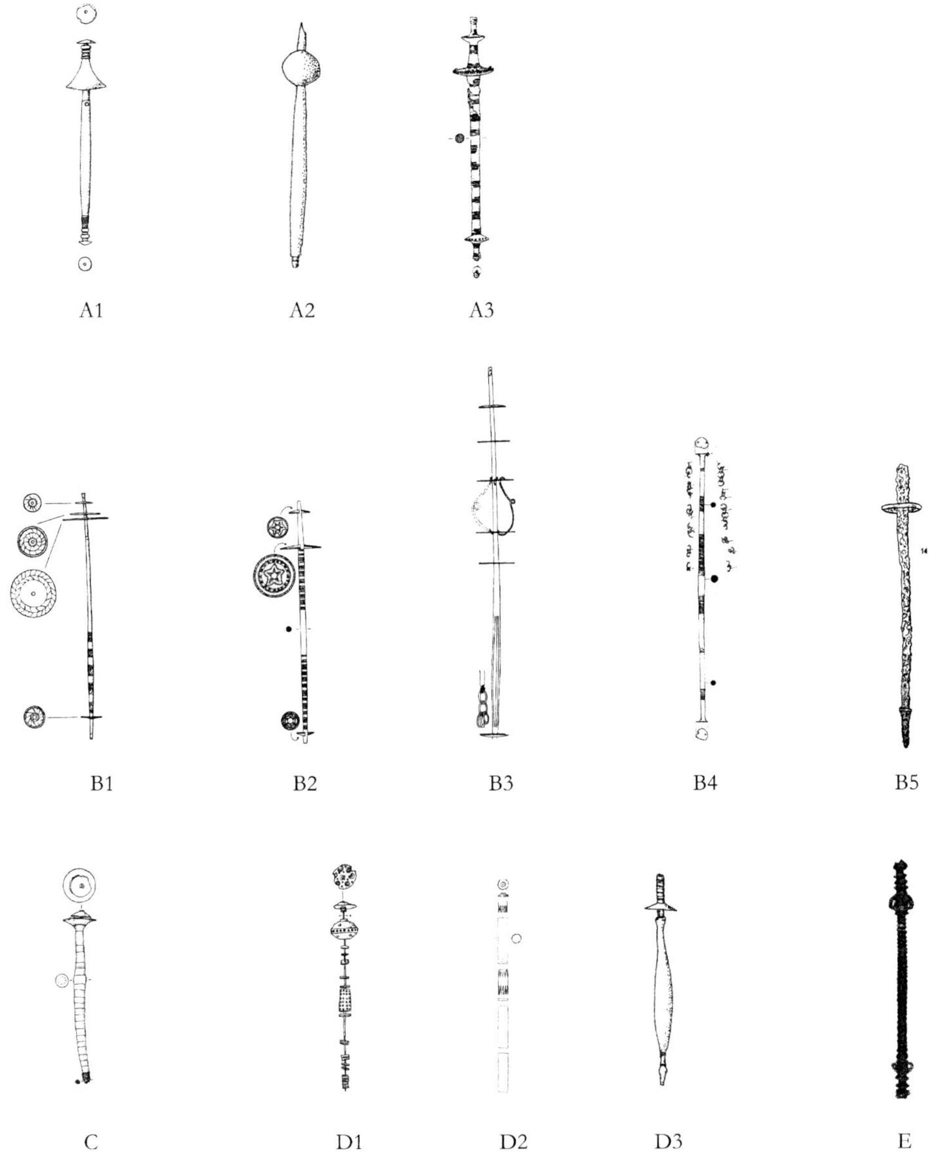

Figure 83. Distaffs of type II: A1 – Tomb HH 11–12, Quattro Fontanili, Veio, 8th century BCE (After Cavallotti Batchvarova 1965, fig. 54); A2 – Siena (Drawing by the author after Cimino 1986, pl. 68); A3 – Tomb 47, Osteria dell'Osa, 8th century BCE (After Bietti Sestieri 1992b, fig. 3a.403 no. 10); B1 – Tomb AA 12A, Quattro Fontanili, Veio, 8th century BCE (After Cavallotti Batchvarova 1965, 71 fig. 12); B2 – Tomb 580, Pontecagnano, 8th century BCE (After Bailo Modesti, d'Agostino and Gastaldi 1974, fig. 22); B3 – Tomb 27, Sala Consilina, 8th century BCE (After Kilian 1970, pl. 190); B4 – Tomb 410, Incoronata, 8th century BCE (After Chiartano 1994, pl. 102); B5 – Grammichele, 8–7th century BCE (After Orsi 1905, 129, 132 fig. 36); C – Tomb 47, Rocca Malatestiana, Verucchio, 8th century BCE (After Forte 1994, 79); D1 – Tomb 25, Caselle di S. Lazzaro, Villanova, 9th–8th century BCE (After Forte and von Eles 1994, 281 pl. XII); D2 – Tomb 89, Este (After Capuis and Chieco Bianchi 2006, pl. 74); D3 – Tomb 704, Locri, 5th century BCE (Drawing by the author); E – Tomb 119, Campovalano, 7th–6th century BCE (After Martelli 1994, 77 pl. II.6).

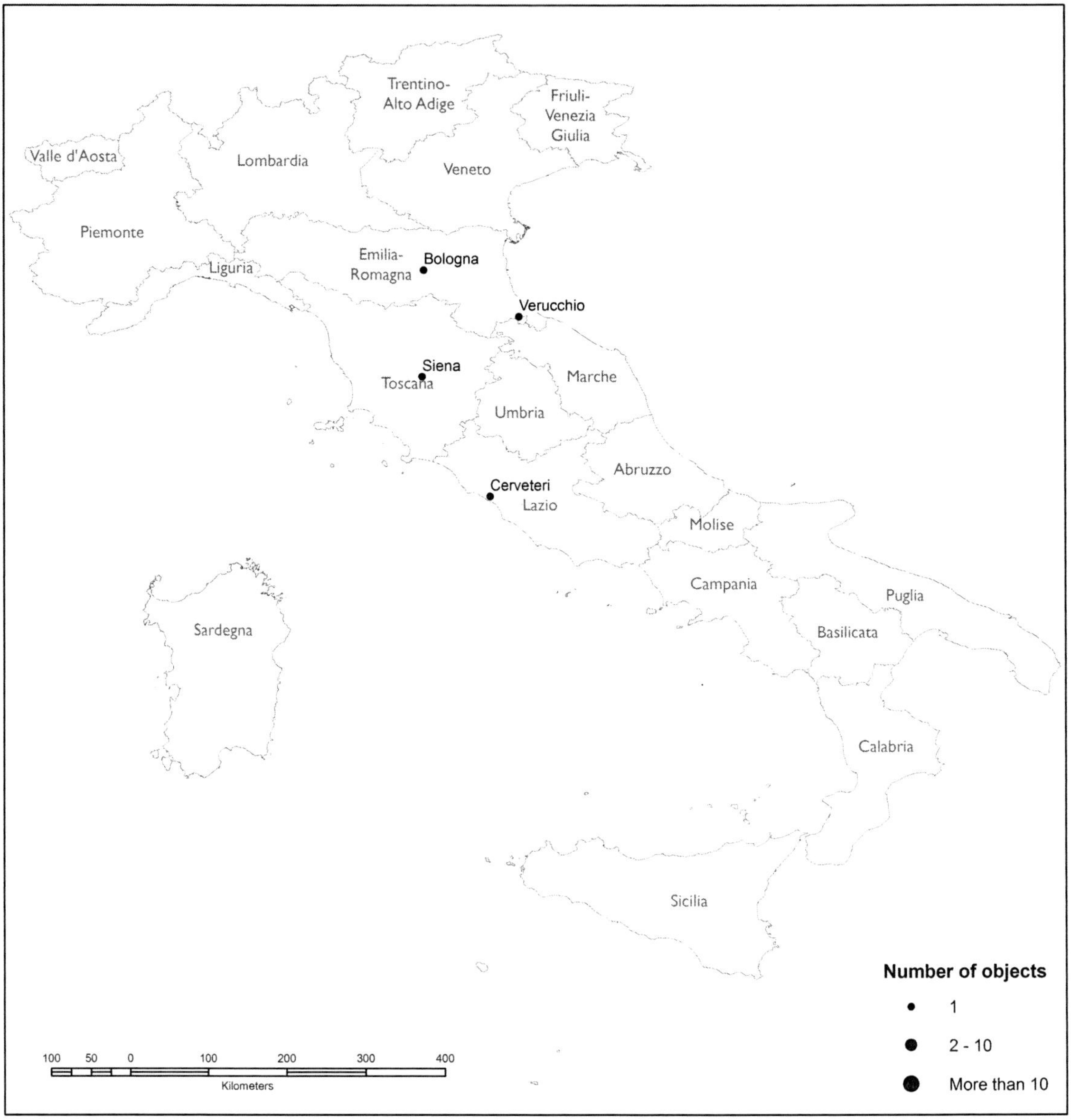

Map 6. Distribution map of distaff type II A2.

at Cerveteri (Montelius 1904, pl. 339.8; Pareti 1947, 217 no. 150).

3. Flat top with attached rings; incised decoration. Length 24–29 cm (Figure 83 A3).

Three items come from Osteria dell'Osa (Bietti Sestieri 1979, 54; 1992b, 755 no. 10), Caracupa (Pasqui 1903, 328, fig. 49), and Sala Consilina (Ruby 1995, 279 no. 7). Another possible example, from Rome, is now lost (Pinza 1905, 72 a). It is hard to judge about the possible centre of production of this distaff type due to few items and their wide distribution (Map 7).

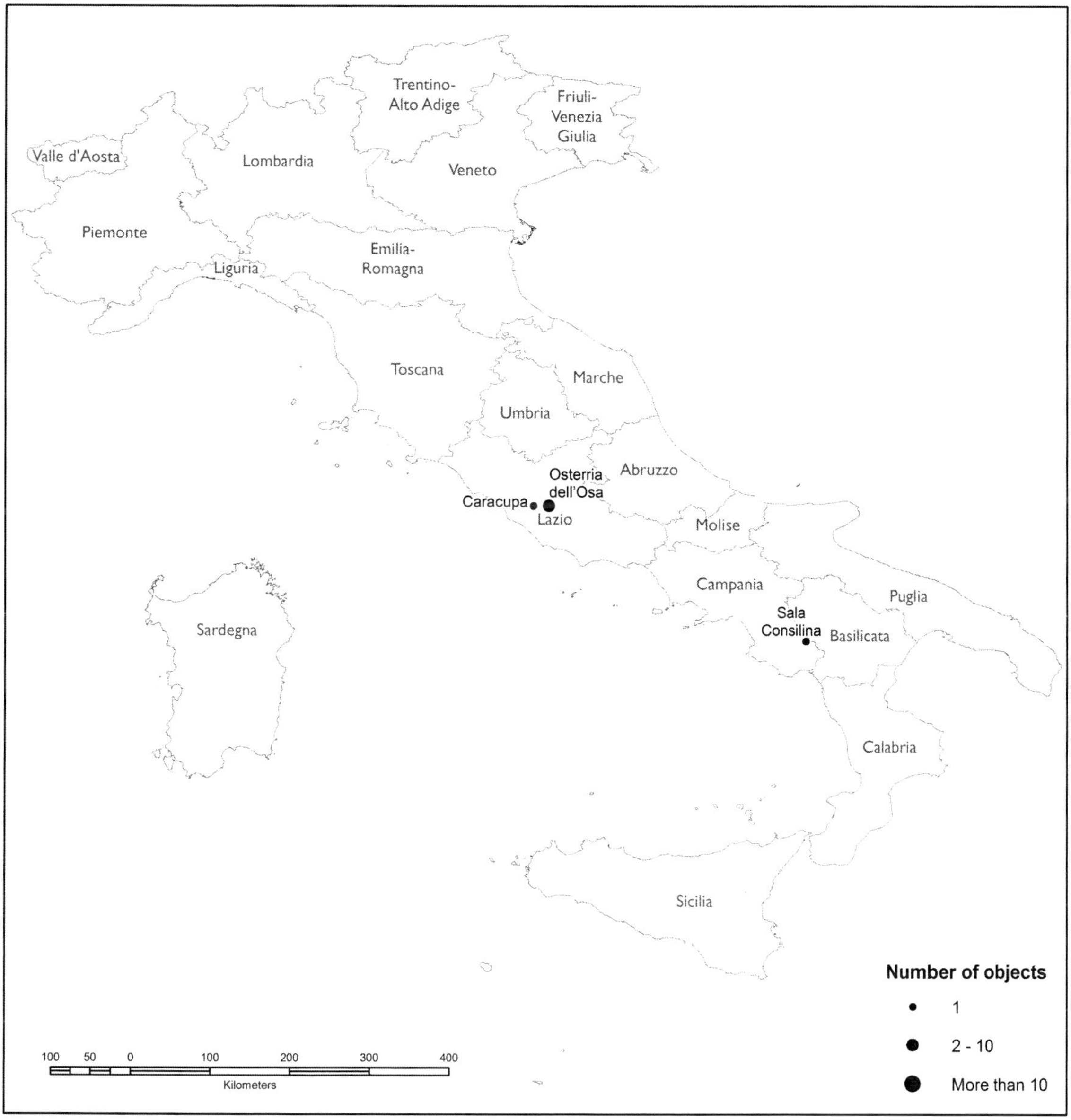

Map 7. Distribution map of distaff type II A3.

B. Solid cast bronze, with disks on each end. Incised decoration on shaft and discs.

1. Three discs on top and one or two disks on bottom, a thickening in the shaft near the lower end. Length 20–30 cm (Figure 83 B1).

Slightly differing examples come from Tarquinia (Montelius 1904, pl. 339.8; Pernier 1907, 71, 73, 241, 245, 2557, 260; Pareti 1947, 217 no. 150; Hencken 1968a, 55, 107, 161, 298, 320), Veio (Cavallotti Batchvarova 1965, 68–69, 220 i; Falconi Amorelli 1967, 217 no. 24; Bedello and Fabricotti 1975, 99 a; Berardinetti and

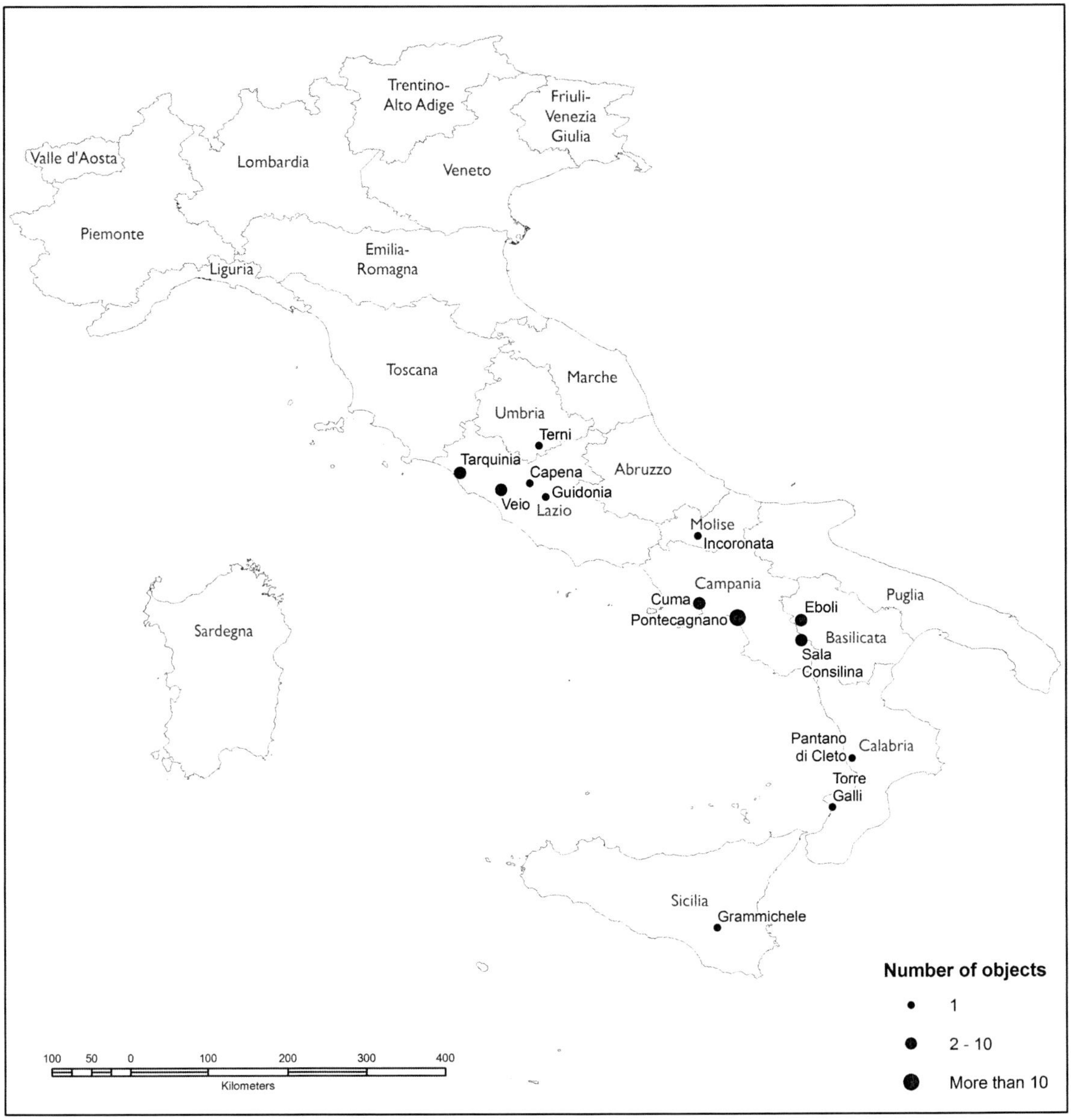

Map 8. Distribution map of distaff type II B.

Drago 1997, 56 fig. 25), Terni (Pernier 1907, 618 hh), and Capena (Stefani 1958, 98 Fig. 26, 100). Three more examples are without provenance (Turfa 1982, 169 no. 13; Neils 1992, 215 no. 49; pers. obs. National Museum in Copenhagen, 2003). The distribution of recorded examples suggests that this type of distaff was popular in southern Etruria, with Tarquinia or Veio as likely production centres (Map 8).

2. Relatively short, with two or three discs on top and one on bottom and often with perforations at one or both ends. Length 18–26 cm (Figure 83 B2).

Numerous examples come from Pontecagnano (Bailo Modesti, d'Agostino and Gastaldi 1974, 98, pl. XXXIX.3, fig. 22; d'Agostino and Gastaldi 1988, 132 no 18–19, 148 no. 70, 179 no. 16; Gastaldi 1992, 89 no. 35 pl. 99, 149 no. 29 pl. 125), Eboli (Cipriani 1995, 557; pers. obs. Eboli Museum, 2002) and Sala Consilina (Kilian 1970, pl. 20 no. 3a, pl. 71 no. 40; pl. 83 no. 1; pl. 88, no. 10c; pl. 113, no. 5; pl. 143, no. 3; pl. 196 no. 4, pl. 241, no. 1h), Campania being a likely area of their production (Map 8). Two examples reportedly come from Cuma (Kilian 1970, pl. 267 nos. 1–2; Albore Livadie 1985, 68 no. 9.14) and one from Pantano di Cleto – Serra Ajello (Luppino 1982, 76, Pl. 9,8 and 10,8; Peroni 1987, 131 and 129 fig. 105 no. 1).

3. Very long shaft with numerous discs and chain attachments, and thickening of the shaft near the lower end. Length *c.* 40 cm (Figure 83 B3).

Several examples have been recovered in Sala Consilina (Ruby 1995, 320 no. 9; Kilian 1970, pls 124 no. 8, 154 no. 4, 190 no. 7h), which must be the centre of production for this type.

4. A shaft with incised decoration and holes at each end, often with attached rings or chains. Length 25–30 cm (Figure 83 B4).

Despite the lack of endings, these objects have been identified as spindles or distaffs. Four have been found at Pontecagnano (d'Agostino and Gastaldi 1988, 149 no. 70 fig. 64, 100 no. 57 fig. 118; De Natale 1992, 65 no. 49, fig. 104; Gastaldi 1992, 141 no. 56), two in Sala Consilina (Kilian 1970) and one example at Incoronata (Chiartano 1994, 155 L).

5. Single disk near the top, cast together with shaft. Length 25–30 cm (Figure B3 B5).

Only two examples are known, from Torre Galli in Calabria (Orsi 1926, 64; Pacciarelli 1999, 165 no. 12) and Grammichele in Sicily (Orsi 1905, 129, 132 fig. 36).

6. Miniature distaff with three disks and a hook at the lower end (Figure 118).

The only example of this type was found in a child burial at Guidonia – Le Caprine together with other miniature textile tools (Giudi and Zaraffini 1993, 191).

C. Composite, made of bronze core with amber elements and conical top.

1. Length 15–20 cm (Figure 83 C).

This distaff type has been found at Verucchio (Forte 1994, 78 nos. 86–87, 106 no. 230; Gentili 2003, 100 no. 51, 178 no. 33, 243 no. 76, 244 no. 77), Bologna (Bartoloni 2000, 278 no. 349) and Populonia (De Agostino 1961, 81, 86 no. 3 fig. 23). In Verucchio, more than 20 examples have been recorded. It appears to be safe to assume that amber distaffs were made in northern Etruria, most likely at Verucchio (Map 9).

In Roman times, amber distaffs were produced in Aquileia and exported to other parts of Italy and Roman Empire (Gottschalk 1996).

D. Composite, made of bone and/or bronze elements.

1. Bronze core with decorated bone elements and a spherical top. Length *c.* 20 cm (Figure 83 D1).

This type is present in several burials at Este (Fogolari and Frey 1965, 277 no. 6 fig. 14, fig. 6.17; Chieco Bianchi and Calzavara Capuis 1985, 89 e, pl. 36; Capuis and Chieco Bianchi 2006, 129 no. 48, 159 no. 19, 299 nos. 55–56), Bologna (Müller-Karpe 1959, pl. 17; Tovoli 1989, 195 no. 6, pl. 86), Villanova (Forte and von Eles 1994, 284 no. 115, pl. XIV), Novilara (Brizio 1985, 423 no. 8) and Tarquinia (Pernier 1907, 338; Hencken 1968a, 349, fig. 347; Camporeale 1985, 78 no. 283;

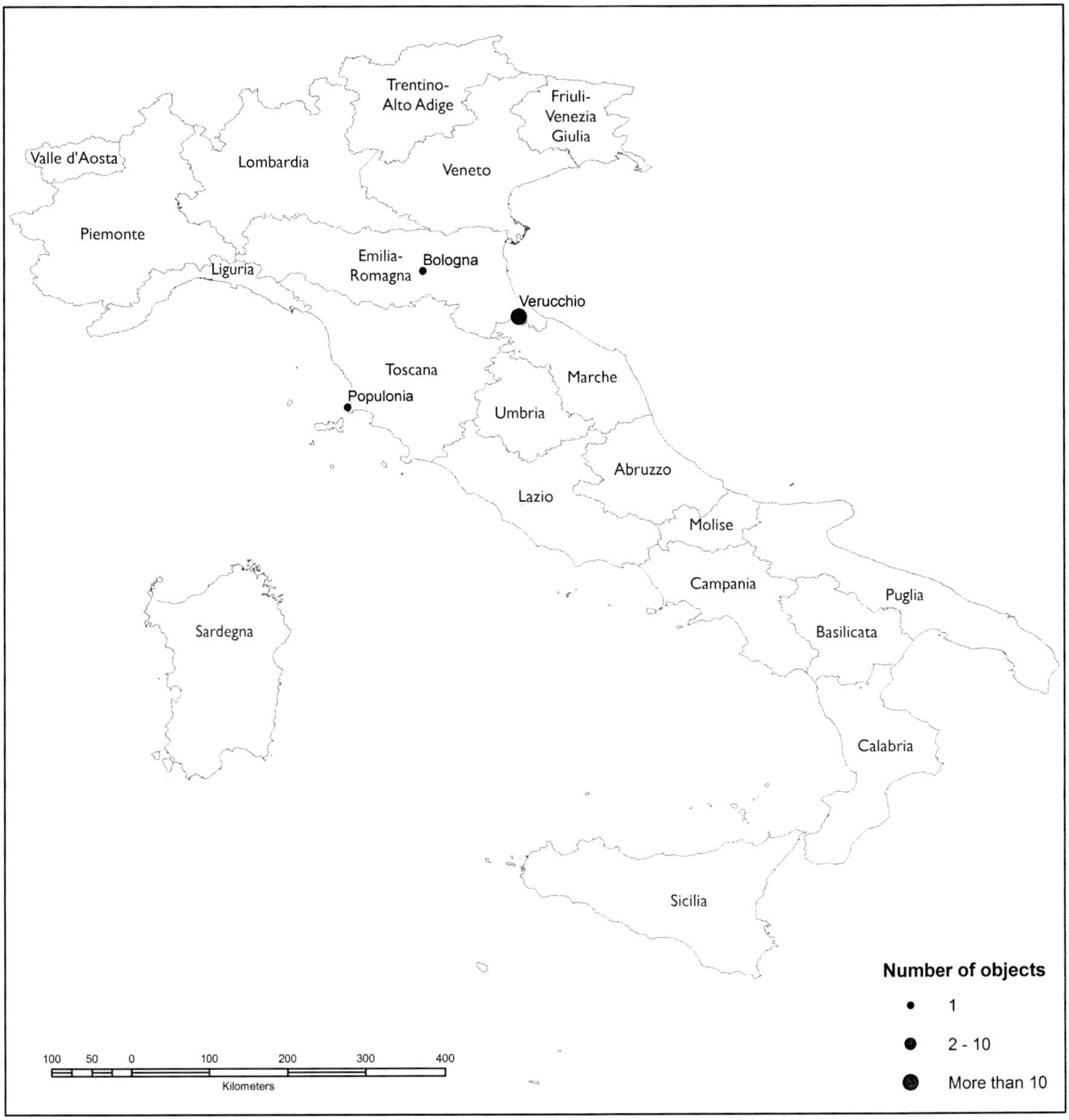

Map 9. Distribution map of distaff type II C.

Bottini *et al.* 2003, 134 no. 76), suggesting that composite distaffs were made in North Italy, possibly Este (Map 10).

2. Cylindrical shaft of bone elements (Figure 83 D2).

Objects of this type identified as distaffs were found in inhumation burials of Monte Tamburino necropolis at Monte Bibele, mostly dating to the 5th–4th centuries BCE (Vitali 2003, 144 no. 12, 153 no. 2, 194 no. 3, 211 no. 15, 216 nos. 3a–b, 220 nos. 15–16).

3. Carved. Length 15–20 cm (Figure 83 D3).

This type of distaff is chronologically later than other types, appearing in burials

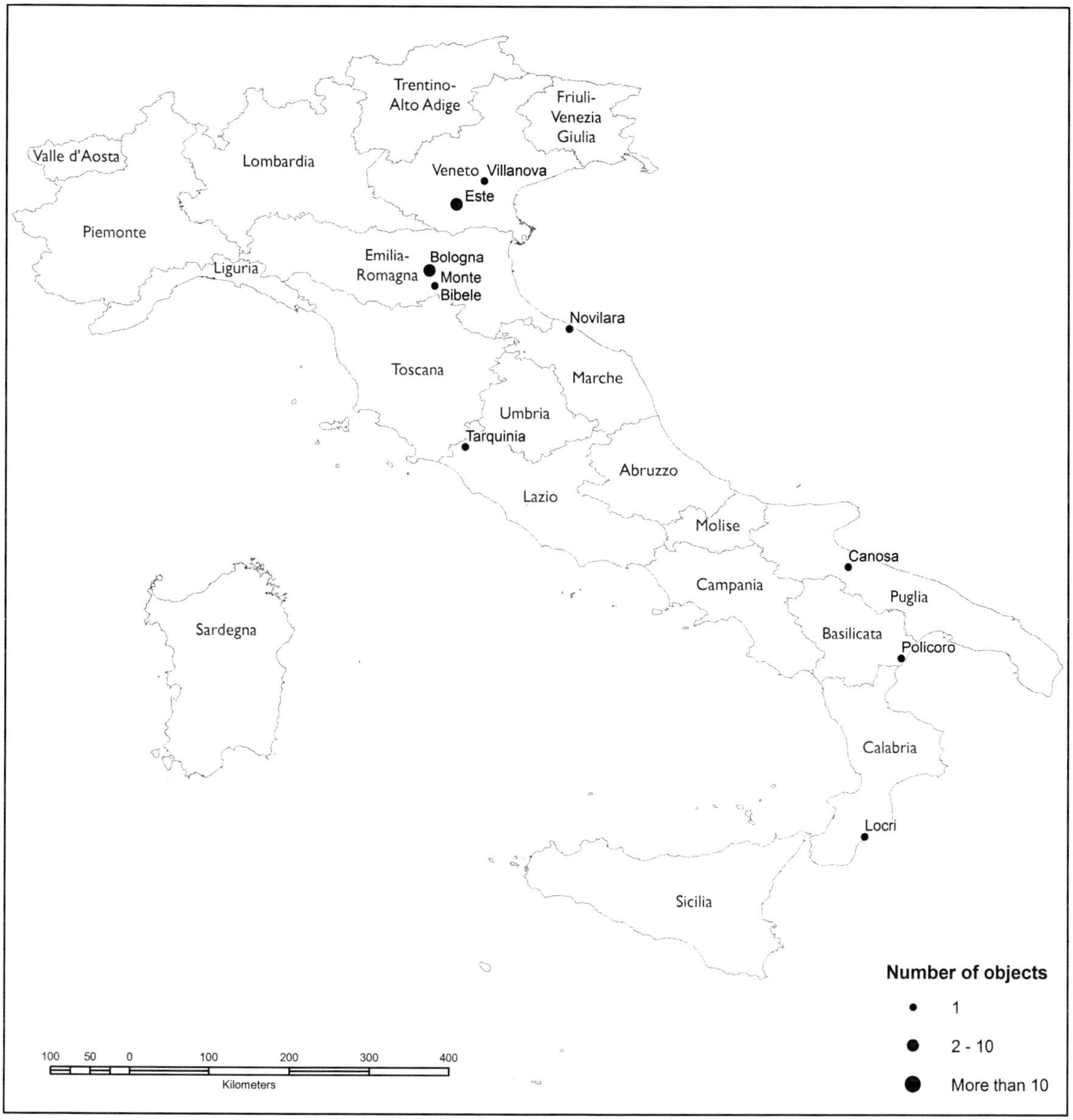

Map 10. Distribution map of distaff type II D.

of late 5th through 2nd centuries BCE, in Hellenic contexts, such as Canosa (De Juliis 1984, 448 no. 33), Policoro Herakleia (pers. obs. Metaponto Museum, 2003) and Locri (pers. obs. Locri Antiquarium, 2003), and is probably Greek in origin (Map 10).

E. Glass with metal rod as a core (Figure 83 E).

The extant examples of this type were studied and identified by Marina Martelli (1994, 75–98) and include two complete items from Cerveteri (Martelli 1994, 75, 77 pl. I no. 1; Bartoloni 2000, 278 no. 347), four from Campovalano (Martelli 1994,

Map 11. Distribution map of distaff type II E.

77 pl. II no. 6; Franchi dell'Orto 1999, 102 fig. 77, 269 no. 556; Bartoloni 2000, 278 no. 348), one from Marsiliana d'Albegna (Martelli 1994, 77), and one fragmentary piece from Castelnuovo Berardenga (Martelli 1994, 76, pl. II no. 4). Another fragment, currently in the collection of the University of Pennsylvania Museum, belongs to a grave group from Ascoli Piceno and probably is a middle element of a distaff similar to the Cerveteri examples. Three fragments of a similar object, the only ones excavated in a settlement context, have been recovered at Poggio Civitate, Murlo (pers. com. Alexis Christensen, 1999). A slightly

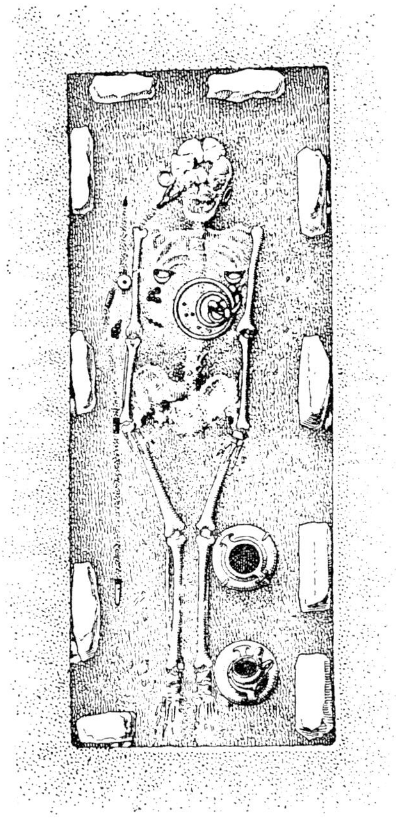

Figure 84 (above left). Tomb Monte lo Greco, Narce, with distaff positioned near belt with wider end pointing towards head (After Cozza 1895, 139 fig. 56).

Figure 85 (above right). Tomb at Nazzano, 7th century BCE, with distaff positioned along the right side of the body and forked end near head (After Stefani 1911, 437 fig. 5).

different glass distaff is noted among the burial goods of a tomb at Monte Soriano, Mazzano Romano (Pasqui 1902, 334 fig. 10). It also has an iron core but its shape is different in that the shaft is smooth and the top is less elaborate. Distribution of glass distaffs indicates that they were produced in Central Italy, possibly in the Picene area (Map 11).

Functional features

It should be noted that a considerable amount of confusion exists in the literature regarding identification of distaffs. Frequently, distaffs of types IIA and IIB are identified as spindles (Amann 2000, 27), while type IA has been called a symbolic staff (Müller-Karpe 1974, 93). My identification of all these implements as distaffs is based on their shape, and location and association with other implements in burials (cf. Brizio 1895, 152–153; Bartoloni 1989, 43). Thus, type IIA and IIB distaffs are frequently found deposited near the head of the deceased, near or directly on the body, with the wider end pointing towards the head (Figure 84). This position is common in Veio tombs (Paribeni 1928, 435; Stefani 1958, 91 fig. 23; Close-Brooks 1963, 237 fig. 195; Cavallotti Batchvarova 1965, 124 fig. 48; Falconi Amorelli 1967, 211 fig. 73). Type IA distaffs were often placed alongside the body of the deceased, with the forked end near the head (Figure 85). Such is the case in some burials at Veio (Close-Brooks 1963, fig. 90), Civita Castellana (Baglione and De Lucia Brolli 1997, 158 note 29), Nazzano (Stefani

1911, 437 fig. 5), and Narce (Barnabei and Pasqui 1894).

All types of distaffs are frequently associated with spindle whorls, suggesting that a spindle complemented the spinning set. For example, in Tomb AA12A at Veio-Quattro Fontanili, a spindle whorl was found near the lower end of type IIB1 distaff (Cavallotti Batchvarova 1965, 67 fig. 9). Bologna's Benacci-Caprara Tomb 56 also provides evidence for this practice: there, a distaff of type IIA1 was deposited with a bronze model of a spindle (Brizio 1889, 329 no. 51; Tovoli 1989, 188 no. 51).

Weaving

Weaving is accomplished on a loom, a special frame that keeps the warp system in place, while allowing the weft to be passed in between warp threads. North of the Mediterranean, and in Italy in particular, a warp-weighted loom (Figure 86) was used in antiquity (cf. Crowfoot 1940; Forbes 1956, 199–202; Hoffmann 1964; Wild 1970, 61–68; Barber 1991, 91–113). The earliest evidence for this type of loom comes from Hungary and is dated to the late 7th millennium BCE (Barber 1991, 93). In Italy, according to Barber (1991, 113), it seems to be in use at least since the Middle Neolithic. The earliest representations of such a loom, in fact, come from Italy: seven depictions were carved into the Great Rock of Naquane in Val Camonica (Figure 5), dated to the 14th century BCE (Anati 1959, fig. 5; 1961, 138–142).

The loom was made up of two upright beams that stood at a slight angle to the vertical plane, and a single horizontal or cloth beam, to which warp was attached. In a warp-weighted loom, as suggested by its name, the warp is kept taut by the weights attached at the bottom to groups of threads. Since weights were made of stone or clay, they survive well in the archaeological contexts and allow us to trace the presence and sometimes even location of a warp-weighted loom on sites.

The frame of the loom was usually leaning against the wall, although occasionally two posts could be set at an angle

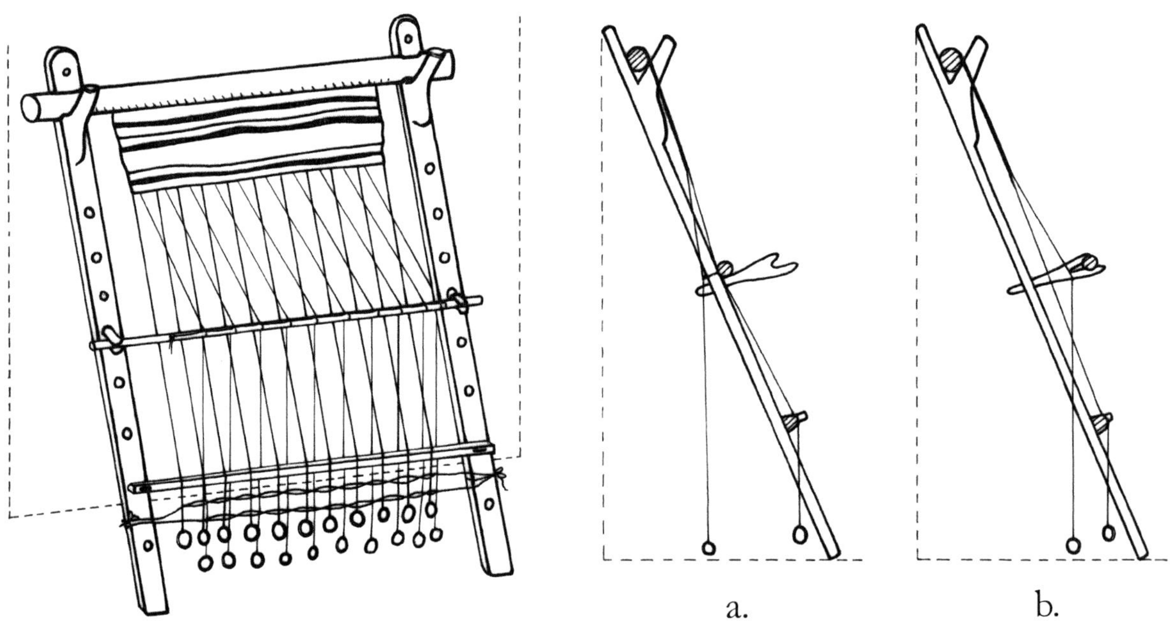

Figure 86. Warp-weighted loom and its position with: a) natural shed; b) artificial shed (Courtesy of Eva Andersson).

to each other (Barber 1991, 103). The warp is divided by pulling every second thread and inserting a rod or shed bar between the two groups in such a way that one of the groups is in front of it. Such a position creates an open, or natural shed through which the weft could be passed all at once. The artificial or counter shed is then achieved by providing the back set of warp threads with heddles, or individual holders, usually made out of string for each individual warp thread, which are attached to a heddle bar. When the heddle bar is lifted, it separates the threads in the direction opposite to the original shed. All of these elements seem to have been in existence at least since the 4th millennium BCE (Barber 1991, 91). This basic system allows a creation of a textile of a simple weave. Additional heddle bars allow multiple sheds and, hence, a mechanised weaving of more complex patterns, such as twills.

The weaving started at the top, hence the weft had to be packed upwards. Before weaving could begin, however, the warp had to be prepared apart from the loom and then attached to the cloth beam. This was accomplished by making a starting border or heading band, the weft of which became the warp of the loom (Forbes 1956, 192–194; Wild 1970, 63–64; Barber 1991, 116–118). This band could be made on a special band loom in order to measure out and organise the warp threads. The only possible representation of such a device is on the Bologna *tintinnabulum* (Figure 8). Another method was tablet weaving (see below).

The weft was kept on a shuttle, which in its simplest form could have been a ball of thread or a simple stick, possibly a spindle shaft on which the yarn had been spun (cf. Barber 1991, 107). Shuttles most likely were made of wood or bone and survive only in exceptional circumstances. The weft was then beaten up with a special tool, possibly wooden or bone comb with short teeth or a weaving knife or sword (Wild 1970, 65–67).

One of the problems with a warp-weighted loom is that its size is limited by the distance that the weaver can reach in vertical and horizontal directions (Barber 1991, 105–106). To increase the width, the weaver can simply walk in front of the loom from one end to another, or multiple weavers can perform the task. The latter situation is depicted on the New York vase. The evidence of such a practice can also be found in cloth itself, as in the case of Danish Early Iron Age textiles, where crossing wefts indicate that in some cases up to four weavers were at work on a single piece of cloth (Hald 1980, 152).

In order to increase the length of the cloth being woven, the weaver had to become 'taller'. Different solutions to the problem developed over time. One way to accomplish the task was by standing on a stool or a bench. A variation of this is depicted on Corinthian *aryballos*, where women wear platform shoes (Barber 1991, 106 fig. 3.24). A more elaborate solution is depicted on the Bologna *tintinnabulum* (Figure 8), where one of the women is shown in a complex two-storied loom. Unusually tall loom-like structures also appear on the Verucchio throne (Figure 87). That such looms were indeed used during the Iron Age is suggested by the Verucchio mantles, which are more than 270 cm wide and were probably woven with the wider dimension as the warp. Another way to increase the length of the cloth could have been accomplished by digging a trench under the loom, as depicted on the famous vase from Sopron (Figure 88).[15] Actual trenches were found in the Hallstatt area in Germany (Vogt 1937, 112, figs. 152–154; Dobiat 1990, 73 Fig. 1b, 84 fig. 6), but, so far, no evidence

Figure 87 (above left). Detail from the Throne of Verucchio showing the tall loom (After Stærmose Nielsen 1999, 60 fig. 33).

Figure 88 (above right). Detail from the urn of Sopron, Hungary, 800 BCE, showing a weaving woman (Courtesy of the Naturhistorisches Museum, Vienna).

for their use has been documented in Italy.

The most practical solution, however, is to make the cloth beam a roller, so that the cloth could be wound on it as it is being woven. Such roller beams were in use at least since the middle of the 1st millennium BCE, as proven by the depiction on the Chiusi *skyphos* (Figure 13).

The various iconographic depictions, thus, show that greater length of the cloth was an important issue and that different solutions were employed for achieving it. Furthermore, the warp-weighted loom was so well adapted to weave large pieces of fabric that it continued to be used in Roman times even after the adoption of the two-beam loom (Wild 1976, 171). In particular, the warp-weighted loom seems to have become specifically associated with linen, possibly for sails, as demonstrated by the following passage in Servius (*Ad Aen.* 7.14): "our ancestors used to weave standing, just like today when we see the linen weavers".[16]

Loom

Occasionally, carbonised remains of the wooden beams are preserved in association with loom weights, which allows speculation about their probable position. For example, the 6th–5th century BCE Structure D10 at Pozzuolo del Friuli preserved a line of weights, concentrated in SW end of the structure, with clear carbon traces of wooden beams in the soil (Vitri *et al.* 1991, 22–25, figs. 8, 13–15). A small room at Santorso had carbonised wooden beams, the largest measuring 85–90 by 10–11 cm, while a smaller one was 40 by 3–4 cm (Figure

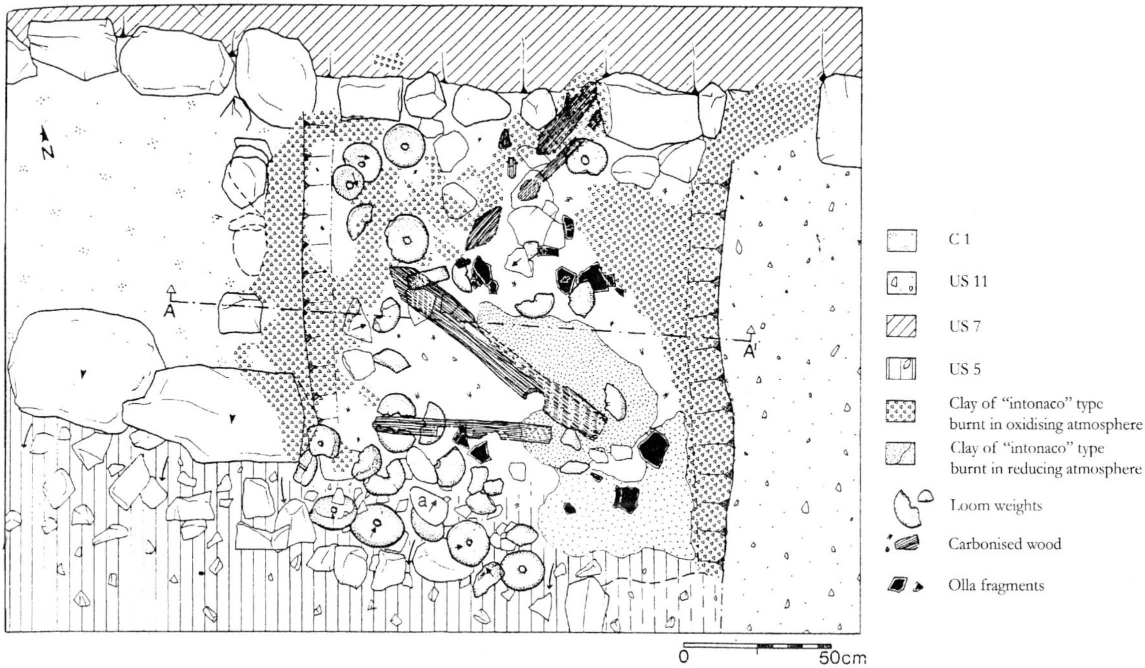

Figure 89. Remains of a loom, Santorso, 6th century BCE (After Balista et al. *1985, 81 fig. 23).*

89) (Balista *et al.* 1985, 79–82, fig. 21, 23, 24). Likewise, at the 4th–3rd century BCE building at La Piana, some of the carbonised wood timbers, found together with an accumulation of loom weights near the north wall in one of the rooms of the house, probably were fragments of a loom frame (Figure 90) (Whitehead 1996, 114, and figs. 6 and 8).

Another kind of evidence for looms consists of postholes, which may indicate the total width of the loom and the angle at which it was positioned in relationship to the wall (Barber 1991, 103). This evidence, however, is more common for prehistoric contexts.

Archaeological and iconographic evidence suggests that the main, upright portion of the loom was probably braced at an angle against a wall or a post, an arrangement that uses gravity, with a shed bar at the bottom, to separate the threads into a shed (Barber 1991, 103). The height is more problematic, since although the normal loom would not be much higher than a human, the Bologna *tintinnabulum* indicates that elaborate two-storied looms also existed. Such tall, vertical loom probably had connections to the Hallstatt practice of placing a loom over a deep pit, as shown in the Sopron vase. Usually, however, the height of the loom probably did not exceed human height, that is 165–180 cm.

Several models of a vertical loom in bronze were found in tombs of Este. A rectangular loom made of two uprights 61.5 cm high, each terminating in two legs, connected by three horizontal strips, all riveted together by nails (Figure 91a), was found in Tomb 23 at Casa di Ricovero, dated to the 4th century BCE (Chieco Bianchi 1987, 211 no. 89, figs. 11, 32, and 58). Next to the loom model was deposited a situla with textile implements: a spindle, a distaff and two combs, all made of sheet bronze, and a glass spindle whorl. Another, contemporary loom

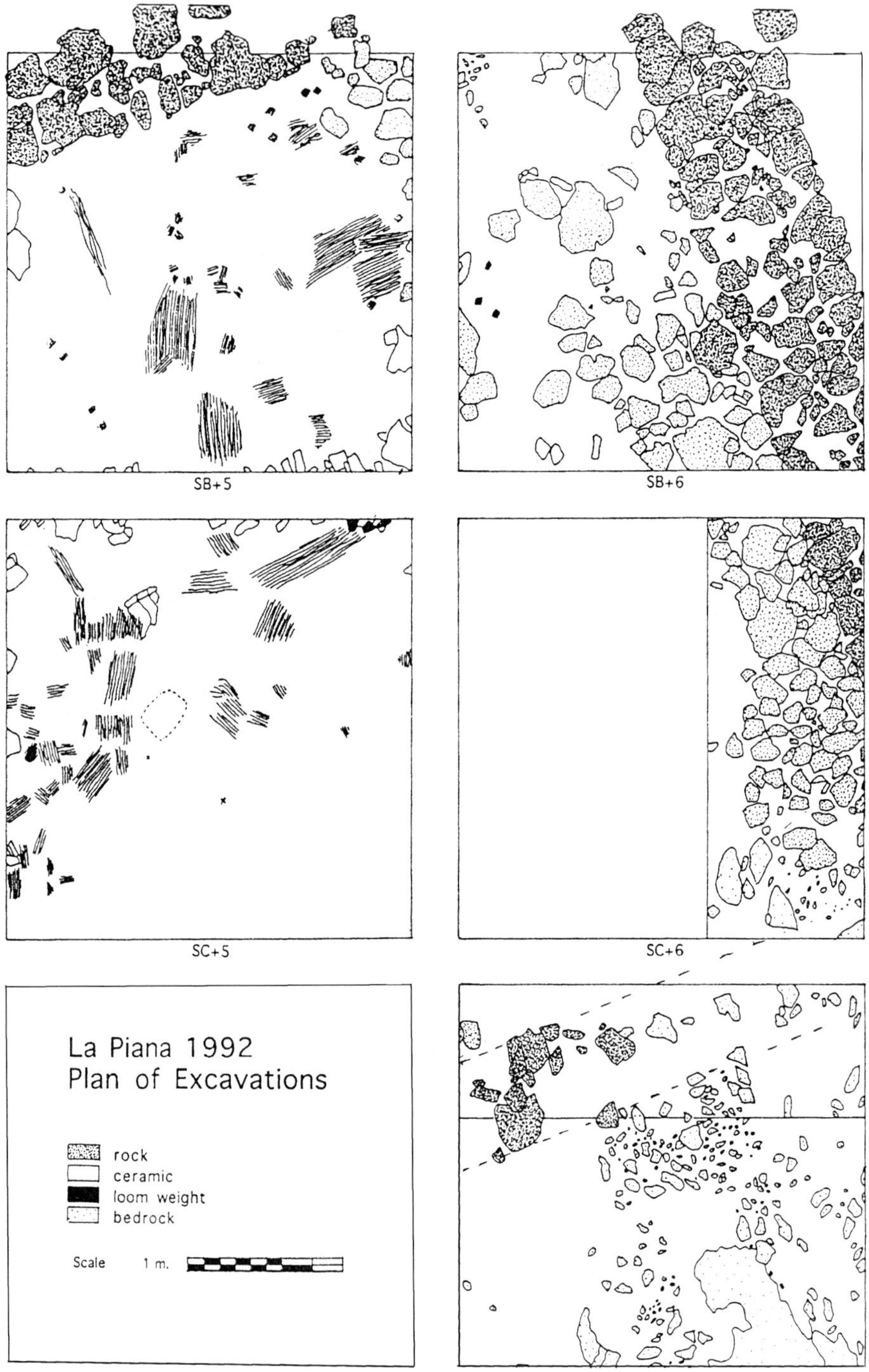

Figure 90. Remains of a loom, La Piana, 4th century BCE (After Whitehead 1996, 111 fig. 6).

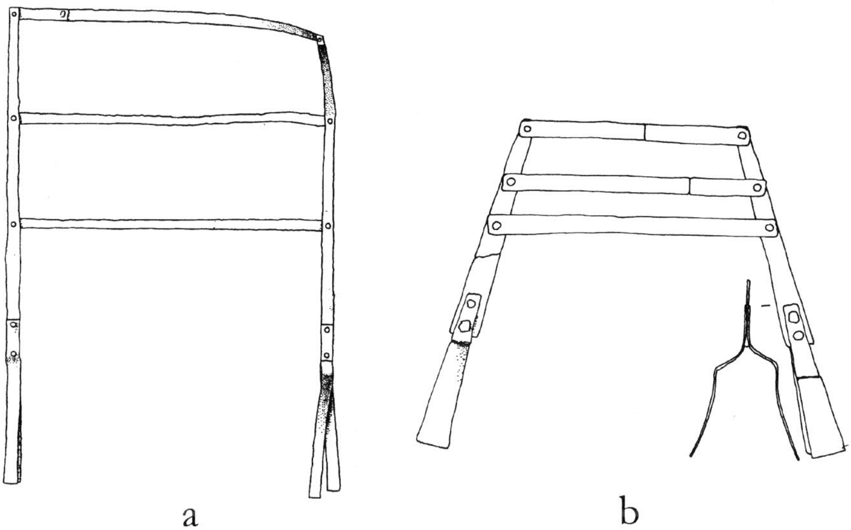

Figure 91. Bronze loom models from Este: a) Tomb 23, Casa di Ricovero, Este, 4th century BCE (After Chieco Bianchi 1987, 212 fig. 32 no. 89); b) Tomb 123 Villa Benvenuti, 3rd century BCE (After Capuis and Chieco Bianchi 2006, pl. 155 no. 58).

model of a trapezoidal shape (Figure 91b) was found in Benvenuti Tomb 123 (Prosdocimi 1882, 33, pl. VIII fig. 47; Capuis and Chieco Bianchi 2006, 282 no. 58). A similar but smaller model comes from Casa di Ricovero Tomb 231 (Chieco Bianchi and Calzavara Capuis 1985, pl. 167). In all cases, loom models were associated with other textile implements.

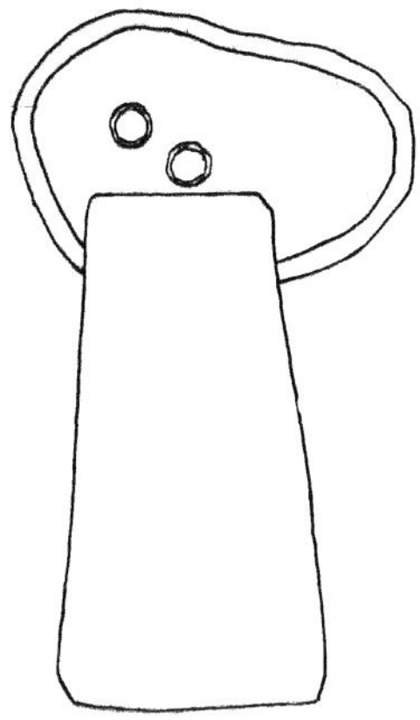

Figure 92. Loom weight from Fratte with a metal ring for the attachment of warp threads, 5th century BCE (After Greco and Pontrandolfo 1980, 162 fig. 55 no. 131).

Loom weight

Function

The function of a loom weight is to keep the warp of a warp-weighted loom taut during weaving. Simple stones could be used for the purpose (Hoffmann 1964), but most cultures developed special implements, made for the purpose. In Italy, the vast majority of loom weights were made of fired clay, although stones could also be used, as in the case of the numerous loom weights in arenaria stone found at Monte Loffa (Battaglia 1934, 122, 137 fig. 20).

The function of the objects in question has been uniformly accepted by now, but as late as 1953 Piero Orlandini argued for their purely symbolic function. It is not always easy to demonstrate that the weights were indeed used for weaving, unless they are found in groups (Wild 1970, 62; Barber 1991, 92–93). Similar objects have been shown to be used as weights for holding down roof thatching, for tying sacks, as fishnet weights, or as supports for holding a spit over fire (Wace and Thompson 1912, 43 fig. 19; Mingazzini 1974, 209–211, 215; Barber 1991, 97 note 11). Another hypothesis is that they were used as supports for the full spindle to facilitate the unwinding of the spun thread onto a spool or into a ball (Buchner and Rittmann 1948, 40, fig. 9).[17] Ring-shaped weights are often associated with hearth and identified as supports for the cooking pots. Despite these other possibilities, most weights appear to be connected to weaving activity.

Attachment to the loom

The manner in which the weights were attached to the warp is still debated. Most likely the threads were tied not to the loom weight itself, but to an intermediary device such as a cord, a metal ring, or a bar, which also attached the loom weights to each other to keep them organised. Barbara McLauchlin (1981) discusses two loom weights from Nemea, Greece, which preserve portions of wooden rods inserted through the holes. Diane Carroll (1983, 96–98) gives a different interpretation of the function of wooden rods. She believes they were used as stoppers to prevent the thread from slipping off the loom weight, which acted as a reel to extend the warp length.

The weight from Fratte (Figure 92) preserves a metal ring on it (Serritella 1990, 164 no. 131). A similar weight is in the collection of the British Museum (Davidson and Burr Thompson 1943, 68, fig. 30). An ancient representation of such loom weights with rings may be seen on the New York *lekythos* with weaving scenes.

In some cases two separate groups of threads were attached to each weight, as shown by the finds of both discoid and pyramidal weights with two holes (Davidson and Burr Thompson 1943, 68).

Chronology

As was the case for spindle whorls, loom weights did not change dramatically in shape, size, or decoration over time and are therefore difficult to date when found out of secure archaeological contexts. In Italy, they are attested from the Neolithic period onward (Bazzanella, Mayr and Rast-Eicher 2003, 92). The main forms of loom weights in ancient Italy were rheniform, cylindrical, ring, trapezoidal, conical and discoid. There does not seem to be a functional difference in this typological diversity (Bazzanella, Mayr and Rast-Eicher 2003, 92). The earliest weights, dated to the 5th–4th millennium BCE, are of cylindrical or conical shape (Barfield 1971, 48; Barber 1991, 100). In the 4th millennium BCE, typology changes: rheniform examples appear on the sites of the Lagozza culture while cylindrical and lenticular forms are common in the Ripoli culture (Bazzanella, Mayr and Rast-Eicher 2003, 92). The trapezoidal shape became prevalent in 1st millennium BCE Italy, and varied from pyramidal to almost cuboid. Conical weights appear in the Archaic period and discoid weights are typical for the Hellenistic times. However, different types frequently coexist even at the same site, to say nothing about different settlements, even located near each other, making chronological comparison difficult.

Typology

The following generalised typology for loom weights is based on material from select archaeological settlement contexts of Italy (Figure 93).[18]

A. Rheniform/crescent/kidney

Objects of semilunate shape with rounded ends, perforated perpendicular to the length of the object seem to be found exclusively in the sites of the Lagozza culture (Cornaggia Castiglioni 1955; 1964; 1965; Barfield 1971, 51, 52 fig. 23; Baioni *et al.* 2003). They have been found at: Lagozza di Besnate (Guerreschi 1967, 185–180; Bazzanella *et al.* 2003, 185), Isolino di Varese (Guerreschi

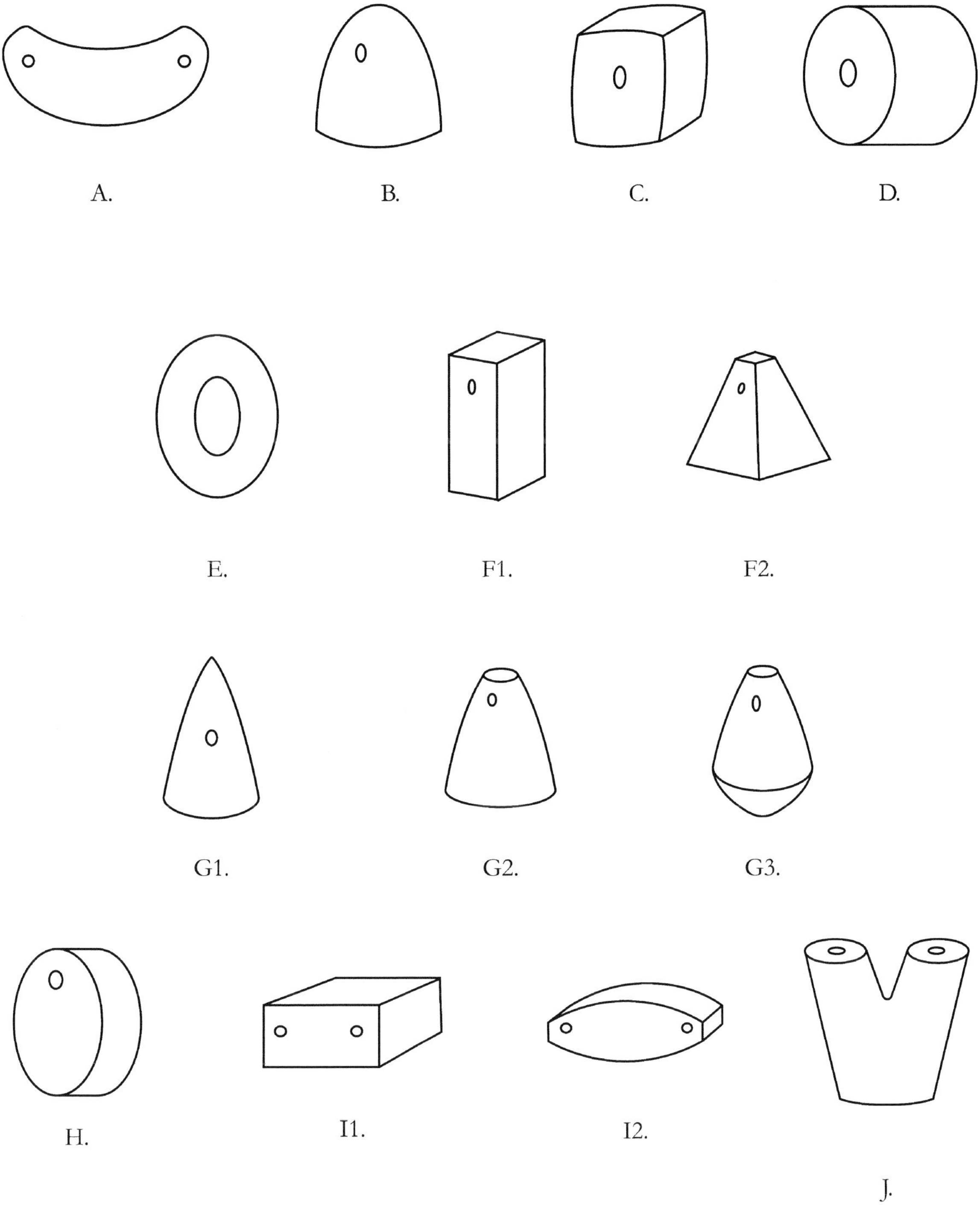

Figure 93. Loom weight types (Drawing by Marianne Bloch Hansen).

1977), Grotta dell'Olivo and Grotta dei Pipistrelli (Almagro Basch, Ripoll and Muñoz 1957, 192; Muñoz 1957, 197), Castelnovo di Teolo (Biagi 1986, fig. 4.3), Pescale (Malavolti 1942, 485), Piacenza le Mose (Bernabò Brea *et al.* 2003), Grotta dei Piccioni (Cremonesi 1976, fig. 51.2), and Podere Casalnuova (Aranguren, Ducci and Perazzi 1987, fig. 2.14). They are usually not well fired and the only decorated example is known from Isolino di Varese. The more numerous examples come from the central area of the Lagozza culture and elsewhere reflect its influence. None are known from contexts later than Late Neolithic (Baioni *et al.* 2003, 101). Rheniform loom weights are also known from contemporary sites of Chassey culture in the south of France and lake dwellings in south-west Germany and Switzerland (Baioni *et al.* 2003, 101). They have also been found in Greece and Anatolia (Lassen 2007).

Their functionality as loom weights for a warp-weighted loom has been a matter of debate. Recently, experiments demonstrated various ways of how they could be used (Baioni *et al.* 2003, 104–106; Feldtkeller 2003).

B. Bell or truncated egg
This type is not very common and appears only at Bronze Age sites, such as Castellaro del Vhò (Frontini 1997, 295, 296 fig. 165 no. 1).

C. Cuboid
Cubical shape with a central perforation. This is another rare type that appears among other shapes at the Bronze Age Castellaro del Vhò (Frontini 1997, 295, 296 fig. 166 no. 1) and Castione dei Marchesi (Mutti *et al.* 1988, 127, figs. 65 no. 6).

D. Cylindrical
Cylinder, centrally pierced along its axis. This shape makes its appearance in the Neolithic period but becomes ubiquitous during the Middle–Recent Bronze Age throughout the *terremare* area in North Italy. Cylindrical weights occur, for example, at Cisano (Fasani 1980, 154 nos. 3 and 4), Castellaro del Vhò (Frontini 1997, 295, 296 fig. 165 nos. 2–3, 297 fig. 166 no. 2), Castione dei Marchesi (Mutti *et al.* 1988, 125–127, figs. 64–65) and Ledro (Rageth 1975, fig. 93.22; Bazzanella *et al.* 2003, 156–158). A few cylindrical weights have turned up in the Late Bronze contexts at Este (Prosdocimi 1887, 186, pl. VII no. 23) and Frattesina (Bietti Sestieri 1984, 422), where they are probably remnants of the earlier period. Cylindrical weights have also been noted in the south, at Broglio di Trebisacce (Peroni and Trucco 1994, 249) and Porto Perone (Lo Porto 1963, 358).

E. Ring or donut
Ring weights (Italian *tarallo* or *ciambella*) may have developed from the cylindrical type. They appear in North Italy in the Bronze Age and become especially popular by the beginning of the Iron Age (Bianchin Citton, Gambacurta and Ruta Serafini 1998, 350). Some of the sites where loom weights of this type have been found are: Castellaro del Vhò (Frontini 1997, 295, 296 Fig. 165 nos. 4–5), Frattesina (Bellintani and Peretto 1984, 19; Bietti Sestieri 1984, 422; Bellintani 1992, pl. 12 nos. 3–10), Villamarzana (Aspes, Bellintani and Fasani 1984, 50, fig. 8), Perteghelle di Cerea (Aspes, Rizzetto and Salzani 1976, 143 nos. 31–32, fig. 8 nos. 9–10), Santorso (Balista *et al.* 1985, 85 figs. 25 and 26), Este (Prosdocimi 1887, 185, pl. 7 nos. 40 and 45; Bietti Sestieri 1980, 30), Montagnana Borgo S. Zeno (Bianchin Citton, Gambacurta and Ruta Serafini 1998, 345–351), and Concordia Sagittaria (Croce Da Villa 1991, 89 nos. 2–3 and 5–6, 90, 94 nos. 19–20; Bianchin Citton 1995, 234, 252 fig. 19 nos. 2–3; *Protostoria tra Sile e Tagliamento* 1996, 231 no. 127, fig. 38 no. 127, 251 no. 149).

Ring weights continue to be used well into the 1st millennium BCE in North Etruria and Veneto, as attested by the finds at Verucchio (Gentili 1988, 83, 84 pl. A.b), Pozzuolo del Friuli (Vitri *et al.* 1991, 22–25, figs. 8, 13–15), Baggiovara (Malnati 1989, 265, 268 fig. 215 no. 5), Forcello di Bagnolo S. Vito (Vay 1988) and Poggio Colla (pers. obs. Poggio Colla excavation, 2006). Occasionally, ring loom weights are found in funerary contexts. Three such weights were found in Tomb 143 of the Casa di Ricovero necropolis at Este (Chieco Bianchi and Calzavara Capuis 1985, pl. 13).

F. Trapezoidal or truncated pyramidal

1. Parallelepipedal
    a. square base
    b. rectangular base
2. Pyramidal
    a. square base
    b. rectangular base
    c. triangular base
    d. polygonal base

This is, by far, the most prevalent loom weight type in the 1st millennium BCE Italy, where trapezoidal loom weights have been found in the majority of the Iron Age and later sites. In North Italy, earliest examples of this type appear during the Bronze Age. Thus, several trapezoidal weights were found at the Middle Bronze Age site of Ledro (Rageth 1974, pl. 93 no. 22), while twelve others have been recovered in Quingento, dated to the Middle–Late Bronze Age (Bernabò Brea and Mutti 1994, 413–414 no. 3; Bernabò Brea, Cardarelli and Cremaschi 1997, 550, fig. 307 no. 7). The shape, however, is not common in the earlier parts of the Bronze Age and becomes diffuse only in the Final Bronze Age–Early Iron Age. Some of the early sites where trapezoidal weights were found, often coexisting with the ring type, are Mariconda (*Antico Polesine* 1986, 108 nos. 7–8, pl. 4 nos. 7–8), Frattesina (Bellintani 1992, Pl. 14 no. 48), Montagnana-Borgo S. Zeno (Bianchin Citton, Gambacurta and Ruta Serafini 1998, 352–361), Este-Canevedo (Prosdocimi 1887, 186, pl. VIII no. 31), Concordia (*Protostoria tra Sile e Tagliamento* 1996, 227 no. 78, 252 no. 164), Como (Casini 1986, 152, nos. 11–12). In Central Italy, the best studied samples are known from Murlo (Gleba 1999), Accesa (Camporeale 1997), Cetamura (Hackworth 1993), Narce (Macnamara 1976, 163–166), Tarquinia (Sartori 2001) and Satrianum (Holloway 1970, 30–32, 113–116, figs. 177–180). Large numbers of loom weights found in excavations in South Italy have been published for the following sites: Broglio di Trebisacce (Peroni and Trucco 1994, 248), Adelfia (Tunzi Sisto 1988, 25–28), Cavallino (Pancrazzi 1979, 188, 190), Monte Sannace (Scarfì 1962; Donvito 1982; Ciancio *et al.* 1989) and Armento (Di Giuseppe 1995, 141–149). The trapezoidal shape is also typical for other areas of the Mediterranean. For example, in Greece, they have been found in Athens (Davidson and Burr Thompson 1943, 73–76, 80–89 nos. 1–115), Corinth (Davidson 1952, 161–162, 170–171 nos. 1192–1204), and Olynthos (Wilson 1930, 123–124 nos. 2–11, 125 nos. 18–20, 126–127 nos. 26–31 and 33–36).

G. Conical

1. Conical
2. Truncated conical
3. Biconical

This type becomes popular during the 6th century BCE, especially in South Italy. It is possible that the shape was brought to Italy by the Hellenic colonists as it is common on Greek sites such as Athens (Davidson and Burr Thompson 1943, 76–79, 89–93 nos. 116–141), Corinth (Davidson 1952, 148–161,164–170 nos. 1083–1191) and Olynthos (Wilson 1930,

124–125 nos. 12–17, 21–23, 126 nos. 24–25, 127 no. 32). In Sicily, conical weights are rare on indigenous sites, especially of the central part of the island (Vassallo 1999, 246). Meanwhile, the shape is ubiquitous in the Greek colonies.

H. Discoid
1. Lenticular
2. Discoid

Disc (*oscillo*) with flat or slightly convex surfaces and suspension hole(s) near the edge.[19] This shape appears by the middle of the 1st millennium BCE but does not become popular until the Hellenistic period. Some of the sites where it has been found are Cozzo Presepe (Morel 1970, 105–107), Torre del Mordillo (Colburn 1977, 512); Francavilla Marittima (Zancani Montuoro 1966, 74–75, 82), Halae (Goldman 1940, 509), Taranto (Wuilleumier 1932), Gela (Orsi 1906; Orlandini 1953, 441–444; Orlandini 1956, 358–359), Rocca d'Entella (Nenci 1995), Montagnola di Marineo (Valentino 1997, 201–202, 205–207, figs. 1–3). Discoid weights are well attested in other areas of the Mediterranean during the same period. For example, in Greece, at Olynthos (Wilson 1930, fig. 285), Corinth (Davidson 1952, 162–163, 171–172 nos. 1205–1209), Athens (Davidson and Burr Thompson 1943, 79, 93–94 nos. 142–149), Perachora (Dunbabin 1962, 331; Tomlinson 1969, 248–250), Delfi (Perdrizet 1906, 197–199), Delos (Deonna 1938, 159–162); on Crete, at Palaikastro (Sackett and Popham 1965, 304 nos. 27–36, fig. 19), Festos (Levi 1966, 569–588); in Asia Minor, at Ephesos (Hogarth 1908, 319–320) and Miletos (pers. obs. Miletos excavation, 2005).

I. Parallelepipedal with two holes through the long side.

1. Rectangular ends: three objects were found at Oliveto Citra Tomb VII, dated to the first half of the 6th century BCE (d'Agostino 1964, 56, fig. 13); four examples come from the settlement of Cavallino (Pancrazzi 1979, 191 3 and 4, fig. 74 nos. 2–3, 9–12); several others are known from Canosa, both from the settlement and from a tomb on Strada Macello Via Cerignola (pers. obs. Canosa Museum, 2003); finally, three objects 4.5–6 cm long and dating to the 4th century BCE have been excavated at Morgantina (pers. obs. Aidone Museum, 2003).
2. Rounded ends: smaller than the rectangular type; at least four implements about 7 cm long, painted with floral decoration, are known from Canosa (pers. obs. Canosa Museum, 2003).

The function of this object type is not quite clear, but genarally it has been interpreted as a loom weight. It is possible, that although it may have been connected with weaving, it did not serve in such a capacity.

J. Double cylindrical attached at one end, with vertical holes passing through each cylinder; single objects of this type have been excavated at Canosa (pers. obs. Canosa Museum, 2003), Satrianum (Holloway 1970, 114 no. 316; fig. 176 no. 316), Monte Sannace (Scarfi 1962, 162, fig. 154; Ciancio *et al.* 1989, pl. 361 no. 2) and Nusco-Serra (pers. obs. Avellino Museum, 2002). This is another type of object usually identified as a loom weight, although its function is not known definitively. Holloway (1970, 114) suggests the object was possibly used for carding wool, although it is unclead exactly how this would have been done. In my opinion, a more likely use of such an object would be for plying, with each single yarn passing through a hole and meeting at one end. Alternatively, the object may not have been connected to weaving at all.

Functional aspects

Number

A set of loom weights is typically composed of six to 30 implements, although it could reach as many as 80 (Barber 1991, 104). Fewer loom weights would be needed if they are heavy and more if they are light.

The evidence for these numbers is provided by ethnoarchaeology and by the configuration of loom weights occasionally found *in situ*, having fallen to the ground when the warp to which they were originally attached was destroyed or deliberately cut. Thus in a Neolithic hut at Fornace Grandi di Bondeno (Ferrara), about 50 loom weights were found on the floor (Cremonini and Steffè 1987). Several sets of loom weights ranging between 12 and 16 in number were found in a Bronze Age habitation of S. Rosa di Poviglio (Rast-Eicher 1997, 550; Bernabò Brea 1997; Bernabò Brea, Bianchi and Lincetto 2003, 112–119). Similar situations are observed at Forno del Gallo a Beneceto (Bernabò Brea, Bianchi and Lincetto 2003, 116), and Fiavè-Carera (Perini 1994, 1019), dated to the Middle Bronze Age.

In a Late Bronze Age hut at Romagnano Loch there were found 24 loom weights in a line two meters long (Perini 1971). Closer to the period of our interest, at Monte Loffa, 49 loom weights were "found in exact position where the loom had fallen" among the concentration of carbon in one of the shallow cellars cut into limestone (Battaglia 1934, 122–123, fig. 6; Barfield 1971, 141). A row of loom weights was found in one of the rooms at Sottosegna (Barfield 1971, 141, fig. 75). Twelve large ring loom weights were found aligned in the corner of Rhaetic House 2 at Montesei di Serso (Perini 1978, 59).

At Francavilla Marittima, in the so-called Weaving Hut, dated to the second half of the 9th and the 8th centuries BCE, two rows of large loom weights were found more or less *in situ* (Attema *et al.* 1998, 337). One of the rows had eleven weights with elaborate incised labyrinth motifs, flanked by two heavier undecorated items. The excavator states that "the intricate and time-consuming decoration of the loom weights with meander and labyrinthine patterns, as well as their disposition and heavy weight (over 800 g), suggest pattern weaving" (Kleibrink 2001, 49).

At Gravina di Puglia, House 2 had 31 loom weights on the floor, assumed to belong to a set (Wilson 1977, 132). At the Hellenistic settlements of Oppido Lucano (Lissi Caronna 1983, 295, 332; Lissi Caronna 1992, 238, 277–285) and Monte Sannace (Scarfi 1962, 160; Donvito 1982, 168–169), in South Italy, numerous houses yielded sets of loom weights, ranging in number from 20 to over 80. In the Casa dei *Pithoi* at Serra di Vaglio, dated to the late 4th–early 3rd century BCE, 108 loom weights were found next to the south-west wall and carbonised remains of large wooden beams were visible during the excavation (Greco 1991, 62). Other loom weight groups found *in situ* have been excavated at Forcello di Bagnolo S. Vito (Vay 1988), Poggio Bacherina at Chianciano (Paolucci 1997, 56–57), Acquarossa (Nylander and Pelagatti 1986, 125 n. 282; Östenberg 1975, 11–12), Pithekoussai – Punta Chiarito (De Caro and Gialanella 1998, 345, fig. 5) and Himera (Allegro 1976, 557).

Loom size

Such loom weight series not only give an idea about the number of loom weights utilised in a loom but also permit reconstruction of the exact orientation and the width of the loom, which, in turn, can be used to speculate about the size and quality of the textile being woven.[20] The number of rows may also indicate the number of sheds, and hence the complexity of the weave (Barber 1991, 110). The pattern of distribution of the loom weights can also be useful in providing information about the location of textile producing activities within the settlement. The degree of uprightness and the height of a warp-weighted loom, as well as the distribution of the weights must be considered when attempting to suggest a possible location for looms in houses.

Weaving was often a seasonal activity in antiquity. Ethnographic and archaeological evidence suggests that warp-weighted looms were generally dismantled and their parts stored when not in use (Barber 1991, 102).[21] Sets of loom weights found in amphorae in Hellenistic Mozia may have been stored there when not in use (Rossoni 2002).

Weight

Weight is probably the most important functional feature of a loom weight. The groups of loom weights from specific sites, which have been weighed, are presented in Table 6. In considering the range of weight only complete examples were used. Overall, they vary in weight from less than one hundred grams to a few kilograms. The number of threads attached to each weight would have varied depending on the weight, type of fibre, thread diameter and type of textile being woven.

Production

The vast majority of loom weights were hand made, resulting in irregularity in their shapes, and the fact that no two weights are completely identical either in shape or in weight. In later periods, the weights were made quickly and by skilled hands, possibly by the same craftsmen who manufactured pottery and/or tile on a site. Thus, 34 weights were found in potters' workshop at Oppido Lucano (Lissi Caronna 1980, 217–218). At Metaponto, pit 7 of a potters' workshop, dated to the 4th century BCE, contained eight misfired discoid weights (Adamesteanu, Mertens and D'Andria 1975, 443 nos. 400–401, 441 fig. 72).

In the Hellenistic period, weights were sometimes mould made. It seems that no single method of producing loom weights ever became standardised. Precision of shape and weight were, evidently, of no great importance since the weights appear to have been made with skill but with not much care. The hole was pierced with some kind of sharp object, possibly a stick, without particular regard to centring. Occasionally, weights with double holes are found, where the second perforation was made because the first was too skewed to the side or upwards. Unequivocal evidence for loom weight production is presented by the rare cases when the objects are found in kilns. Thus, in the S. Domenico area of Orvieto, two kilns of Etruscan-Roman period contained trapezoidal loom weights (Minto 1936, 263, fig. 16). Other loom weights were found in kilns at Cetamura (de Grummond 2001) and Pisticci (*Metaponto II* 1977, 382). Although loom weight production was, in most cases, a local affair, de Grummond recently suggested a possibility that, occasionally, they may have travelled considerable distances. Two weights, found in the Hellenistic context at Cetamura, may have originated in Cosa and were made by the workshop of the Sestii, which also produced tiles, amphorae and red-glazed ware (de Grummond 2005). The weights are made of clay different from other materials at the site and bear S and asterisk stamps connected to the workshop of the Sestii family.

Decoration

Weights often bear simple markings on the top surface, including lines, crosses,

Site	Date	Number	Weight (g)	Bibliography
Lagozza di Besnate	4th mil.	-	690–730	Baioni *et al.* 2003, 101.
S. Rosa di Poviglio	MBA	28	230–360, 450–760, 980–1300	Bernabò Brea, Bianchi and Lincetto 2003, 114.
Fonte Tasca	FBA	72	20–110	Di Fraia 1995, 460.
Francavilla Marittima	9th c.	-	600–900, 2 over 1000	Maaskant Kleibrink 2000, 174.
Incoronata	8th–7th c.	-	18–80	Franchi 1986, 178.
Narce	8th–7th c. 4th–3rd c.	3 3 4	*c.* 50 300–350 *c.* 2000	Macnamara 1976, 163–66 and 172–73.
Acquarossa, Zone L Zone B	7th–6th c. 3th–1st c.	16	400–425 300–350 and 1000–2000	Wikander 1986, 112 and 125 no. 282.
Murlo	7th–6th c.	69	37–795	Gleba 1999, 74; Gleba 2000.
Accesa	7th–6th c.	89	125–370 and one 450	Camporeale, 1997, 199
Ficana Zone 4a	7th–5th c. 950–500	 29	250–350 147–2000 most 180–350	Brandt 1980, 118; Lipponen 207, 3.
Forcello di Bagnolo San Vito	6th–5th c.	105	300–800	Vay 1988, 171–72.
Monte Savino	6th–5th c.	80+	500–750; over 1500	Macellari 1983, 123–126.
Cavallino	6th–4th c.	42	27–270	Pancrazzi 1978.
Adelfia	IA–4th c.	358	20–419	Tunzi Sisto 1988.
Tarquinia	6th–3th c.	58	200–800+ most 300–410	Sartori 2001.
Satrianum	5th–4th c.	57	52–117, 380, 480, 580	Holloway 1970.
Heraion alla Foce del Sele	5th–4th c.	266	100–150, 200, 350–425	Zancani Montuoro 1966, 61–82.
Paestum, Santa Venera	5th–3rd c.	50+	50–310	Pedley 1993, 19.
Locri	5th–3rd c.	300?	70–200	Dotta 1989; Arboletti 1992.
Cetamura	5th–2nd c.	18	90–1000+	Hackworth, 1993, 10.
La Piana	4th–3rd c.	44	68–975	Hackworth 1993, 13; Whitehead 1996, 105–46, 114.
Pomarico Vecchio	4th c.	50+	50–200	Sartoris 1997, 228.
Mozia	4th c.	241	30–40, 75–85	Rossoni 2002.
Oppido Lucano	4th–3rd c.	86	30–465 (medium 150–200)	Lissi Caronna 1992, 460.
Gravina di Puglia	4th–1st c.	500+	50–100, 175–200, 250+	Tatton-Brown, 1992, 218.
Armento	3th–2nd c.	27	44–234	Di Giuseppe 1995, 142.

Table 6. Weighed loom weight groups.

zig-zags and round finger impressions (Figure 94a). Sigla (graffiti and stamps of single letters) appear during the Orientalising period and, starting with the Archaic period, full or abbreviated words are occasionally inscribed or stamped on loom weights (Figure 94b).[22] Gem impressions also become popular around the same time (Figure 94c).[23] Impressions of various objects are also common as a means of decorating the loom weights. Fibulae are among the objects most frequently used for this purpose.[24] Occasionally, tweezers were impressed on the side of a loom weight (Figure 94d), such as at Pomarico Vecchio (Sartoris 1997, 228, pl. 92 and fig. 100 no. 18) and Oppido Lucano

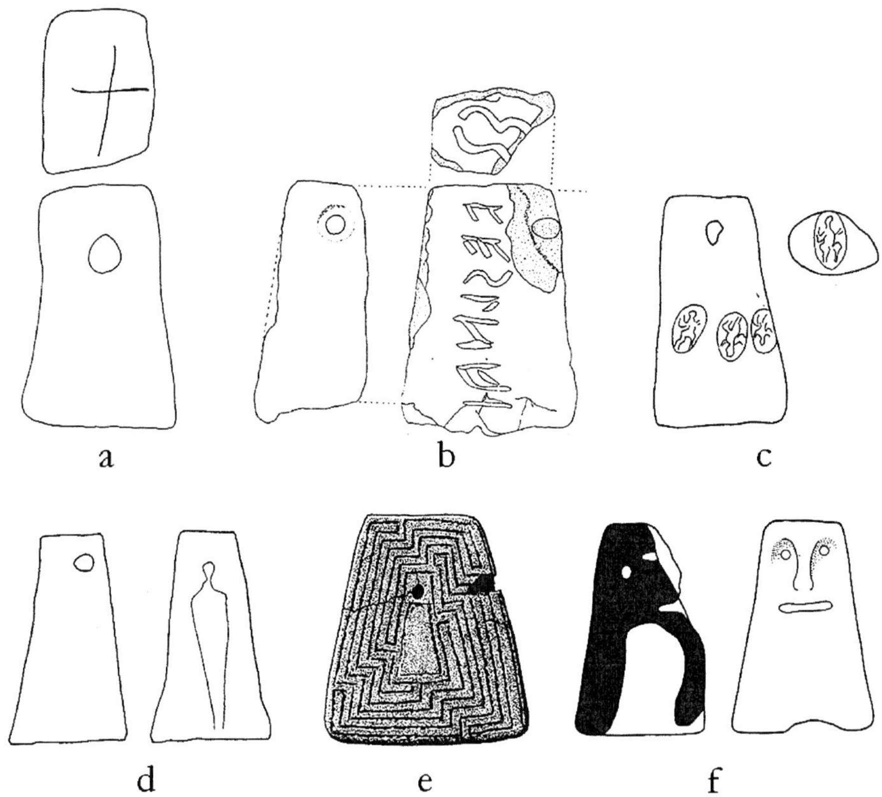

Figure 94. Decoration on loom weights: a) loom weight with incised cross on top from Cetamura (After de Grummond 2000, pl. XX no. 36); b) loom weight with Etruscan inscription, Museo Nazionale Romano (After Ambrosini, 141 fig.3); c) loom weight with impressions of gems from Pomarico Vecchio (after Sartoris 1997, 100 fig. 92); d) loom weight with impression of tweezers from Pomarico Vecchio (after Sartoris 1997, 100 fig. 92); e) loom weight with labyrinth decoration from Francavilla Marittima (Courtesy of Marianne Kleibrink); f) anthropomorphic loom weight from Pomarico Vecchio (After Barra Bagnasco 1996, 214 fig. 52 no. 137).

(Lissi Caronna, Armignacco Alidori and Panciera 1992, 468 no. 83, fig. 247–248). A combination of impressions of a fibula and tweezers occur at Pomarico Vecchio (Barra Bagnasco 1996, 383 fig. 94); Oppido Lucano (Lissi Caronna, Armignacco Alidori and Panciera 1992, 277–279 nos. 1–3, fig. 105, 284 nos. 33, 35 fig. 108), Montescaglioso (Lo Porto 1988–1989, 383, fig. 94). Other objects, impressions of which appear on loom weights include knuckle bones or *astragalli* (Lissi Caronna, Armignacco Alidori and Panciera 1992, 224 fig. 47),[25] cords (Lissi Caronna, Armignacco Alidori and Panciera 1992, 465 no. 36, figs. 244, 246), keys (Lissi Caronna 1983, 287 no. 1, 338 no. 11 and 343 fig. 140), sheaves of grain and other objects.

Loom weights decorated with labyrinth and swastika motifs have been found in northern Calabria (Figure 94e). In particular, at Francavilla Marittima numerous weights incised all over with labyrinth patterns and dated to the 8th century BCE, have been excavated (Stoop 1972, 65–66, pl. XXVI; Attema *et al.* 1998, 337; Maaskant-Kleibrink 2000, 174, fig. 92). Similar examples have been found in contemporary burials of Canale and Ianchina near Locri (Orsi 1912a, 48, figs. 52–55; Orsi 1926, pl. XVII; de la Genière 1968, 67). They have been interpreted as having mythological meaning. Lissi Caronna (1972) reads the decoration as representation of Labyrinth of Knossos and suggests Greek influence. Zancani Montuoro (1975) further connects the iconography with the Cretan palaces. Godart (1992) argues that the labyrinth motif goes back to the Bronze Age Aegean presence in Calabria (cf. Mertens-Horn 1994). The labyrinth motif, however, remains an intriguing question (Maaskant-Kleibrink 1993, 22).

Anthropomorphic or zoomorphic motifs, such as a deer on a loom weight from Burial 94 at S. Maria d'Anglona, also occasionally appear (Frey 1991, pl. 3A no. 6). An interesting comparison may be noted with a loom weight from Hallstatt settlement Smolenice-Molpír, located in the north-eastern Alpine region in Slovenia. Its decoration consists of a human figure and an animal (Stegmann-Rajtár 1998, 273 fig. 6a). In rare instances loom weights have been fashioned into anthropomorphic shape (Figure 94f). Such is the case of a weight from Pomarico Vecchio with one of the sides shaped into a human face (Barra Bagnasco 1996, 214 fig. 52 no. 137, 215 no. 137, fig. 53; Sartoris 1997, 230).

Occasionally, loom weights were painted. This type of decoration occurs in Hellenic settlements Gravina di Puglia (Tatton Brown 1992, 224) and Incoronata (Franchi 1986, 184 no. 141c) in the south and seems to be of Greek derivation. In Greece, painted weights have been found in Athens (Davidson and Burr Thompson 1943, 74, 82 fig. 33).

Various interpretations have been offered for the decoration on ancient loom weights (Mingazzini 1974, 202). The markings can represent one of the following:

1. Weight measure (Davidson and Burr Thompson 1943, 74). However, they do not appear to represent numerical values since there is no consistency in weight corresponding to a particular mark.
2. Maker's mark (Davidson and Burr Thompson 1943, 76). This interpretation has not been widely favoured by scholars. Recently, however, de Grummond has suggested that S and asterisk stamps found on loom weights in the Hellenistic context at Cetamura may be stamps from the factories of the Sestius family, which also appear on amphorae, lamps, tiles and red-glazed pottery produced by the Sestii (de Grummond 2005).
3. Owner's mark. Names or their abbreviations, sometimes amounting to a single initial, were incised, stamped or painted on the weights. A weight with stamped inscription **ΣΩΚΡΑ** from Lariano has been interpreted as owner's/maker's mark (Ceccarini *et al.* 1989, 125 V33). Weights with the same inscription have been found in Taranto. Another weight, from Policoro, had a name **ΙΣΟΔΙΚΗΣ'ΗΜΙ** painted in red (Orsi 1912b, 61 and fig. 63). A number of others, found in Taranto, are currently conserved in the Provincial Archaeological Museum of Bari.[26] The majority of the names are feminine but occasionally male names also occur (Mingazzini 1974, 220). Most recently, an analysis by Laura Ambrosini of loom weights bearing Etruscan inscriptions has demonstrated that all the inscriptions are of personal (mainly feminine) names in a possessive case, indicating ownership (Ambrosini 2002, esp. 156). A different take on the meaning of owner's mark is proposed by Mingazzini, who even suggests that the last weight on the left remained attached to the cloth so that the merchant or the buyer knew who made the cloth (Mingazzini 1974, 202). A loom weight with owner's mark could also serve to guarantee the quality and authenticity of the product. Gem impressions in particular would have made especially meaningful owners marks.
4. The order in which loom weights were attached to specific sets of warp threads. This interpretation was proposed by Ferrandini Troisi (1986, 93–94) for the weights inscribed with single letters.[27]
5. Type of cloth. Abbreviated incriptions **ΤΕΤΑΡΤΙ(ον)**, **ΤΡΙΤΑΙΑ**, **ΗΜΙC(υ)**, **ΔΥΟΤΡΙ(ταια)**, have been interpreted by Mingazzini as indicative of the size of the cloth: a quarter, a third, a half, and two-thirds piece of the normal size (Mingazzini 1974, 202–203).[28]

6. The deity's name to which the weight is dedicated. Votive inscriptions on loom weights seem to be a practice common in the South Italy, most likely adopted from Greece, where the phenomenon is also documented. Only two possible cases of the same practice have been documented in Etruria and in the absence of clear context for either loom weight, it is impossible to make judgments about the destination of these objects (Ambrosini 2002, 159). As far as the deities to whom such votives were dedicated, ubiquitous is Athena, the guardian of the feminine crafts. In addition to the inscriptions with deity's name, loom weights decorated with the owl of Athena recall the connection of the goddess with the art of weaving. Thus, 4th–3rd century BCE weights from Puglia have an owl with human hands in the act of spinning (Wuilleumier 1932, 47 no. 104; Di Vita 1956, 43; Barber 1992a, 106–107, 151 no.12).[29] Greek terracotta of Athena Ergane with a dressed distaff from Scornavacche in Sicily, dated to the 5th century BCE, also illustrates the goddess at the task of spinning (Figure 15). Depiction of Athena Ergane with spindle and distaff was typical, but Pausanias (7.5.9) notes that Athena Polias of Eretria also had these attributes. Other gods, whose inscribed names or stamped attributes appear on loom weights, are Heracles, Demeter, Hera and the Fates (Wuilleumier 1932, 38–39, no. 20 a–e; Orlandini 1953, 442; Ferrandini Troisi 1986, 97; Ambrosini 2002, 156 note 149).

Figure 95 (top). Beating knife from Fiavè-Carera, Early–Middle Bronze Age (From Bazzanella et al. *2003, 141).*

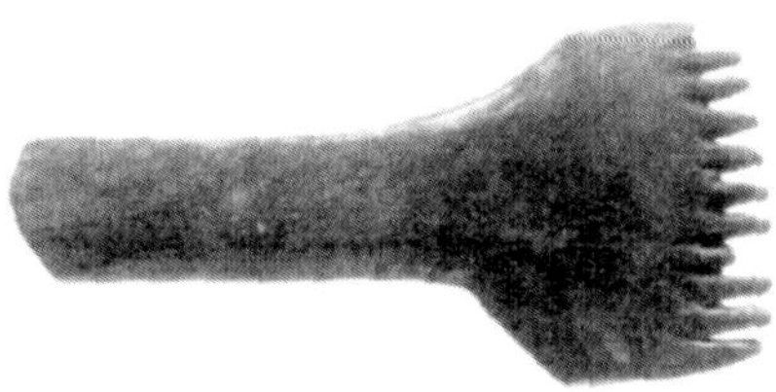

Figure 96 (bottom). Beating comb from Fiavè-Carera, Early–Middle Bronze Age (From Bazzanella et al. *2003, 141).*

It is apparent that depending on the type of decoration or inscription, as well as context, cultural and geographical location and chronological period, the signs and impressions on loom weights could have diverse meanings and functions and it is hardly possible to find a single explanation and interpretation.

Beating tools

The tools used to beat up weft against the cloth already woven are not well known from the Italian archaeological record, partly because they were made of perishable materials. Throughout Europe, a variety of implements were utilised for the purpose, including combs and special weaving swords and knives, made of bone or wood.[30] Large bone needles could also be used for for distributing the weft evenly, particularly in fine and/or open weaves. The Middle Bronze Age settlement of Fiavè-Carera produced a wooden weaving sword, 52 cm in length, as well as bone weaving knives (Figure 95) and numerous horn combs (Figure 96) (Bazzanella *et al.* 2003, 138, 141–142). Similar implements have been found in the Neolithic Swiss lake dwellings (Bazzanella *et al.* 2003, 229, 234, 246).

Tablet weaving

Tablet weaving involves passing threads through holes in the corners of (usually) square tablets, which, when rotated forward or back, force the threads to form different sheds (Figure 20). By rotating cards in different combinations, it is possible to achieve numerous patterns. This method is suitable for weaving narrow bands, such as belts, heading bands for the warp of a warp-weighted loom, or decorative borders for the base textile. The later use of the tablet technique is aptly demonstrated by the Verucchio mantles where 36 tablets with four holes were utilised to create a triangular pattern (Figure 30). One of the fragments of the third garment found in the burial had a tablet-woven border of a simpler kind, obtained using 13 tablets (Ræder Knudsen 2002, 222–225). Fragments with tablet-woven bands from Sasso di Furbara (Cat. No. 46) and Palestrina (Cat. No. 61) also show that the method was used since the Iron Age in Italy, paralleled by the frequent finds of tablet borders in Western and Central Europe (Bender Jørgensen 1992, 123; Ræder Knudsen 2002, 232). The more significant finds come from Hallstatt in Austria, the Kerameikos in Athens, Greece, Apremont in Luxemburg, Hochdorf and Hohmichele in Germany (Banck-Burgess 1999, 80–82; Ræder Knudsen 2002, 233–234; Grömer 2005b).

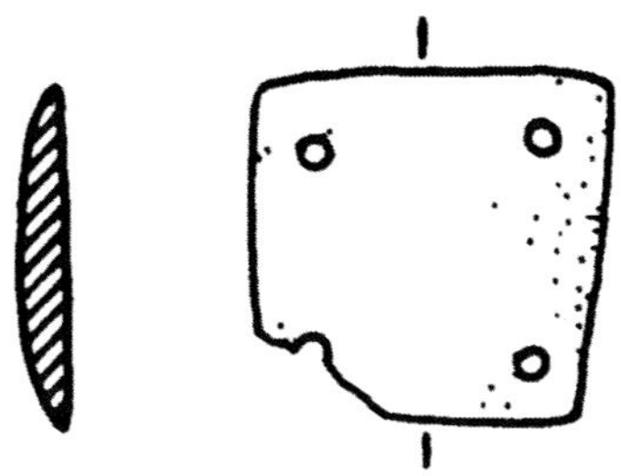

a

b

Figure 97. Tablets from a) Ficana, 8th century BCE (Courtesy of Eero Jarva); b) Poggiomarino, 9th century BCE (After Pizzano 2005, fig. 16).

Among the surviving tools that may have been utilised in tablet weaving are tablets, terracotta spools, metal clasps and bone spacers with pegs.

Tablet

Tablets themselves, having been made of wood, bone or hardened leather, rarely survive. Even larger problem however is that of identification: few excavators are familiar with weaving technology and even fewer still are aware of tablet weaving, thus hampering the identification and publication of tablets. Recently, tablets from two sites have been identified. At Ficana, an almost complete small bone tablet, 2 × 2 cm (Figure 97a), was found in a context of the late 8th century BCE (Lipponen 2007, 4). Other important finds are from Longola di Poggiomarino, an Iron Age settlement, recently identified and excavated north-east of Pompeii. Here, objects in organic materials have been preserved in unique waterlogged conditions. Among these are small tablets, also *c.* 2 × 2 cm (Figure 97b), in both wood and bone of domestic and/or wild pig (Pizzano 2005, pl. 10). Bone tablets have also been observed among the burial goods at Santa Maria d'Anglona (pers. com. Marianne Kleibrink, 2007).

It has been suggested that certain clay tablets with holes in them may have been utilised for some sort of cord-making technique, similar to tablet-weaving. For example, 28 round and triangular clay tablets with three holes each, 1.5–2 cm in diameter, have been found in the 6th century BCE Tomb 309 at Alianello-Cazzaiola, in Basilicata (Bottini 2000; Russo 2006, 144 note 65). Three oval terracotta tablets, 2–3 cm long and with two holes each were found in the 7th century BCE Tomb 6 of M. Ripa at Eboli (pers. obs. Eboli Museum, 2003).

Spool

Function

Spools are small cylindrical objects, often with concave shaft and flaring ends. They are made of terracotta and range in length between 3 and 10 cm. Spools (Italian *rocchetti*) are sometimes also called bobbins in the archaeological literature. They are known in Italy since the Final Bronze Age and their function has been widely debated since the first examples were found. No representations of spools exist to aid in understanding their function. For the most part, scholars have agreed that these objects are in some way associated with textile production. There are, however, other interpretations. For instance, one scholar suggests that spools were ancient yo-yos (Gould 1973). This seems to me the most implausible hypothesis, based on the fact that spools frequently appear in funerary contexts, associated with adult burials, and often occurring in multiples. Further, the sheer number of spools found throughout Italy belies their use as mere toys. Another theory suggests that spools were used as spacers in pottery kilns, to support the weight of and provide stability for pottery or tile stacked inside (Hackworth 1993, 34). Alternatively, they could have been used as testers in kilns during firing. Both interpretations are, again, unlikely due to the context, large number and high quality of many examples.

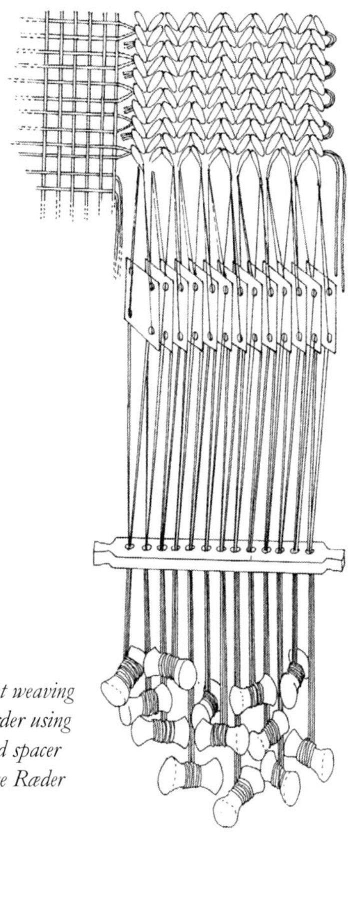

Figure 98. Tablet weaving of a garment border using tablets, spools and spacer (Courtesy of Lise Ræder Knudsen).

Most excavators agree that spools functioned simply as holders for yarn. I disagree with this theory as well, for the following reasons. First of all, most spools have only very slight taper and would not have provided an efficient way of storing yarn. Made of terracotta, they are also heavy and prone to break easily. For instance, at Murlo, 73 percent of spools are broken. Their production requires the expenditure of time, energy and resources that are not really justifiable by the storage function. The most efficient way to use spun yarn would have been without rewinding it from the spindle, which could then also serve as a shuttle during weaving, and was cheap and easy to make – an important factor when a large quantity of yarn is to be produced and utilised. The yarn was stored in baskets as depicted on some Greek vases, or wound in balls, as is still done by modern weavers (Hoffmann 1964, 289).

For these reasons, I have argued that spools were, in fact, used as small weights (Gleba 1999, 92; Gleba 2000, 79). The possibility has been previously suggested by Edith Hall Dohan (1942, 17) who stated: "that these objects are weights, not spools, is probable because similar specimens have a central perforation". Lauren Hackworth (1993, 47) proposed that spools were used as lighter loom weights for more delicate fibres such as cotton and silk. Recently, Lise Ræder Knudsen (2002, 228–229) has demonstrated quite convincingly that they could have been utilised as weights for the sets of threads passing through the tablets used for making the borders on the Verucchio textiles (Figure 98). Their weight supports such an interpretation. A sample of spools from Verucchio varied in weight between 8 and 55 g, with concentrations around 20–30 g and 35–45 g (Ræder Knudsen 2002, 228). Spools from two Early Iron Age burials at Narce have also been weighed: 5 spools from Tomb 23F weighed 21–34 g, while 8 spools from Tomb 105 ranged 19–49. At Murlo, more than 580 spools had a weight range of 8–112 g, with a mean around 50 g and a median at 48 g. Each spool would thus provide the optimal amount of tension for four fine threads passing through a tablet.[31] Swiss Bronze Age spools have been found to weigh 46–244 g at Eschenz-Werd and 10–190 g at Ürschhausen-Horn (Brem, Bolliger and Primas 1987, 129; Nagy 1999, 75).

Alternatively, spools could have been used for other types of weaving, resulting in narrow strips of patterned textile. Surprising parallels are found in Japan, where a cord-making technique called *kumihimo* involves the use of a spool-shaped weight that provides tension and stores the excess thread in the initial stages of production (Kinoshita 1980; Martin 1984; Moss 1987, 20–21, 29; Berlin 1999). Depending on the pattern, 4 to 36 threads may be used, each wound on its own spool. One of the variations of the technique can utilize more than 100 such spools (Berlin 1999, 31). The pattern is formed by a process akin to braiding. Interlacing threads in different patterns can produce round, square, or flat bands. Barber has proposed that Neolithic inhabitants of Greece, where similar objects are found, utilised a technique similar to *kumihimo*, which then spread east with the Indo-European migrations (Barber 1999, 60). Thus, the cord belt found on one of the Ürümchi mummies in the Tarim Basin of Chinese Turkestan appears to have been made using this method around 1000 BCE (Barber 1999, 58–59). It is not unlikely that from there, the technique spread even further east, all the way to Japan, where it survives up until this day.

Funerary contexts also provide some indication of the function of spools since, in burials, they usually occur in multiples. The numbers range from 1–3 to over 40. In one extraordinary case, the Isis Tomb at Polledrara necropolis in Vulci, dated to the 7th–6th century BCE, 130 spools have been found (Haynes 1977, 29). Single spools may have had a *pars pro toto* significance in burial contexts.

In cases when an entire necropolis has been published, the material can be subjected to simple statistical analyses (Table 7). Six Iron Age Italian necropoleis have been selected: Este-Casa di Ricovero (Chieco Bianchi and Calzavara Capuis 1985), Verucchio-La Rocca (Gentili 2003), Veio-Quattro Fontanili (Close-Brooks 1963; Cavalotti Batchvarova 1965; Falconi Amorelli 1967; Franco *et al.* 1970; Fabricotti, Buchanan and Paton 1972; Bedello and Fabricotti 1975; Toms 1998), Osteria dell'Osa (Bietti Sestieri 1992a, 1992b), Pontecagnano–Picentino (d'Agostino and Gastaldi 1988) and Torre Galli (Pacciarelli 1999). Spools are almost exclusively associated with female burials. In the case of Osteria dell'Osa, the number of female burials is based on osteological analysis: 545 individuals were studied, of which 99 had undetermined sex (Bietti Sestieri 1992a, 99). In other

Site	Date	Total no. of burials	Nos. of ♀ burials (% of total)	Nos. of ♀ burials with S (%)	Burials with 1 S (%)	Range in nos. of R	Mean no. of S per burial (*) [**]	Median no. of S per burial (*) [**]	Standard error (*) [**]
Este	8th–5th c.	171	60 (35)	13 (22)	4 (31)	1–31	7 (10) [3]	2 (4) [2]	2.82 (3.72) [1.06]
Verucchio	9th–7th c.	190	NA	39	5 (13)	1–62	13 (15) [8]	10 (12) [8]	2.19 (2.37) [1.06]
Veio	9th–7th c.	651	NA	32	6 (19)	1–56	12 (15) [6]	6 (8) [6]	2.51 (2.85) [0.92]
Osteria dell'Osa	9th–7th c.	595	258 (43)	24 (9)	6 (25)	1–13	4 (5)	4 (4)	0.62 (0.66)
Pontecagnano	9th–7th c.	99	34 (34)	13 (38)	3	1–16	7 (9)	7.5 (8)	1.34 (1.3)
Torre Galli	9th–8th c.	277	138 (50)	37 (27)	5 (13)	1–13	5 (5)	5 (5)	0.48 (0.48)

* Recalculated with burials with one S removed from the sample.
** Recalculated with burials with more than 20 S removed from the sample.

Table 7. Statistical analysis of spool numbers in selected Iron Age necropoleis.

necropoleis, the numbers are based on sex attribution through burial assemblage analysis.

Although it was not always possible to calculate the total number of female burials, the percentages of female burials containing spools at these sites range between 9 and 38 per cent. At Este, where 60 of 171 burials were identified as female, 13, or 21 per cent, contained spools. All burials with spools also contained spindle whorls, often multiple, although given that the depositions were often multiple, a more careful examination is necessary to correlate the data on the co-occurrence of these two types of textile implements. The data from Verucchio-La Rocca indicate that at least 39 female burials had spools among their burial goods, complemented by usually multiple spindle whorls. At Veio, 32 of the total 651 burials yielded spools. Only about half of these burials also included spindle whorls, which in three cases were associated with bronze distaffs. At Osteria dell'Osa, 24, or 9 per cent, out of 258 female burials contained spools. Twenty of these burials also contained spindle whorls, in the majority of cases multiple, ranging between 2 and 15 in number. Pontecagnano yielded 34 burials identified as female out of 99, of which 13 contained spools, comprising the highest percentage among the examined necropoleis, 38 per cent. At Torre Galli, 138 out of 277 burials were identified as female and 37 of these, or 27 per cent, contained spools. Twenty-nine of these burials also contained single spindle whorls.

One spool is present in over 10 per cent of female burials in each site, reaching 30 per cent at Este. Removal of burials with just one spool from the sample raised both the mean and the median number of spools per burial, especially in the cases where the range was high, although the increase was not significant given the standard error values. Removal of the burials with high numbers of spools (over 20), on the other hand, lowered significantly the mean and the median, while resulting in much lower standard error.

Regarding the chronological distribution, the data suggests that burials of earlier date contain fewer spools, as shown by the ranges in the earlier necropoleis of Osteria dell'Osa and Torre Galli. The larger numbers in the later burials could also be culturally conditioned as it is the Villanovan sites of Verucchio and Veio, which yielded the largest numbers of spools per tomb. More complex analysis is necessary to see the correlation of textile tools with other objects of the *corredi* (*i.e.* wealth) and chronology.

Spools do not change dramatically over time in shape and size, and style may not be used to date examples found without provenance. The decoration is often helpful, but many patterns go back to the Proto-Villanovan period and so are not useful in dating.

The following typology is intended only as a general guide (Figure 99):[32]

Typology

I. Symmetrical
- A. Cylindrical body
  - 1. flat ends
  - 2. concave ends
  - 3. convex ends
- B. Concave body
  - 1. flat ends
  - a. cross-shaped ends
  - 2. concave ends
  - 3. convex ends
  - 4. domed
  - 5. conical/truncated conical ends
- C. Dumbbell
- D. Disc ends
- E. Domed ends

II. Asymmetrical
- A. Cylindrical body
- B. Concave body
  - 1. flat and convex ends
  - 2. flat and conical ends

III. Reel: body very short, length is smaller than diameter.

Perforation

Frequently, spools have a perforation, which may be in one of the following positions (Figure 100):

1. Oblique
   The most common type of perforation, present to a varied extent in spools from sites throughout Italy from the Final Bronze Age on, but especially dominant in Picenum and South Italy. This type of perforation has been noted in Luni sul Mignone (Hellström 1975, 63 no. 12, pl. 51 no. 12), Narce (Macnamara 1976, 169, 171–172 nos. 198–203, 170 fig. 55), Sorgenti della Nova (Negroni Catacchio 1992, 11, fig. 3 no. 11), Gran Carro (Tamburini 1995, 245, figs. 58, 48, 61A), Montagnana Borgo S. Zeno (Bianchin Citton, Gambacurta and Ruta Serafini 1998, 349 no. 21, 348 fig. 204 no. 21), Villamarzana (Aspes, Bellintani and Fasani 1984, 49, fig. 7), San Giovenale (Olinder and Pohl 1981, 50–51 no. 638, pl. 29), Murlo (Gleba 1999, 85), Ascoli Piceno (Silvestrini, Lavagnoli and Cazzella 1981, 154, 159 fig. 43), Tarquinia (Mandolesi 1999, 92 no. 10 and fig. 40 no. 10), Veio (Cavalotti Batchvarova 1965), Ficana (Stopponi 1985, 169 no. 5, fig. 5–5), Cavallino (Pancrazzi 1979, 191, fig. 74 nos. 1, 5–8, and 16), Capua (Johannowsky 1983, 138 nos. 8–18, pl. XXXVI), Pontecagnano (d'Agostino and Gastaldi 1988; De Natale 1992, 63 nos. 10–19, fig. 104, 98 nos. 8–11, fig. 118), Torre Galli (Pacciarelli 1999); Calanna (Procopio 1962); Locri-Canale/Ianchina (Orsi 1926).
2. Transverse central
   Second most common type of perforation; present in North and Central Italy. Some of the sites where this type of perforation is present include Este (Prosdocimi 1887, 187 pl. VII fig. 22), Padova (*Padova preromana* 1976, 86 no. 97, pl. 8 no. 97), Montagnana Borgo S. Zeno (Bianchin Citton, Gambacurta and Ruta Serafini 1998, 346 nos. 7–9, 347 nos. 17–18, 349

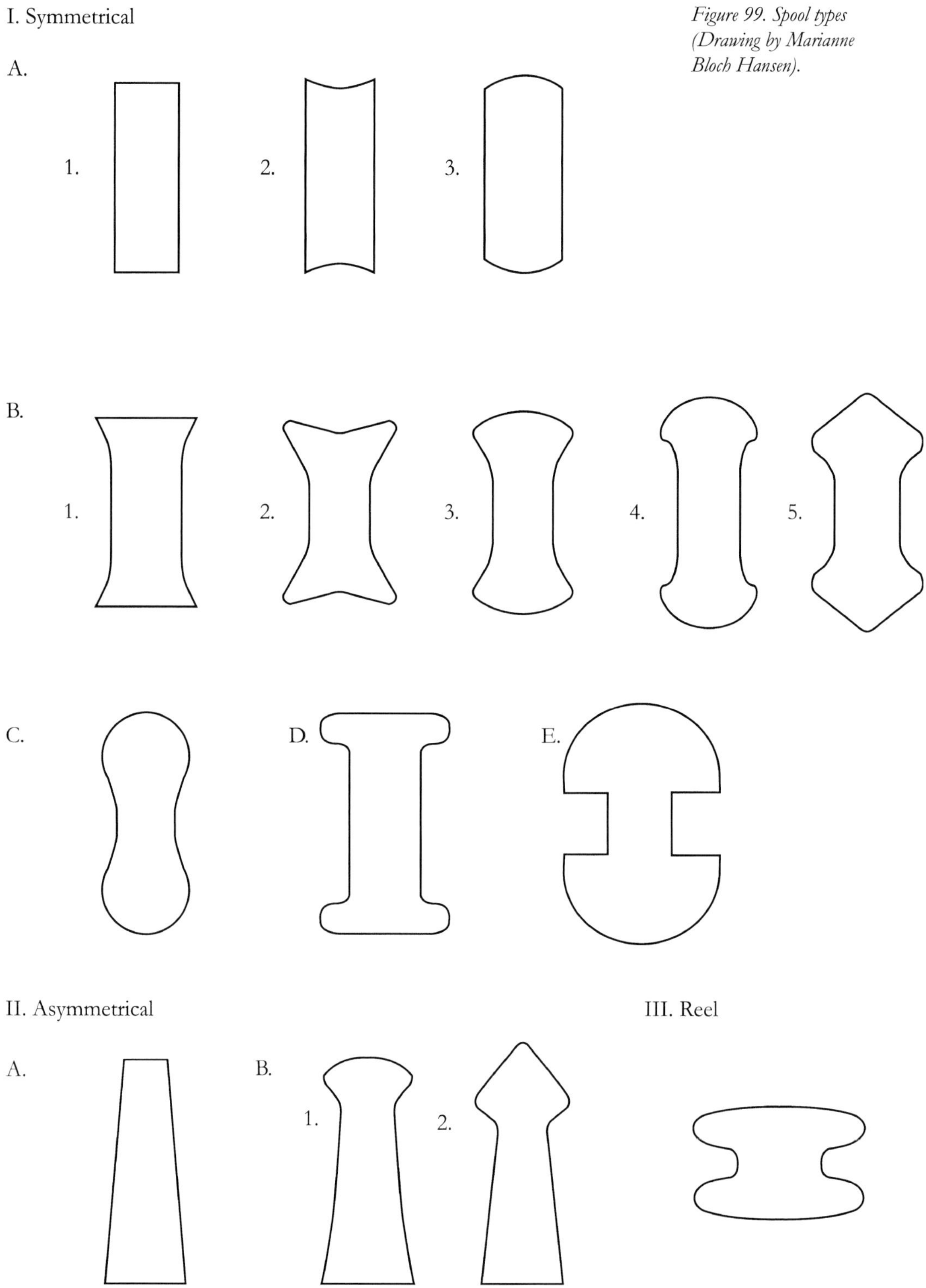

Figure 99. Spool types (Drawing by Marianne Bloch Hansen).

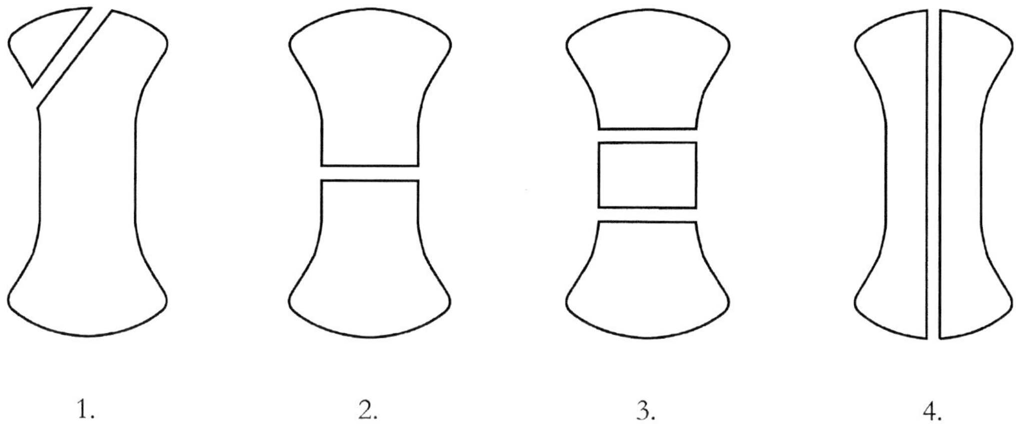

Figure 100. Spool perforation types (Drawing by Marianne Bloch Hansen).

nos. 19–20, 348 fig. 204 nos. 6–9 and 17–20), Verucchio (Forte 1994, 142 nos. 399–402, 411–412 and 415), Roselle (Donati 1994, fig. 8 no. 71, Fig. 27 no. 279), Murlo (Gleba 1999, 84), Manganello (Hackworth 1993, 54), Sorgenti della Nova (Negroni Catacchio 1981, 433 no. 77, pl. 125 no. 77), Luni sul Mignone (Hellström 1975, 63 nos. 15–16, pl. 51 nos. 15–16), Veio (Cavallotti Batchvarova 1965, 128 n, 135 fig. 55; Falconi Amorelli 1967, 213 no. 4, 215 fig. 76; Bedello and Fabricotti 1975, 93 no. 11, fig. 18), Osteria dell'Osa (Bietti Sestieri 1992b, 315 types 34 a, c, e, f, g, h), Cerveteri (Arancio and D'Erme 1992, 307, fig. 4 nos. 1–2, 5–7, 9–10, 14–15 and 18), Rome (Gjerstad 1956, 244 no. 3, Gjerstad 1966, 143 no. 19, 181 no. 64), Ficana (Brandt 1997, 270 cat. 260, fig. 162 nos. 258c, 260a, 260b), Capua (Johannowsky 1983, 138 no. 15, pl. XXXVI no. 14).

3. Double transverse
   The example known so far comes from Murlo and may be a production mistake (Gleba 1999, 85).
4. Longitudinal
   Only three examples are known: Capua (Johannowsky 1983, 138 no. 17); Pianella di Monte Savino (Marchi 1983, 118 no. 32); Canosa (pers. obs. Canosa Museum, 2003).

The function of the perforation is not clear but most likely it served to tie the yarn to the spool. The vast numbers of spools without perforation, however, indicate that it was not a necessary feature.

Chronology

Chronologically, spools appear in the Late Bronze Age, although they do not become widely diffused until the Final Bronze Age (Marzatico 1997, 146). The fact that they appear earlier in the north seems to indicate that the technique arrived from Europe via the Alps. And, in fact, similar spools are common in Swiss Bronze Age contexts, while tablet-woven textiles are frequent in the Hallstatt area. In Italy, in addition to being common on settlement sites, they are frequent in burials of the 9th–6th century BCE in areas associated with Villanovan culture.

Spools with perforations of types 1 and 2 are common in both the Late Bronze Age and the Early Iron Age. In terms of chronological development, early types are cylindrical with oblique holes, while later types tend to be concave with no holes (*e.g.* Macnamara 1976, 171; Bianchin Citton, Gambacurta and Ruta Serafini 1998, 345). This change may be explained by the fact that cylindrical spools need a hole in order for the thread to stay on, while in the concave spools the thread will not slip off as easily. Concave types are also more useful for long textiles since excess yarn can be wound around them.

In North Italy, spools have been found since the Recent Bronze Age at Como (Casini 1986, 152 nos. 7–9), San Brizio (Mancassoola *et al.* 1999, 53, 56 figs. 14 and 15, 57), Verona (Aspes, Rizzetto and Salzani 1976, 143, 152, 158, 170), Este (Prosdocimi 1887, 186, pl. VII fig. 22), Montagnana – Borgo S. Zeno (Bianchin Citton, Gambacurta and Ruta Serafini 1998, 345–349), Padova (*Padova Preromana* 1976, *passim*), Concordia (Croce Da Villa *et al.* 1991, 91; Croce Da Villa and Mastrocinque 1995, 234, 252 fig. 18 no. 12), Frattesina (Bellintani, Perretto and Perretto 1984, 19; Bellintani 1992), Case Nuove di Siccomonte at Cabriolo di Fidenza (Rebecchi 1994, 250), Villamarzana (Apes, Bellintani and Fasani 1984, 49), Verucchio (Gentili 1988: 83, 84 fig. A–b), Forcello di Bagnolo San Vito (Vay 1988, 173–174), Pianella di Monte Savino at Monterenzio (Marchi 1983, 118 nos. 27–32), all settlements of Bologna (Morigi Govi and Vitali 1979, 467; Forte and von Eles 1994, 16, 88, 89, 204, 209; Taglione 1999, 159 pl. XXVIII nos. 1–3, 222 pl. LI), Baggiovara (Malnati 1989, 265, 268 fig. 215 nos. 3–4), and others.

In Central Italy, spools appear during the Final Bronze Age at the sites of Livorno (Zanoni 1997, *passim*), Gran Carro (Fioravani and Camerini 1977, 82 nos. S1–17, 9, 7, 19, 90, 1; Tamburini 1995, 242–245), Norchia (Mandolesi 2000, 144), Monte Bisenzio (Fugazzola Delpino 1982, pl. LXII.2), Sorgenti della Nova (Negroni Catacchio 1981, 247, pl. 8., 431 nos. 71–72, 433 nos. 73–77, pl. 125 nos. 71–77; 1982, 105 nos. 71–77; 1995, 372, fig. 146 nos. 79–83), San Giovenale (Pohl 1977, 93, pl. 21; Berggren and Berggren 1980, 18 no. 179; 1981, 16 note 15, 43 nos. 16–22, pl. 36 nos. 16–22; Olinder and Pohl 1981, 50–51 nos. 631–645, pl. 29, 79–80), and Luni sul Mignone (Hellström 1975, 63, 65, pl. 51 nos. 7–18, pl. 53 nos. 32–35).

Even more conspicuous numbers have been excavated at the following Etruscan sites: Fiesole, Volterra (Cristofani 1973, 92 fig. 64 no. 21, 103 fig. 70 no. 144), Casale Marittimo – Casalvecchio (Esposito 1999, 28 fig. 15), Roselle (Colonna 1985, 55, 56 B, figs. 2.1 B1 and B4), Murlo (Gleba 1999; 2000), Acquarossa (Wendt 1982, 65 nos. 386–402, 47 fig. 36, pl. 3), Narce (Macnamara 1976, 169–172, fig. 55), Vulci (Fugazzola Delpino 1984, 89), Monte Rovello (Allumiere) (Toti 1964, 25, fig. 17i), Santa Severa (*Pyrgi* 1970, 543 no. 4 and fig. 399 no. 1), Veio (Murray Threipland 1963, 66), Osteria dell'Osa (Bietti Sestieri 1979, 42.34, pl. VII), Satrianum (Holloway 1970, 114 nos. 317–318, fig. 176 nos. 317–318), Rome (Dondero Bricchi 1953, 132 fig. 7 no. 3; Gjerstad 1960, *passim*), Crustumerium (Quilici and Quilici Gigli 1980, 78 no. 23, pl. XXV no. 23), and others. At Tarquinia, an extensive survey recently conducted on Civita di Tarquinia yielded numerous spools among the Early Iron Age materials from Pian di Civita, Pian della Regina and Poggio Cretoncini (Mandolesi 1999, *passim*; Bonghi Jovino 2001, 65, pl. 54).

At Ficana, spools of type B are present in all phases of the settlement, although type B1 was more prevalent in earlier levels, while type B2 was more common in later periods (Brandt 1997, 271–272, 270 fig. 162). The lower and upper levels of the hut excavated in 1909 on the Acropolis of Ficana, dated to the 8th–6th centuries BCE, also contained numerous spools of types B, C and D (Ginge 1996b, 132, 133 nos. HLS20–24, 150 nos. HUS 31–32, figs. 41 and 47). At Borgo le Ferriere (Satricum), a large number of spools has been recovered from strata I and II of the habitation, chronologically corresponding to the period between the

Early Iron Age and the transitional phase to the early Archaic period (Maaskant-Kleibrink 1987 and 1992, *passim*). The same spool types appear in foundation and votive deposits of the Southwest Sanctuary and Iron Age burials of West Necropolis (Ginge 1996b, 88 nos. SWT23 and PER9, 108 no. SWS?7, fig. 35).

In Umbria, spools turned up in Terni (Giontella 2001, 368), while in the Picene area, they were found at Ancona – Colle dei Cappuccini (Lollini 1956, 258, figs. 10.7 and 11.7; Percossi Serenelli 1998, 35), Osimo and Moie di Pollenza (Percossi Serenelli 1998, 39, 40), Cartofaro (Ascoli Piceno) (Silvestrini, Lavagnoli and Cazzella 1981, 154, 159 fig. 43) and Fonte Tasca (Di Fraia 1995, 458–459, 469, fig. 8 nos. 10–11).

In South Italy, spools were found in Villanovan sites in Campania (Capua, Salerno – Grotta di Pertosa, Pontecagnano and Eboli), in burial contexts of Calabria, and in the following sites in Puglia: lower levels at Cavallino (Pancrazzi 1979, 189, 190–191, fig. 74 nos. 1, 5–8 and 16), Late Apennine levels at Leporano – Porto Perone near Taranto (Lo Porto 1963, 358, 357 fig. 68 nos. 14–15), and in an Iron Age context at Satyrion, Taranto (Lo Porto 1964, 204, 209 fig. 29 no. 3). Surprising and isolated discovery of spools was made at Incoronata (Metaponto) where they appeared in habitation contexts of trenches I, B, C, D, O, S, *fossae* C and D and inside houses H and S (Franchi 1986, 178–179, 184 nos. 142–143; Castoldi 1995, 110 nos. 12–15, 112 fig. 124–126;

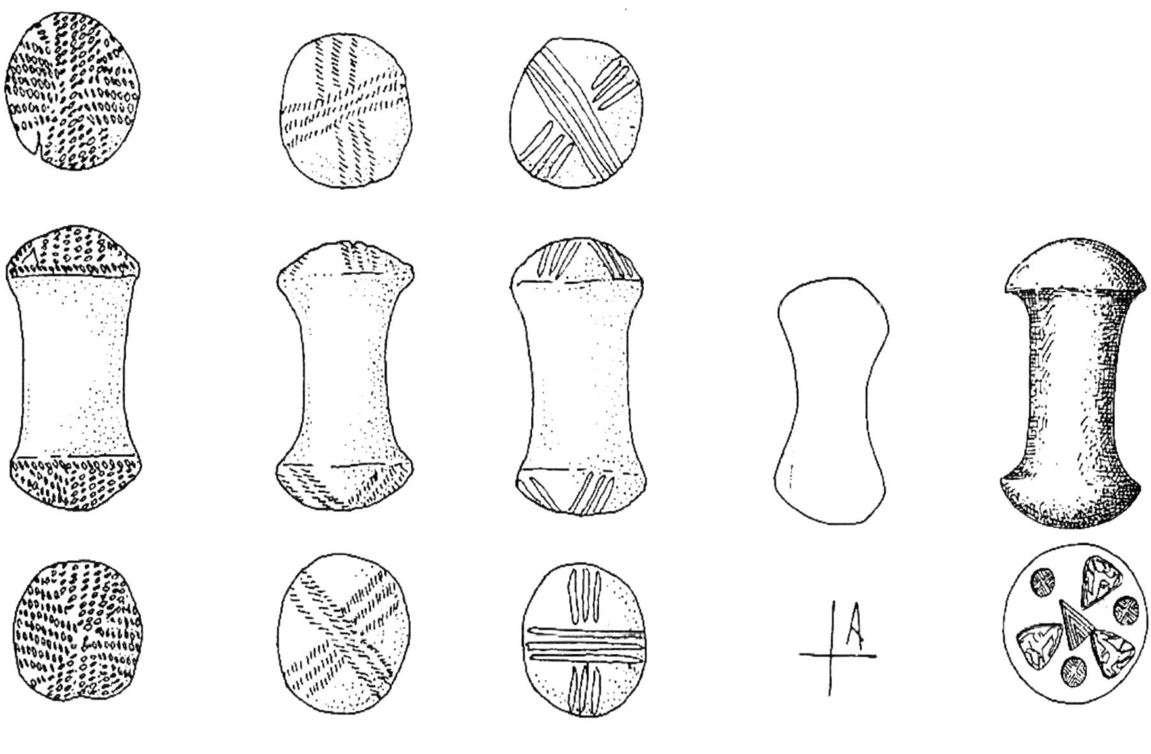

a b c d e

Figure 101. Decoration on spools: a–c) spools with ends decorated with incised crosses and patterns in false cord technique, Tomb 4, Le Pegge, Verucchio (After Gentili 2003, pl. 11 nos. 8–9); d) spool with alphabetic sign, Tomb 870, Casale del Fosso, Veio (After Bagnasco Gianni 1998, 88 fig. 2); e) spool with stamped decoration from Populonia (after Minto 1925, 361 firg. 18).

Stea 1997, 87 no. 1, 90 fig. 97, 149 fig. 261).

Further studies are needed to understand the distribution and chronology of spools and technology they might represent throughout Italy. However, this quick overview shows that spools are particularly common on sites associated with Villanovan culture.

Decoration

Decoration was applied to spools most likely for aesthetic reasons (Figure 101). None of the patterns seem to convey any sort of technical information. Decoration occurs on the ends and never on the shaft of the tool, confirming that the later was its functional part. Simple incisions or impressions are common, the most frequent decoration being a cross on one or both ends (Figure 101a–c); the cross can be simple or with elaborately filled-in quarters (*e.g.* Bocci Pacini 1985, 56; Marchi 1983, 118 no. 30; Tovoli and Vitali 1979, 55 fig. 32 nos. 12–13; Close-Brooks 1963, 152 fig. 49; de la Genière 1977, fig. 25 no. 15). In some instances, alphabetic signs have been incised on one of the ends of a spool (Figure 101d). Decoration is frequently done by impressing a pattern with a cord, or imitating a cord pattern with small diagonal strokes incised with a sharp instrument (*e.g.* Gentili 2003, pl. 5 nos. 19–20, pl. 11 nos. 8–9; Bonghi Jovino 2001 nos. 3/82, 295/5, pl. 54; Close-Brooks 1963, 238 fig. 106, 251 fig. 115). Sometimes, decoration consists of any number of impressed points, which can form a pattern or be random (Marchi 1983, 118 no. 31; Negroni Catacchio 1981, pl. 125 no. 74).

There are also numerous spools with stamped decoration (*e.g.* Tovoli and Vitali 1979, 55 fig. 32 nos. 14–15; Minto 1917, 85 fig. 18; Phillips 1994; Warden and Kane 1997, 177; Mangani and Pacciani 1992, 71 nos. 206–238, fig. 73; Franco *et al.* 1970, 262 fig. 49). The use of stamped decoration on pottery in Etruria can be traced back to the beginning of the Villanovan period. The technique became very popular during the second half of the 7th century BCE in Veneto and northern Etruria, particularly in Bologna, Vetulonia, Roselle, Populonia (Figure 101e) and Chiusi, although the practice is also documented in southern Etruscan sites Tarquinia and Cerveteri (Berkin 2003, 86). One of the earliest and most common stamp types consists of concentric circles. Figural stamps are less frequent and seem to be typical of Etruscan material. Other motifs include asterisks, rosettes, 'S' shapes, ribbed triangles, palmettes *etc.* (Morigi Govi 1982, 224; Berkin 2003, 87–91). Often, the false cord and stamped decoration are combined (Minto 1921b, 306 fig. 5; Bocci Pacini 1985, 56; Phillips 1994, 33 no. 14, fig. 4.21). In Veneto, spools are occasionally decorated with bronze studs (pers. obs. Este Museum, 2002).

Comparanda

Outside Italy, similar types of spools have been found in Bronze Age contexts in Switzerland (*e.g.* Brem, Bolliger and Primas 1987; Steinhauser-Zimmermann 1989; Nagy 1999). Earliest fragments of spools come from Lake Zürich and have been dated to *c.* 1600 BCE. In terms of geographical distribution, spools have only been found in Eastern Switzerland and exclusively in settlement contexts. In Austria, on the other hand, spools are common in Late Urnfield and Hallstatt periods in graves and settlements (*e.g.* Lochner 1986, 350; Lochner 1991, 249ff; Smolnik 1994, 91; Klemm 1992).

Spools more similar to the Italian examples have been recovered in significant quantities on Greek sites,[33] such as Olynthos (Robinson 1930, 128, fig. 296) and Corinth (Davidson 1952, 178 nos. 1279–1281, pl. 79). Objects of similar size and shape have been noted

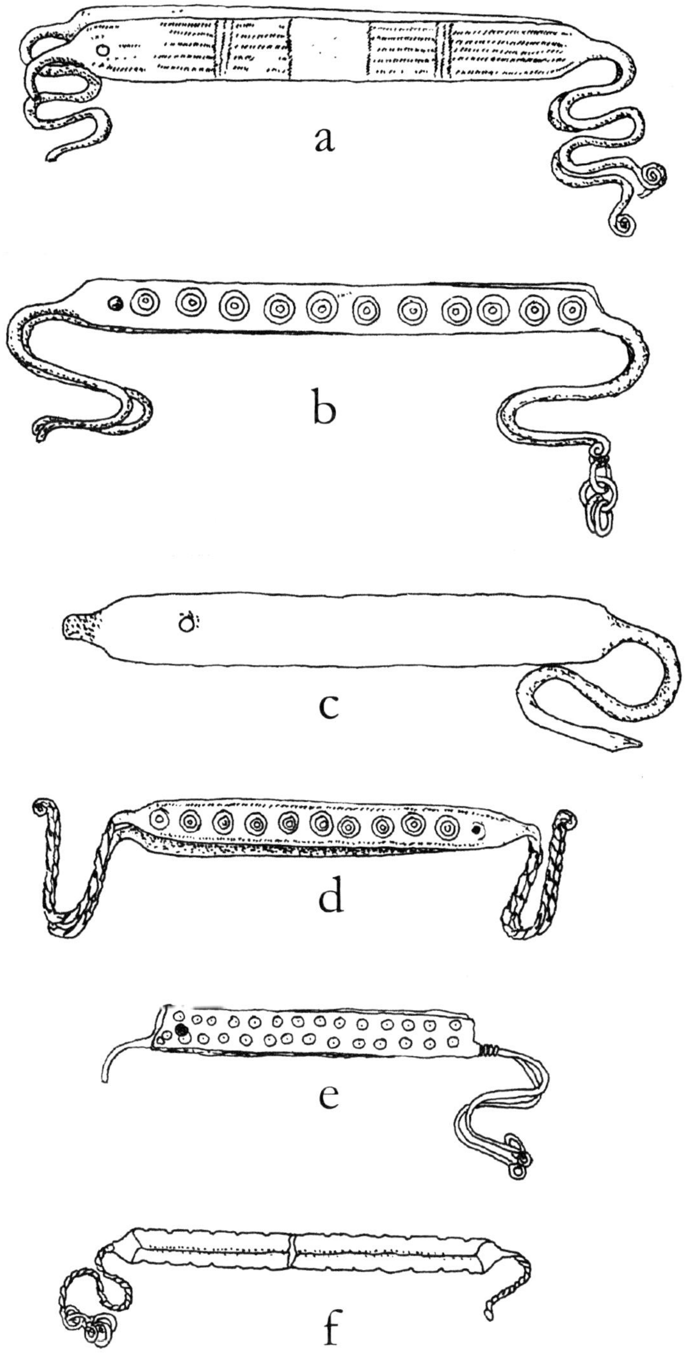

Figure 102. Bronze clasps, 8–7th century BCE: a) Marsiliana d'Albenga (After Minto 1921, fig. 23), b) Vetulonia (After Minto 1921, fig. 23bis), c) Vulci (After Falconi Amorelli 1983, fig. 86), d) Capena (After Paribeni 1906, fig. 35), e) Veio (After Fabricotti et al. *1972, fig. 38), f) Capua (After De Juliis 1996, fig. 17).*

in Albania (Ugolini 1927, pl. VI) and Croatia (pers. com. Antoinette Rast-Eicher, 2007). Interesting parallels can be found in Ukraine, where spools have been found in the 9th–8th century BCE contexts of Zhabotin Culture (Daragan 2004, 57, 59 fig. 12 nos. 1–6). I have also observed a spool from the 4th century

Map 12. Distribution maps of clasps.

BCE settlement of De Panne in north-west Belgium on display in the Musées Royaux d'Art et d'Histoire in Brussels.

Clasp

The use of another metal implement *(forcella da telaio)*, possibly associated with textiles, is uncertain but may be related to tablet weaving. It consists of two long, thin rectangular pieces of bronze, fixed together with a rivet on one end and fastened together with a clasp on the other, and ending in thin coiling strips, sometimes with short chains attached (Figure 102). Although only one of the implements has a clasp, given the bad state

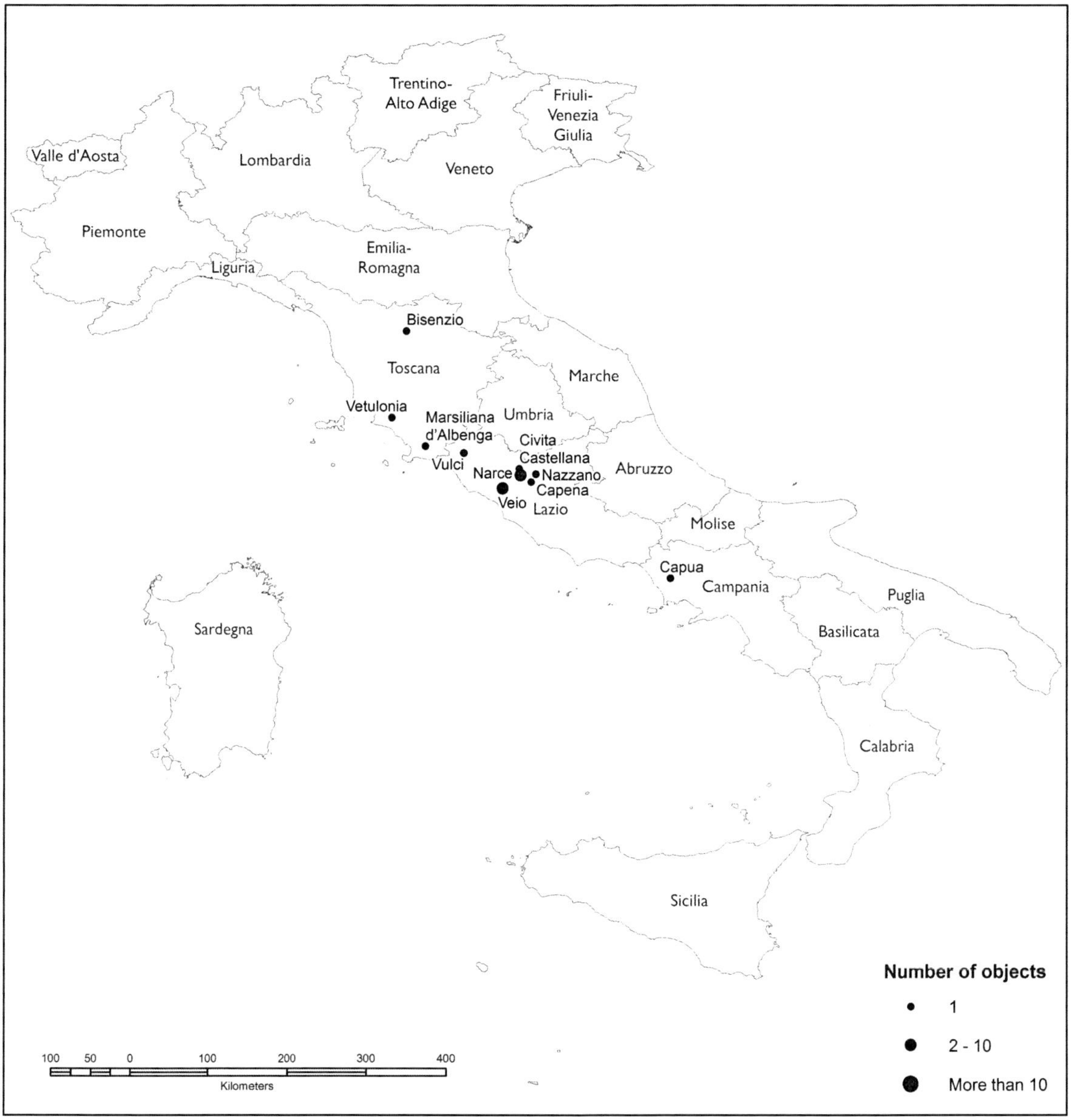

of preservation of the rest, it is probably safe to assume that similar clasps were originally present in all items but have not survived.

The extant examples come from burials at Marsiliana d'Albegna and Vetulonia (Minto 1921a, 257), Vulci (Falconi Amorelli 1983, 196 no. 354, fig. 96), two tombs in Veio (Cavallotti Batchvarova 1965, 126, 128; Bartoloni 2001, 100 no. I.G.5.17; Fabricotti *et al.* 1972, 250 no. 7; Ræder Knudsen 2002, 231), Bisenzio (Paribeni 1928, 438–439), Civita Castellana (Barnabei and Pasqui 1894, 390–391), two tombs at Narce (Pasqui 1894, 441 no. 15, pl. XII no. 19; Dohan 1942, 44 nos. 35–36, pl. XXI; Turfa 2005, no. 141), Nazzano (Stefani 1911, 435), Capena (Paribeni 1905, 346 no. 19, 347; Paribeni 1906, 419–420) and Capua (Johannowsky 1983, 110 no. 9, pl. XVI no. 9, pl. 4b; De Juliis 1996a, 34). Excepting the Capua example, all objects come from southern Etruria, with three items from the Faliscan-Capenate area (Map 12). All items are dated to the 8th–7th centuries BCE.

Bronze objects from Molino della Badia near Grammichele (Orsi 1905, 128, fig. 36) and Nazzano (Stefani 1911, 435 o, fig. 3d) cited as comparanda in the early works (cf. Minto 1921a, 257 note 2, and Paribeni 1906, 419), have a different shape in that they consist of one piece of metal, perforated with large regular holes, and their ends rise perpendicular to the main body of the instrument. Identical objects have since been found in the indigenous Sicilian graves in association with bronze tubes that fit into the perforations and have been identified as musical instruments (*e.g.* Militello and La Piana 1969, 225 fig. 14, 254 fig. 39).

Most of these clasps are decorated with incised or impressed geometric designs. The Marsiliana d'Albenga and Narce examples have alternating parallel or/and perpendicular lines made in false cord technique. Vetulonia, Capena, Bisenzio and one of the Veio examples have one or two rows of concentric circles.

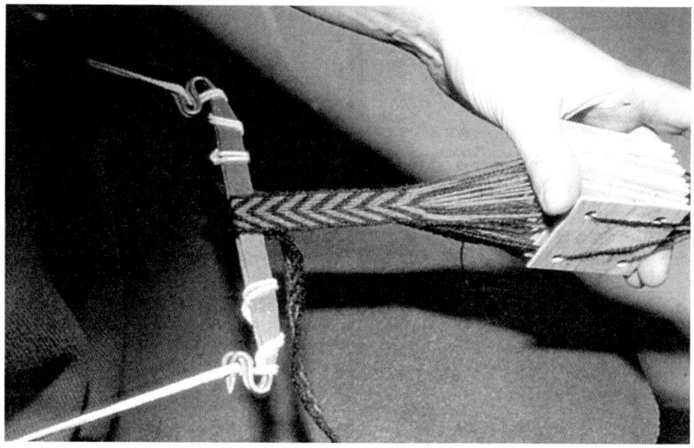

Figure 103. Tablet weaving using a clasp (Courtesy of Lise Ræder Knudsen).

It has been suggested that these objects either functioned as small embroidery looms (Paribeni 1906, 419–420) or were used as clasps grouping together warp threads on a warp-weighted loom (Hackworth 1993, 54). Their use for textile activities was deduced from the fact that the Falerii example had textile remains between the two bronze plates (Barnabei and Pasqui 1894, 390, fig. 180). Textile traces were also noted inside the implements from the Quattro Fontanili Tomb YZ5 at Veio. Lise Ræder Knudsen, in her analysis of the tablet-woven borders of the Verucchio mantles from Tomb 89, suggests that, rather than 'looms', such instruments could have been much more practical for fixing a narrow cloth band in place, as, for example, during tablet weaving (Figure 103) (Ræder Knudsen 2002, 231). In the process, the coiling ends function to wind a string which goes around the weaver's waist, while the finished end of a tablet-woven textile is held firmly between the two rectangular plaques, which keep it taught and prevent coiling. The weaver, thus, has both hands

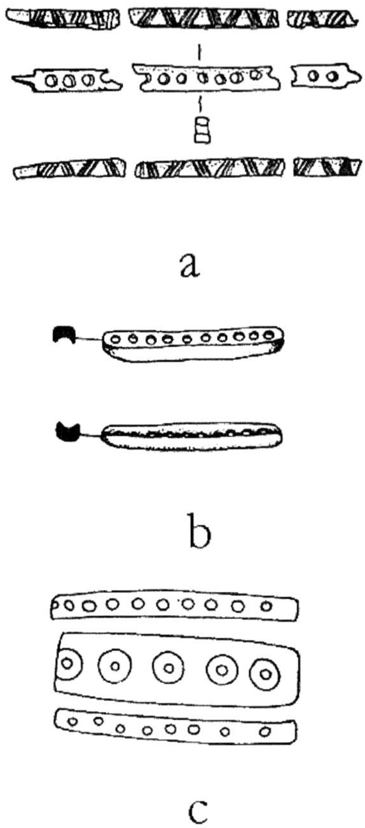

Figure 104. Spacers: a) Tomb 55, Rocca Malatestiana, Verucchio, 8th–7th century BCE (After Gentili 2003, pl. 124 no. 17); b) Tomb 683, Pagliarone, Pontecagnano, 8th century BCE (After Gastaldi 1998, pl. 100 no. 32); c) Tomb 343, Incoronata, 8th century BCE (After Chiartano 1994, pl. 81).

free to rotate the tablets and pass the weft back and forth. The object from Tomb HH 11–12 at Veio was, in fact, found near the waist area. Grave good assemblages with clasps, for which information is available, also contained spools.

Similar instruments, made in wood or metal, were used until recently in North Africa (Ræder Knudsen 2002, 230). It is likely that, in ancient Italy, these implements were usually made of wood or bone and hence did not survive. The metal variety, like bronze distaffs and spindles, were made as objects of status, thus accounting for their rarity and elaborate decoration. The Capua example, in fact, though superficially similar to the rest, is constructed in a different way, having being made of one piece with top and bottom edges bent instead of two plaques, and may be a non-functional imitation of the actual tool.

Spacer and pegs

Other objects possibly related to textile production are spacers and pegs. Spacer *(distanziatore)* is usually a small narrow bone piece with a rectangular section, numerous holes along its length and, sometimes, narrowing at each end (Figure 104). The most common interpretation for the function of this implement is that it was used to keep distance between warp threads in order to avoid entanglement. Given its length, it may have been used in tablet-weaving or any other kind of band-weaving: threads of one tablet would pass through one hole of a spacer to be subsequently wound on a single spool (Ræder Knudsen 2002, 229).

The largest number of complete or fragmentary spacers has been identified in the burials of Verucchio. Tomb 33 had 2 examples 2.8 and 4 cm long (Gentili 2003, 221 no. 6, pl. CXCIX.5). Tomb 36 had three spacers ranging in length from 3.8 to 6 cm (Gentili 2003, 226 nos. 33–35, pl. 106, pl. CCIII). Tomb 55 had one spacer 4.4 cm long, as well as eight spools and three spindle whorls (Gentili 2003, 255 no. 17, pl. 124). Tomb 102 yielded two examples 4.3 and 7 cm in length respectively and the burial goods also included 14 spools and one spindle whorl (Ræder Knudsen 2002, 229 fig. 103; Gentili 2003, 325–326 nos. 33–34, pl. 167, pl. CCCI.2) Finally, Tomb 118 also had two spacers, 5 and 4.9 cm in length (Gentili 2003, 344 nos. 16–17, pl. CCCXVIII).

Several similar examples have been found at Longola di Poggiomarino in Campania (Pizzano 2005, pl. 10, figs. 11–12). They are about 10 cm long and most bear incised decoration. Another group of spacers comes from burials at

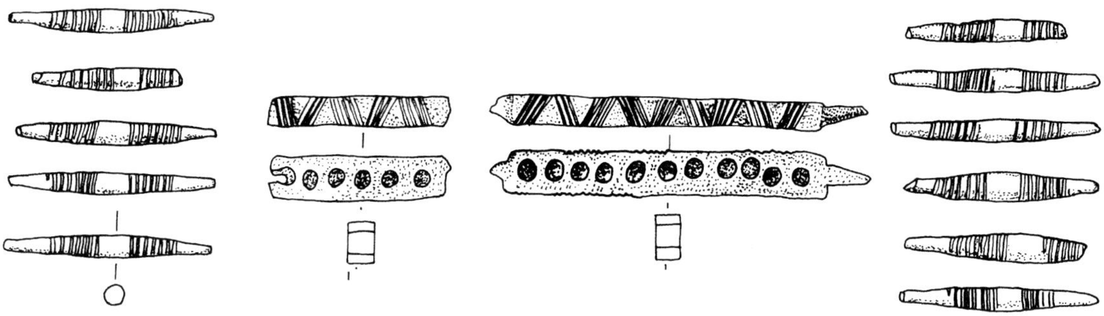

Figure 105. Spacers and pegs from Tomb 102/1972 Lippi, Verucchio (After Ræder Knudsen 2002, 229 fig. 103).

Pontecagnano. Pagliarone Tomb 683 had one implement 3.8 cm long (Gastaldi 1998, 89 nos. 32–33, pl. 100). Tomb 899 from the same necropolis had four fragmentary instruments, ranging in length between 1.7 and 2.2 cm (Gastaldi 1998, 130 nos. 10–11, pl. 116). In the Picentino necropolis, Tomb 2057 had one spacer 4.8 cm long (d'Agostino and Gastaldi 1988, 179 no. 15, pl. 153). Tomb 2106 had one fragmentary implement 1.8 cm long (d'Agostino and Gastaldi 1988, 196 no. 33, pl. 161). Other examples have been noted in Picentino tombs 496, 558, 683, 889 (d'Agostino and Gastaldi 1988, 70).

Two items, about 8 cm long, were included among the burial goods at Canale/Ianchina necropoleis of pre-Hellenic Locri, while one example was identified in Tomb 343 at Incoronata (Chiartano 1994, pl. 81). All examples date to the 9th–8th centuries BCE and most are decorated with incised lines or circles.

Pegs – small slightly biconical cylinders with transverse incisions – have been found at Verucchio (Figure 105). They were interpreted as stoppers for the holes in the spacers, based on the fact that in Tomb 102/1972 Lippi 11 pegs were found together with a spacer with 11 holes (Ræder Knudsen 2002, 229).

Other textile techniques

Weaving on a loom is not the only way to create a web-like structure. One technique, sprang, was known throughout Europe. In Scandinavia and Egypt, there are actual examples of sprang textiles (Hald 1980, 245ff), while in Greece we have evidence for it in the vase paintings (Figure 106) (Clark 1983). Sprang permitted the creation of highly flexible and stretchable cloth, useful for various items of clothing. Unfortunately, small wooden frames used for sprang weaving do not survive in the archaeological record in Italy and the only evidence for the use of sprang consists of iconographic depictions of hair nets.

Finishing

Once the textile has been woven, it had to be finished. Different finishing techniques can be used for different purposes. Thus, bleaching or smoothing bring out certain inherent qualities of linen, while fulling gives woolen cloth a different surface or makes it more water resistant. Pleating, on the other hand, would arrange the fabric in specific shapes, giving it a very particular look. All these techniques further increased the variety of textiles available for consumption, both in terms of their qualities and appearance.

Figure 106. Woman with a sprang frame. Drawing after a red-figure kalathos, c. *440 BCE (After Clark 1983, 94 fig. 2).*

Linen: bleaching and smoothing

Linen cloth could be subjected to various rough treatments to make it softer or it could be rubbed with a special stone or glass piece to give it extra lustre (Wild 1970, 84–85). While no evidence of this practice exists in pre-Roman Italy, smoothing or slick stones made of glass are known from Roman contexts (Wild 1970, 84–85; Wild 2002, 22) and Late Iron Age Scandinavia (Andersson 2003, 33). Subsequently, linen could be bleached and/or decorated in some way. Bleaching could be done naturally, by exposing cloth to the sun, or by using the sulphur treatment. The later was used in Roman times, but may have originated in the earlier period (Plin. *NH* 35.175). Wild (1970, 83) describes the process of sulphur-bleaching as involving "a semi-circular cage over which the cloth was laid and a pot, probably containing lump-sulphur, which burnt slowly inside the upturned cage". Such equipment seems to be represented on a wall painting from Pompeii (VI.8.20). In Roman times, the linen of Faenza (ancient Faventium) was preferred for its whiteness to that of Alia, which was never bleached (Plin. *NH* 19.9).

Wool: fulling

Woollen cloth could be subjected to fulling, a treatment with water and sometimes soap that produces a very tight fabric, although no evidence for the practice in Italy exists before the Roman period. Fulling establishments used very large basins. The surface of a woollen cloth could also be raised to produce a nap, which could have been accomplished with teasels (Goldmann 2007).

Pleating

Pleating could be applied to both linen and wool. During Roman period, both literary and archaeological evidence exists for clothes-presses (Granger-Taylor 1987, 120). While no such sources are available for the earlier times, depictions of Etruscan garments in art indicate regular and often elaborate pleating of certain garments, particularly mantles. Verucchio textiles may provide first hard evidence of such practice (von Eles 2006).

Dyeing

The main material to be dyed was wool. Although linen is notoriously difficult to dye, occasionally it was coloured (Goyon 1996). Pliny the Elder (*NH* 19.22–24) mentions that extravagantly decorated awnings were commissioned for special occasions and red covers were used in the inner courtyards of houses to keep the sun off the moss growing there.

The simplest way of imparting colour onto any material is staining, which could be accomplished using mineral pigments or with plant extract and does not require any special equipment. True dyeing, on the other hand is more complicated. Unfortunately, while botanical, zoological and geological studies have demonstrated the potential sources of textile colouring substances, very little is known about the tools and methods used by ancient dyers. Raw materials for the dyestuffs have to be ground in order to be used, making mortars, pestles, grinders *etc.* necessary equipment for the process. Then, relatively large pots or vats are needed for dissolving or fermenting the colouring substance, as well as for immersing the fibres or cloth to be dyed. In addition, "some means is needed to extract (and preferably to salvage) the excess dye", especially in the case of more expensive dyes (Barber 1991, 240). Channels, drains or sieves may be used for the purpose. Finally, areas for rinsing and open spaces for drying are needed to complete the process, neither requiring any sort of special installation. Barber brings in examples of dye-works identified at various sites in Greece, Cyprus and the Near East, but no such installation has been definitively identified for pre-Roman Italy so far (Barber 1991, 239–243).

Dyeing installations

Dyeing facilities, especially the large-scale ones, tended to be located on the outskirts of settlements, due to unpleasant smells produced in the process of dyeing. Large quantities of water required for rinsing also necessitated that dye-works be located near a body of water. Very little work has been carried out on the functional identification of such installations, especially in the pre-Roman period. A major problem is presented by the fact that similar structures were used for washing wool and cloth, fulling, and other purposes not related to textile manufacture, such as tanning (*e.g.* Sagona 1999). Recently, an important study on basin typology and chronology was carried out at Pompeii by Philippe Borgard and Marie-Pierre Puybaret (2004), who suggest that tanning vats are differentiated by their cylindrical shape. Maria Emanuela Alberti attempted to define criteria for washing and dyeing installations in Minoan Crete (Alberti 2007).

Grinding and pounding tools

Grinders, pestles, mortars and pounders constitute a category of tools, which could have been used by a dyer in order to prepare dyes by crushing or grinding the raw materials. These objects, although frequently found at archaeological sites, are rarely recognised as significant to the textile industry because they could have been utilised for other purposes, such as cooking. Presently, the possibility of mortars and pestles having been used in textile industry has been recognised only at the Hellenistic site of Cetamura, where these items were excavated in the area of the North-West Complex in Zone II (Hackworth 1993, 38–43)[34] and on Crete (Alberti 2007, 60).

Purple

The dye that left the most evidence is the Royal purple, mainly because its large scale production required enormous numbers of molluscs whose shells leave clear physical traces, as at Coppa Nevigata (Figure 107). The manufacturing process was governed by access to the sea, to salt and vats for heating the mixture, thus restricting production to coastal regions. It also produced a strong stench, forcing

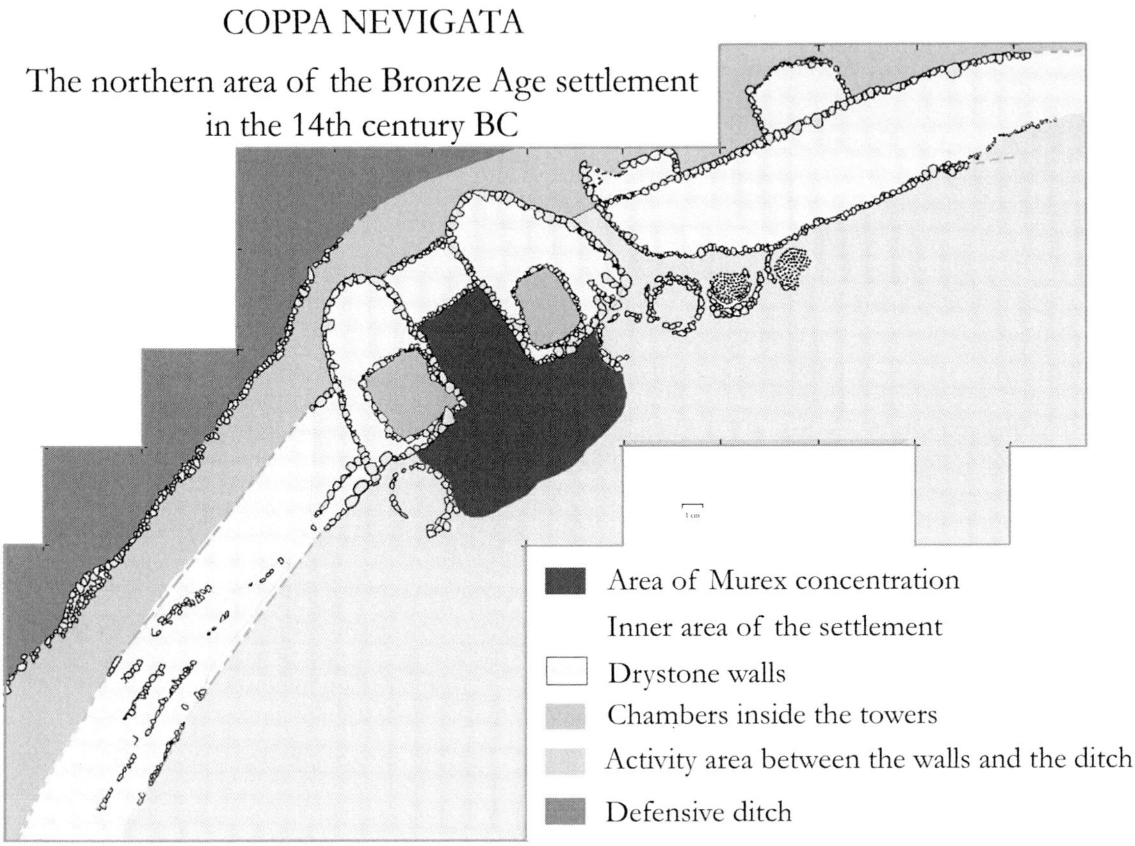

Figure 107. Purple production remains at Coppa Nevigata, Bronze Age (Courtesy of Alberto Cazzella).

the production outside settlement limits. The four stages of production involved collecting the shellfish, extracting the dye by breaking the shells, preparing the dye bath and dying, although not all aspects of the process are clear (Sagona 1999, 37–50; Haubrichs 2005). Only the evidence of extracting the dye survives in the form of the heaps of broken shells.[35] These have been found at Mozia (*Mozia* 1989, 40; Reese 2005). Here, too, were found Sperm whale vertebrae, which have been interpreted as platforms for breaking the shells (Reese 2005).

Large piles of *Murex* shells were still visible on the Monte dei Coccioli of Taranto during the 19th century, remnants of the dyers' establishments on the Mar Piccolo (Forbes 1956, 136, 162; Morel 1978, 104ff; De Juliis 2000, 81). Vats that may have been used in the purple dyeing process have been identified in Roman Pompeii, but no evidence has yet been recognised for the earlier periods.[36]

Sewing

Textile work involves sewing to construct new or mend already existing garments, and there is evidence for both in the Vedretta di Ries and Verucchio textiles. Leather could also be sewn to make garments, as in the case of the Similaun Man (Bazzanella and Mayr 1995a, 109). The tools used for the process are needles (Figure 108).

Needle

Needles, characterised by one pointed end and an eye at the other, were used to pass thread through fabrics and leather to reinforce, mend or join articles together. Most extant needles found in ancient Italian sites are made of bronze. Bronze gives strength and thinness, which are advantageous for fine stitching of cloth and leather. In terms of size, bronze needles usually do not exceed 10 cm in length. They could be manufactured by bending a thin bronze wire or by piercing it to create an eyelet. Another, more popular, method was to cast a needle with an eyelet (Macnamara 1976, 146). More often, however, needles were probably made of bone, a material that is cheaper, but also less likely to survive. Bone needles are often shorter and thicker that the bronze ones, most likely to prevent breakage. They were in use much earlier than the bronze ones. Numerous examples have been found at the Early–Middle Bronze Age site of Fiavè-Carera (Bazzanella *et al.* 2003, 139–140).

Needles are not always mentioned among materials recovered from settlement sites. Being small, these tools were probably often lost in past excavations, despite the fact that they often survive intact. Two bronze examples have been found at the Montagnana – Borgo S. Zeno settlement, dated to the Late Bronze Age (Bianchin Citton, Gambacurta and Ruta Serafini 1998, 293 nos. 18–19, fig. 164 nos. 18–19). Bronze and bone needles were found at the contemporary site of Fonte Tasca (Di Fraia 1995, 459 fig. 8 nos. 20–24, 461, 465 fig. 11 nos. 10–12) and in Structure IV D at Castenaso (Bologna) (Forte and von Eles 1994, 211). Bronze needles have also been recovered at the 11th–10th century BCE site of Sorgenti della Nova (Negroni Catacchio 1981, 247, 435 no. 83; 1982, 108 no. 83, fig. 67A–83; 1995, 378 fig. 249 nos. 127–128). Twelve bronze needles and one made of bone were recovered from the Orientalising contexts at Poggio Civitate, Murlo (Gleba 1999, 95–98), and 15 bronze examples were found at Narce in contexts dated 10th–4th centuries BCE (Macnamara 1976, 144 fig. 46, nos. 91–99). Numerous bronze needles come from the Iron Age and Archaic levels in Rome (Gjerstad 1966, 172 fig. 68 no. 5, 449 fig. 123 no. 1). In the south, one bronze needle was found in the Late Bronze Age stratum at Broglio di Trebisacce (Peroni and Trucco 1994, 572 pl. 120 no. 11). Numerous bone needles, mostly made of pig bones, have been found at the Iron Age Longola di Poggiomarino (Pizzano 2005, pl. 10 figs. 4–5).

A surprising number of needles, single or multiple examples, is known from burial contexts, mainly in North Italy (*e.g.* Brizio 1895, 152; Chieco Bianchi and Calzavara Capuis 1985, *passim*; Franchi Dell'Orto 1999, 52) and Sicily (Frasca 1982; Denti 1982, 632; Meola 1996, vol. 1, 263). In addition to bronze examples, silver needles have been found in two burials in Taranto (De Juliis 1984, 331) and probably refer to the social position of their owners as well as their wealth. Needles also occasionally turn up in votive deposits. Thus, two unusual gold

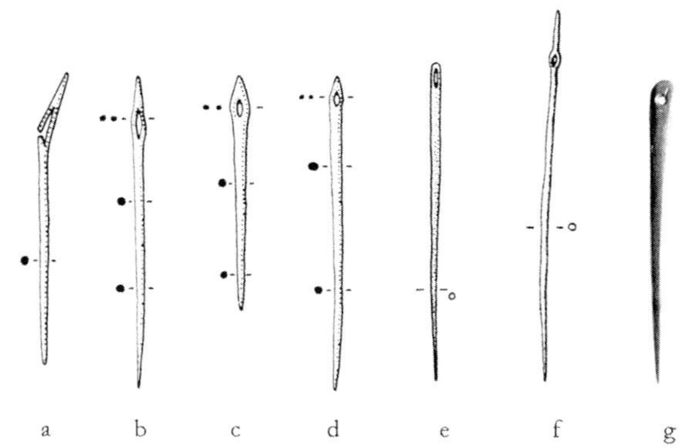

Figure 108. Needles: a–d) Narce, 8th–6th century BCE, bronze (After Macnamara 1976, 144 fig. 46 nos. 92–95); e) Tomba 44, Este, 7th century BCE, bronze (After Bianchin Citton, Gambacurta and Ruta Serafini 1998, 123 fig. 59 no. 19); f) Broglio di Trebisacce, Late Bronze Age, bronze (After Peroni and Trucco 1994, 572 pl. 120 no. 11); g) Fiavè-Carera, Middle Bronze Age, bone (After Bazzanella et al. *2003, 139).*

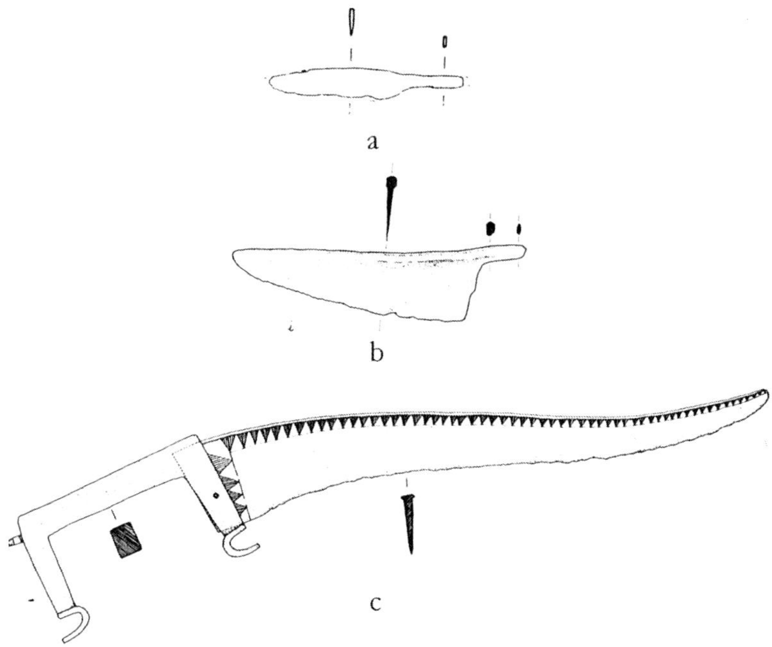

Figure 109. Knifes: a) Fondo Rossi, Cerveteri, Final Bronze Age (After Arancio and D'Erme 1990); b) Tomb 838, Casale del Fosso, Veio, Late Bronze Age (After Vianello Cordova 1967); c) Tomb XCI, Sermoneta-Caracupa, 8th century BCE (After Bedini 1976, 262 no. 1).

needles have been found in a votive deposit of a sanctuary at Saturo in Taranto, dated to the 4th century BCE (De Juliis 1984, 331, 366 nos. 337–338). They were probably precious versions of an everyday item made to be used as dedications.

Knife

One last tool should be mentioned in connection with textile making activities. Knives have occasionally been found in female burials, often in association with spinning and weaving implements (Figure 109). Thus, in the Benacci Tomb 888 of Bologna, a small knife was deposited next to a spindle and a distaff (Bianco Peroni 1976, no. 372, pl. 71.1). A Late Bronze Age cremation burial at Cerveteri had 18 spools and a small bronze knife (Arancio and D'Erme 1990, 307 no. 6, 309, 312). Contemporary Tomb 1 at Grotte Barche of Sasso di Furbara had a spindle whorl associated with a bronze knife (Brusadin Laplace 1984–1987, 369 no. 17, fig. 34). The 8th century BCE Tomb L at Casale Marittimo had an iron knife and a spindle whorl (Esposito 1999, 65, fig. 60). Similar miniature bronze knives have been found in Tomb 126 of Osteria dell'Osa (Bietti Sestieri 1979, 54 no. 58, pl. IX), Tomb 5 at Le Caprine, Guidonia (Damiani *et al.* 1998, 206; Guidi and Zaraffini 1993), and Tomb 838 at Casale del Fosso necropolis at Veio (Vianello Cordoba 1967, 299, pl. LVI). The association with textile tools and the small size (less than 15 cm long) of many knives suggest that they, too, may have been connected with weaving activities, possibly utilised to cut off the warp after the cloth was finished (Chieco Bianchi and Calzavara Capuis 1985, 54; Bartoloni 1988, 46; Amann 2000, 43).[37]

Larger knives are known from the 8th century BCE Tomb CXXXII of Castel di Decima (Bedini 1976, 258 no. 12 and pl. LIX, 262 no. 1 and pl. XCVIII), and Tomb C at Torrino (Bedini 1985, 48 no. 15).[38] Knives of large dimensions are more likely to represent another aspect of female role in the society, that of carving and distributing meat for the community or in cultic activities (Bartoloni 1988, 44–45; Bietti Sestieri, De Santis and La Regina 1991, 82; Bietti Sestieri 1992a, 228; Amann 2000, 41–42). However, since a knife may be used in many different contexts for many different functions, it is impossible to assign it a specific significance, especially when two or more knives of different dimensions are found in burials (Amann 2000, 43).

Conclusions

This chapter has hopefully demonstrated that textile tools from archaeological contexts represent one of the most abundant and richest sources of information about ancient textile production. We have evidence almost for every stage of the textile making process (Table 8), albeit not in every chronological period.

While some tool types are not preserved, we know that they existed because they were necessary for the various stages textile production. Thus, the investigation of textile tools covers the entire *chaîne opératoire*. Textile production requires many specialised tools, such as spindles, distaffs, looms, which are of use only for making textiles. At the same time, some textile tools are non-specialised and multifunctional implements. Multifunctional tools like knives, shears, grinders and mortars, tell us that textile production was embedded in the household and shared tools with agriculture and, particularly, cooking – activity that also belongs to the female sphere of the household. In order to understand the organisation of textile production, next we must examine the different contexts, in which textile tools have been found.

Process	Tool	Presence in BA	Presence in IA	Presence in Roman period
Obtaining fibre	Shears		X	X
Processing fibre	Water installation			X
	Comb	X	X	X
Spinning	Spindle	X	X	X
	Spindle whorl	X	X	X
	Spinning hook		X	
	Distaff		X	X
Weaving	Loom			
	Loom weight	X	X	X
	Beater	X		
	Shuttle	X		
Tablet weaving	Tablet		X	X
	Spool	X	X	
	Clasp		X	
	Spacer		X	
	Pegs		X	
Sprang	Frame			
Finishing	Needle	X	X	X
	Smoothing stone			X
Dyeing	Mortar		X	X
	Basin			X
	Water installation			X

Table 8. Archaeological evidence for textile tools from Bronze Age through Roman times.

PART 5

Contexts

Tools, as indicators of production, although ubiquitous, are not always recovered *in situ*. Yet, our identification of the function of textile implements is often determined by the context in which they have been found. The main archaeological contexts for textile implements are: settlements, burials and votive deposits. Each provides information about the role of tools and of textile production in society. A better understanding of the craft can only be gained from the combined study of the material from all of these contexts.

Settlement context

Implements found in settlement contexts are invaluable for the understanding of the production process. In fact, the primary, technological function of spinning and weaving tools can best be inferred from habitation sites. A study of the number, distribution and morphology of these and other textile implements provides information about the technology and scale of production, as well as the raw materials and final products. Moreover, we can begin to reconstruct the history and the economic role of the textile craft in ancient Italian communities.

Artefact distributions are especially important for such an analysis since concentrations of implements reflect the areas of activity (Hodder and Orton 1976). When found *in situ*, textile implements can provide specific information about the location and the dimensions of tools, such as looms, parts of which were made of perishable materials and thus do not survive in the archaeological record. Furthermore, distributions of textile implements can help to identify where the textile-making activity took place and what was its scale.

Following is a sample of sites where textile production can be demonstrated unequivocally. It is more than probable, however, that spinning and weaving took place in most ancient Italian settlements, whether large urban sites or small farmsteads.

Bronze Age settlements

The North Italian Bronze Age habitation of S. Rosa di Poviglio yielded several areas of loom weight concentration (Bernabò Brea, Bianchi and Lincetto 2003, 112–119). In the Middle Bronze Age *Villaggio piccolo*, one group consisted of 16 loom weights and another one of 12. The finds indicate that at the time of destruction of the village, looms were set up in several houses. The Late Bronze Age *Villaggio grande* yielded at least 17 loom weights in a corner of House A (Figure 110), and a smaller set of at least ten weights in the same general area. Another small concentration of loom weights, consisting of five complete and ten fragmentary elements, was found in

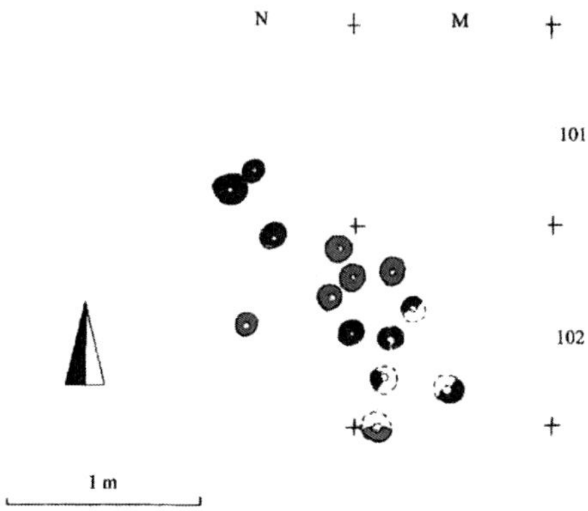

Figure 110. Loom weights in situ, Vilaggio grande, *S. Rosa di Poviglio, Late Bronze Age (After Bernabò Brea, Bianchi and Lincetto 2003, 117 fig. 6).*

House C. In their comparison of the two sets of loom weights dated to the Middle Bronze Age with three Late Bronze Age sets, the excavators note that earlier examples are smaller and weigh less. However, the size of the loom remains the same, with the width of about 170–180 cm and the loom weight set consisting of sixteen or a few more elements. The main difference between the two periods of the settlement seems to be the scale of textile production: weaving seems to have been practiced on a larger scale in the Late Bronze Age. In addition, a large number of spindle whorls were recovered from both phases of the site, demonstrating substantial homogeneity in terms of shape and weight, with items of middle and middle–heavy weight predominating (Bernabò Brea, Bianchi and Lincetto 2003, 111). Such consistency indicates either that only one type of fibre was spun at the site or that specific type of yarn was sought.

Late Bronze Age–Early Iron Age settlements

Late Bronze–Early Iron Age textile production at Sorgenti della Nova, in the future territory of Vulci, is attested by a relatively homogeneous distribution of spindle whorls, spools and loom weights in the areas of domestic activities (Negroni Catacchio 1982, 49). Specialised production, however, is suggested for Hut I in Sector I, where numerous loom weights have been found, and Sector III, which also had a high concentration of spools (Negroni Catacchio 1981, fig. 8). Contemporary habitation at Luni sul Mignone to the south produced spindle whorls in large quantities, dated to all periods of occupation, but no loom weights have been found on the site (Östenberg 1967, 127).

The Faliscan town of Narce, located south-east of Viterbo, and occupied since the Late Bronze Age, not only produced a fair number of weaving implements, but also shows a change in their forms in successive phases (Macnamara 1976, 172–173). Early loom weights, for example, are small, weighing only 50 g, while the later, larger pieces are between 1000 and 2000 g in weight. Also, the discoid form of spindle whorl is found only in the early strata. The earliest examples of spools do not appear until about the 9th century BCE; this is paralleled by their absence from contemporary levels at Luni sul Mignone. As on the majority of other sites, spindle whorls and spools outnumber loom weights. Numerous bronze needles were also found at Narce, indicating that some sewing activity was taking place on the site.

Another Iron Age site in the area of Viterbo, San Giovenale, produced enough implements to indicate extensive textile production. Area B yielded 18 spindle whorls and 15 spools (Pohl 1981, 50–51, 79–80); Area D produced three spindle whorls, one fragmentary spool, and two loom weights (Malcus 1984, 43); and in Area E many spindle whorls were found, but there were also three spools and one loom weight.

In South Italy, one of the few well-excavated Late Bronze Age sites is Broglio di Trebisacce. Here, the concentration of spindle whorls and loom weights in certain areas of the acropolis has been interpreted by the excavators as an indicator of a greater intensity of spinning and weaving in the Late Bronze Age, particularly in some habitations or groups of them, where inhabitants may have had a more direct access to raw material (Peroni 1993, 116).

Orientalising and Archaic period settlements

During the Orientalising and Archaic periods, implements associated with spinning and weaving are more numerous on the settlement sites. Sometimes, as in the case of Poggio Civitate, a workshop mode of production may be suggested on the basis of tool quantity, distribution and standardisation.

In North Italy, at Verucchio, part of the 8th–7th century BCE hut identified in the Pian del Monte area, yielded about 30 loom weights and numerous spools (Gentili 1988, 83–84). It is tempting to interpret this find as direct evidence for the textile production depicted on the Verucchio throne.

Structure R18 at Forcello di Bagnolo San Vito, located north of the Po river, has been dated to the 6th–5th century BCE; it has produced large quantities of loom weights concentrated in two main areas along the north-west wall of the building (De Marinis 1988c, 151–154 fig. 70; Vay 1988, 170–174). The loom weights, which carry evident signs of wear, represent a wide range in weight and shape, indicating that several looms were in operation and several types or qualities of textiles were produced. A large concentration of sheep bones, of adult animals, on the site suggests that wool may have constituted the main fibre in this settlement (Scarpa 1988, 192). Structure R18 has been interpreted as a workshop for textile production. Only weaving was carried out here, though, for to judge from the nearly complete absence of whorls, spinning took place elsewhere. Twenty-one spindle whorls have been found at Forcello, only one of which came from structure R18 (Vay 1988, 174). Also, 15 spools were recovered from the site, but their specific find area is, unfortunately, not reported.

A structure destined for weaving activities has also been excavated at Pozzuolo del Friuli, although the excavation report claims the function of the structure as unknown (Vitri *et al.* 1991, 22–27). Here, in a small rectangular house D10, measuring 2 by 2.9 m, carbonised remains of a wooden structure and a large number of complete and fragmentary loom weights were found, the latter concentrated in the south-west part of the structure (Figure 111).

Another small room with a loom, which was still set up at the time the settlement was destroyed, has been identified at Santorso (Balista *et al.* 1985, 79–82). Here, parts of carbonised beams were noted during excavation and a significant number of complete and fragmentary loom weights were found concentrated in a small area (Figure 89).

The site of Monte Savino in the territory of Monterenzio, yielded over 80 loom weights, distribution of which (Figure 112) suggests that almost every household on the site produced textiles (Macellari 1983, 119–126). Spindle whorls and spools were also common at the site (Marchi 1983, 114–118).

In the Etruscan territory, at the habitation site of Lago d'Accesa in the area of Massa Marittima, large concentration of loom weights (24) indicates the position of a loom in room III of complex X in area A. The house has been dated to the 6th century BCE. In rooms I and III of complex VII in area B, dated to the 7th–6th century BCE, 17 and ten

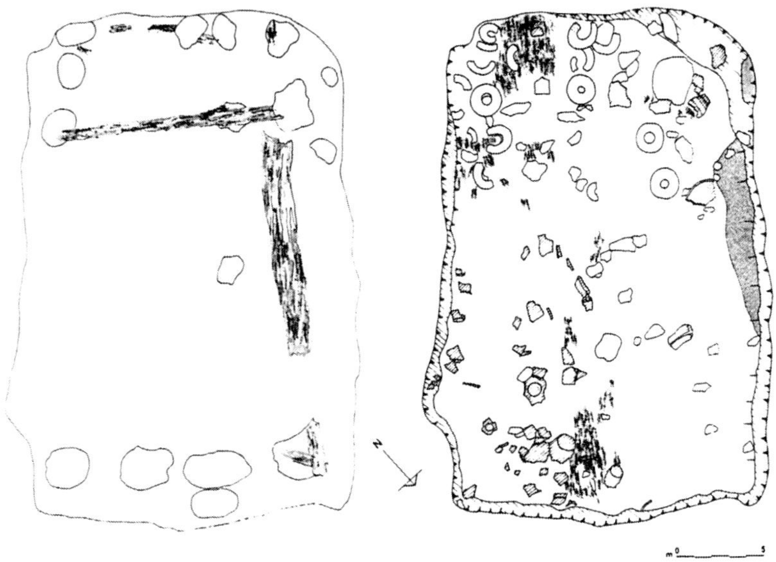

a b

Figure 111. Loom remains, Struttura D10, Pozzuolo del Friuli, 6th century BCE: a) wooden beams; b) ring loom weights (after Vitri et al. *1991, 27 figs. 13 and 15).*

loom weights were found, respectively (Camporeale 1997, 209–211). Apart from one specimen, all loom weights are of the same weight of about 800 g. Room I of complex VII in area B had the largest concentration of spools, a total of eight. The majority of the spindle whorls were also recovered from area B.

The small concentration of loom weights at the Casa d'Impluvio at Roselle is not sufficient to support speculation as to the location of a loom, but five weights in Room VII and another five in the adjacent courtyard IX may indicate that weaving took place in the central area of the house (Donati 1994, 139–140). The small courtyard also produced a large number of spools and spindle whorls. The area to the north of another house at Roselle, dated to the Orientalising period, yielded large quantities of loom weights and spools (Bocci Pacini 1985, 54–57). Forty-six of the total of 146 loom weights were unfired and were found together in a corner formed by the northern wall of the house and the precinct wall. It has been suggested that these loom weights may have been placed there to dry on a wooden board, burnt traces of which were visible nearby (Bocci Pacini 1985, 55).

At Acquarossa, an important habitation site of the Orientalising period in the area of Viterbo, weaving implements have been excavated throughout the site. In Structure A of Zone B, 16 loom weights were found on the ground in a row (Figure 113), probably where they fell from the loom when the site was destroyed *c.* 500 BCE. The weights, together with some spools, were found in the largest room in the house, Room 2, which has been identified as the kitchen (Östenberg 1975, 12, 78–81, 96–97). Another concentration of loom weights was recovered from the courtyard of Zone L (Östenberg 1975, 34; Rallo 1989b, 18–19). Their number, weight and position indicate a presence of a loom in the southern room of Structure C. The Acquarossa loom weights vary in weight; those from Zone B weigh approximately 400 g each, while the ones found in Zone L weigh about 300 g each (Wikander 1986, 112). This suggests

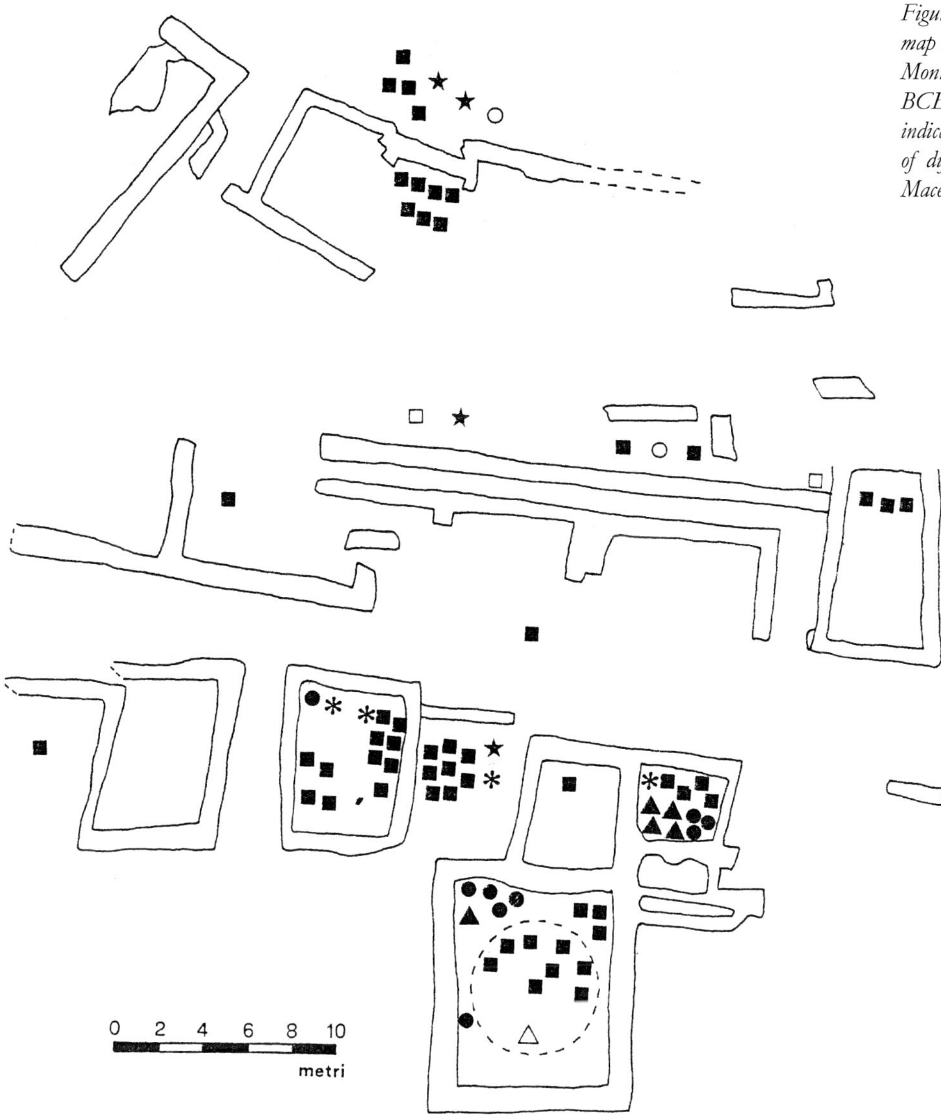

Figure 112. Distribution map of loom weight groups, Monte Savino, 6th century BCE; different shapes indicate loom weights of different types (After Macellari 1983, 120).

that different looms and different sets of weights were used to produce fabrics of varied fineness. Spindle whorls are also numerous at Acquarossa, and range in size from earlier, large ones to later examples, which are smaller and more refined in shape. Large numbers of spools were found at Acquarossa but have not been studied.

Among the southern sites, Sibari is a site of particular interest, since literary sources speak highly of its textile industry. Unfortunately, the stratigraphy of the Archaic Greek city excavated so far has been very unclear. However, stratum E at Parco del Cavallo, associated with a 6th century BCE settlement phase, yielded significant quantities of loom weights. These have been identified as votive but more likely they come from the Archaic

Figure 113. A group of loom weights found in situ, *Casa A, Zona B, Acquarossa, 6th century BCE (From Östenberg 1975, 79).*

houses in this area and document the celebrated Sybarite textile industry (Galli 1930, 93–94; *Sibari* 1969, 48 no. 139, 69, 89, 88 fig. 78, 94 nos. 206–207, 125 no. 109, 129 nos. 136–137). Furthermore, a notable number of loom weights often found in groups of 10–12 have been excavated at the nearby indigenous settlement of Amendolara (de la Genière 1969, 87; Guzzo 1996, 70). Loom weights inscribed with names in the Achaean alphabet indicate a connection with the Greek city (de la Genière 1992, 119). It is likely that the famed textile industry of Sibari depended on its countryside satellite settlements not only for the raw material but for the final product as well.

Finally, in Sicily, at the site of Monte Saraceno, at least 57 loom weights were found in room 6 of House D, dated to the second half of the 6th century BCE (Calderone 1982, 610 pl. LXXXIX.3). In Himera, room 23 in the eastern quarter of the city yielded 19 loom weights, dated to the 5th century BCE (Allegro 1976, 557, 559). The excavator suggested that the room may have functioned as a workshop for textile production and metal working (Allegro 1976, 494).

Post-Archaic period settlements

A careful analysis of textile implements found at Cetamura once again demonstrates the potential of this material for drawing conclusions about the economy of ancient sites. Here, Lauren Hackworth looked at loom weights, spindle whorls, spools and grinders (44 objects; more have been excavated since). All of these tools, although not numerous and lacking primary context, indicate that facilities for textile production were set up in the NW Complex of the site (Hackworth 1993). A loom weight found in the nearby kiln shows that the tools for the textile industry were produced locally (de Grummond 2001, 60).

Another site with evidence of textile manufacture is La Piana, located in the territory of Siena. The settlement is dated to the Hellenistic period and is important because it yielded significant finds associated with weaving. In 1992, a concentration of 23 loom weights was uncovered in a destruction layer in the corner of one of the rooms of a house (Figure 90) (Whitehead 1996, 114, figs. 6, 8). From the scatter pattern, the excavators concluded that a loom leaned against the north wall. It is also possible that some of the carbonised wood timbers found in the same area and situated near the north wall, were fragments of a loom frame. The weight of loom weights excavated throughout the site ranges between 68 and 975 g, with the majority of objects weighing 200–500 g. The numerous spindle whorls (40) vary in terms of shape and size, their weight ranging between 6 and 56.6 g. Only two spools have been found at the site, which is consistent with the chronological span of this implement's use.

A small farm at Poggio Bacherina, near Chianciano Terme, was also producing textiles. At the base of the western wall of room 5, numerous loom weights have been found where they fell from a loom at the time of the destruction of the site (Figure 114) (Paolucci 1997, 56–57 fig. 47). Excavation of small farm sites like Poggio Bacherina is important for the assessment of the scale of textile production in settlements of different size or social status.

Figure 114. Loom weights found in situ*, Room 5, Poggio Bacherina, 5th century BCE (After Paolucci 1997, 55).*

In South Italy, at Monte Sannace in Puglia, many loom weights were found in numerous rooms of the Hellenistic houses, especially in the western zone. Thus, in House I1, Room b had 70 loom weights, while Room c had 48 loom weights. House I2 had a set of 48 weights in Room d (38 in one group), and additional 27 weights in Room e. Isolated room in House I7 had 67 loom weights. Eighty-eight weights were found in the courtyard of House II1. Room c of House II7 yielded 22 loom weights. It appears, thus, that every family engaged in textile production and not one but several looms were set up simultaneously in each household (Scarfi 1962, 160; Donvito 1982, 168–169). It is possible that textiles were made not only for household needs but for market and exchange as well.

A similar situation seems to have occurred at Oppido Lucano in Basilicata, where several 4th–3rd century BCE houses evidently had looms set up when the settlement was destroyed. Room 2 of House A had a concentration of 21 loom weights (Figure 115a); other 21 weights were found in Room 2 of House B; and Room 3 of House D yielded 79 weights, 58 of which were in the north-east corner and 21 in the south-east corner of the room (Figure 115b), suggesting that at least two looms were set up there simultaneously (Lissi Caronna 1983, 295, 332; Lissi Caronna, Armignacco Alidori and Panciera 1992, 238, 277–285). Smaller groups of loom weights were excavated in other parts of the site as well.

By contrast, at the nearby Serra di Vaglio, the 4th–3rd century BCE Casa dei *Pithoi*, yielded 108 loom weights along its south-west wall, where the loom may have been set up at the time when the house was destroyed (Greco 1991, 62). Compared to the data from the contemporary Monte Sannace and Oppido Lucano, this set has a much larger number of weights, suggesting that a much larger and/or more complex textile was being produced here.

The Centocamere area of Locri in Calabria yielded large numbers of loom weights, most, unfortunately, from secondary contexts. Several groups, however, have been noted: sets of 16 and 12 from the southern part of the wall; 11 and 15 from *Isolato* I2 (Dotta 1989, 199). Although these groups are too small to represent entire looms, such concentrations may still indicate the areas of textile production on the site.

Figure 115. Plans of the houses with loom positions at Oppido Lucano, 4th–3rd centuries BCE: a) House A; b) House D (After Lissi Caronna 1984, 195 fig. 1, 207 fig. 4).

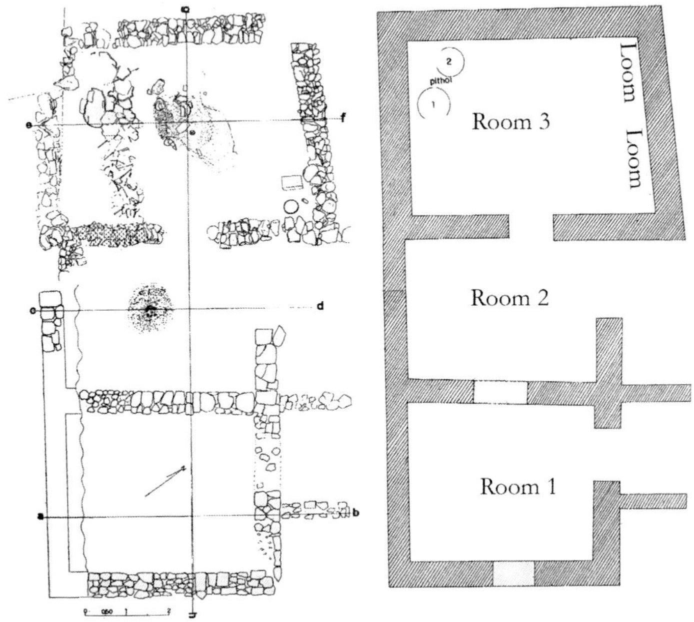

a

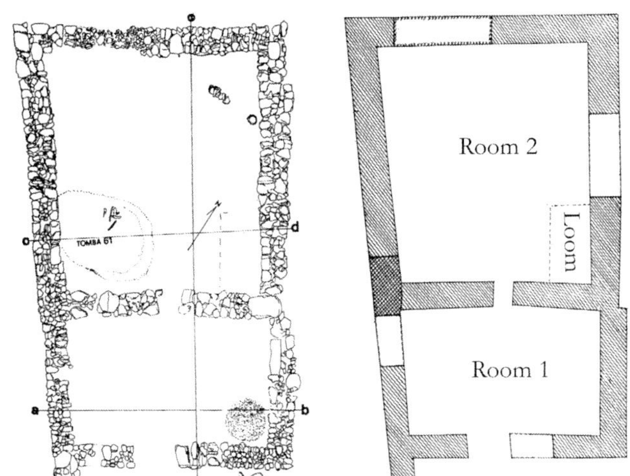

b

The analysis of over 240 loom weights excavated in the Isolato I of Zone A at Mozia in Sicily indicates extensive textile production at the site (Rossoni 2002). The weights were often found in groups of similar size; thus room 41 of the 4th century BCE Casa delle Amphore yielded 31 weights, weighing 75–85 g, while room 39 of the same house had 23 loom weights weighing 30–40 g. The excavated sets seem to have been at different stages of their use when the settlement was destroyed. The sets from rooms 39 and 41 were found in amphorae, where they were probably kept when the looms were not in use. In Room 12, on the other hand, 72 loom weights were dispersed near the east and north walls, suggesting that the loom was set up at the time of abandonment.

The above examples show how, despite the total absence of textiles at a site, it is possible to draw some conclusions about their role in the economy of the settlement. A tool-based approach to textile studies can be used to start reconstructing this craft, which until now has been missing in the context of the study of pre-Roman economy in Italy. The feasibility of such an approach hinges on the publication of necessary data by the excavators and I would like to underline the importance of publishing the physical parameters of excavated objects in archaeological reports (Gleba 2004a). Until now, the weight values have been published only rarely for loom weights and almost never for spindle whorls and spools. Just as important is the stratigraphic information since the most reliable way of dating implements is through their association with other material. Artefact distributions are also essential for a study of this type since concentrations of implements reflect the areas of activity. Finally, a good understanding of the archaeological context is necessary in order to draw conclusions about the production of textiles at a given site.

A case study: Poggio Civitate (Murlo)

The site of Poggio Civitate (Murlo) furnishes excellent material for a case study, demonstrating the kind of information that can be retrieved from textile implements found in settlement contexts. Judged solely by the number of spinning and weaving implements, Poggio Civitate appears to have been a significant textile-producing centre. In 1998, the catalogued finds included 441 spindle whorls, 69 loom weights, 580 spools, and 13 needles (Figure 116). The results of the analyses of these implements support some generalisations about textile production on the site.[1]

The great number of spindle whorls at Poggio Civitate, especially when compared to other sites for which similar information is available, indicates that the scale of yarn production was large, well beyond the requirements of domestic needs. The mean weight of spindle whorls at Poggio Civitate is 9.6 g (median 8, range 2–48), which is appropriate for spinning very fine yarn, probably in wool. The number of spindle whorls further suggests that a large quantity of this very fine yarn was manufactured at the site. Ninety-three per cent of the whorls are of truncated conical shape, suitable for spinning yarn of medium twist. The very uniform proportions of the whorls suggest that spinning was both organised and specialised. The concentration of the large number of very uniform implements in specific areas of the site suggests a workshop situation: spinning seems to have been one of the major crafts on the site.

The number of loom weights at Poggio Civitate seems to be small in comparison to the number of spindle whorls and

Figure 116. Textile tools from Poggio Civitate, Murlo, 7th–6th century BCE: a) spindle whorls; b) loom weights; c) spools (Courtesy of Anthony Tuck).

spools. The weight of the Murlo loom weights ranges from 34 to 795 g. This is a very broad range in comparison to loom weight size at contemporary sites (cf. Table 6), which suggests that textiles of various qualities and complexities were being produced at Poggio Civitate.

Although not all parts of the site have been excavated, the numbers of spindle whorls with respect to loom weights suggest that the amount of yarn spun significantly exceeded the quantity of fabric that was being produced at the site. This inconsistency may be explained by the extraordinary quantity of spools on the site. Their number (over 580) is higher even than the quantity of spindle whorls and, if the hypothesis that they were used in tablet weaving and other kinds of band weaving is correct, it may indicate that the weaving workshop at Murlo specialised in the production of textiles with fine patterned strips. This interpretation is suggested by the popularity of patterned borders on garments depicted in contemporary figurative representations (Bonfante 1975) and actual borders found at Verucchio (Stauffer 2002) and Sasso di Furbara (Mamez and Masurel 1992). Such a situation would not be surprising, especially since these elements, in addition to being decorative, were also indicators of status (Stauffer 2002, 192, 208).

Finally, 13 fine bronze and bone needles have been found at Murlo. Not only do they indicate that sewing was practiced on the site, they also suggest that several workers could have been sewing simultaneously, because in normal domestic contexts needles are quite rare.

The distribution and style of decoration of the spinning and weaving implements at Poggio Civitate may be taken to indicate that most of them were probably associated with the Orientalising phase, and hence, were manufactured and/or used over the period of about a quarter to a half of a century. Although scattered by the destruction and levelling of the site, the tools for textile production appear to be concentrated in just two areas, both associated with structures of the late 7th century BCE. One is in the northern part of the Lower Building/Orientalising Complex 1. Another area in which implements are concentrated is around, and to the north of, the so-called Southeast Building/Orientalising Complex 2. This structure has been identified as a workshop, because of deposits of the debris of ivory and bronze-working, as well as the presence of unfired tiles (Nielsen 1998a, 98–99). The distribution of spinning and weaving implements not only confirms the functional identification of this structure but also suggests that textile manufacture was as important as the production of other types of materials and goods, mostly luxury in nature.

From this brief summary it is clear that textiles were manufactured at Poggio Civitate on a scale significantly greater than what would be considered production for domestic consumption. In fact, the evidence of spinning and weaving implements points towards Poggio Civitate having played a significant role in production and possibly even textile exchange in the Orientalising period.

Burial context

Burials and funerary gift assemblages represent an important source of information about a society and its hierarchical and ideological aspects. The last thirty years of research have advanced significantly the analytical methodology of burial studies. The publication of important Early Iron Age necropoleis at Tarquinia (Iaia 1999), Veio-Quattro Fontanili (Guidi 1993), Osteria dell'Osa (Bietti Sestieri 1992b), Pontecagnano (d'Agostino and Gastaldi 1988), Sala Consilina (Ruby 1995) and Torre Galli (Pacciarelli 1999) have demonstrated the complexity of early societies in Italy through studies of burials. Comparisons based on spatial analysis within the same necropolis or across necropoleis of the same site, allow scholars to characterise the chronological, gender, familial and rank distinctions between members of the community (Bergonzi 1981; Zifferero 1995).

Many spinning and weaving tools have been recovered from tombs (Figure 117). Shears, spindle whorls, distaffs, loom weights, spools, and needles are the types of textile tools most frequently deposited in graves beginning in the Late Bronze Age. The practice was especially common during the Early Iron Age and the Orientalising period, but continued well into Roman times as well.

Every region and even every community had its own rules for the type and number of implements chosen to be deposited with the dead. In North and Central Italy, especially the areas of Villanovan culture and its influence, spindle whorls, spools and distaffs are frequent finds in female burials. In South Italy, there is more diversity in the ritual. For instance, in Campania, Villanovan necropoleis yield assemblages similar to those of Central Italy, but textile implements are absent from the *Fossakultur* sites. In Basilicata and parts of Calabria, spindle whorls are present but spools seem to be substituted by loom weights. In Puglia, loom weights begin to appear in burials only as the region becomes increasingly hellenised. In Sicily, we find an association of needles, spindle whorls and knives in female tombs (Frasca 1982, 30; Leighton 1999, 197–198). The only tool found in association with male burial assemblages, shears, appears exclusively in the Gallo-Roman necropoleis of North Italy and in Punic areas on Sicily.

One could argue that objects deposited in graves belong to a symbolic universe and do not reflect the daily reality of pre-Roman Italy. After, all, funerary rituals constitute an autonomous variable of a social system and constitute only one of the archaeological sources of information about a society. I believe, however, that differences in the numerical and chronological distributions of spindle whorls, distaffs, spools, loom weights and other textile tools and their combination in a given context reflect, to some extent, textile production process and justify the use of grave material as a source for

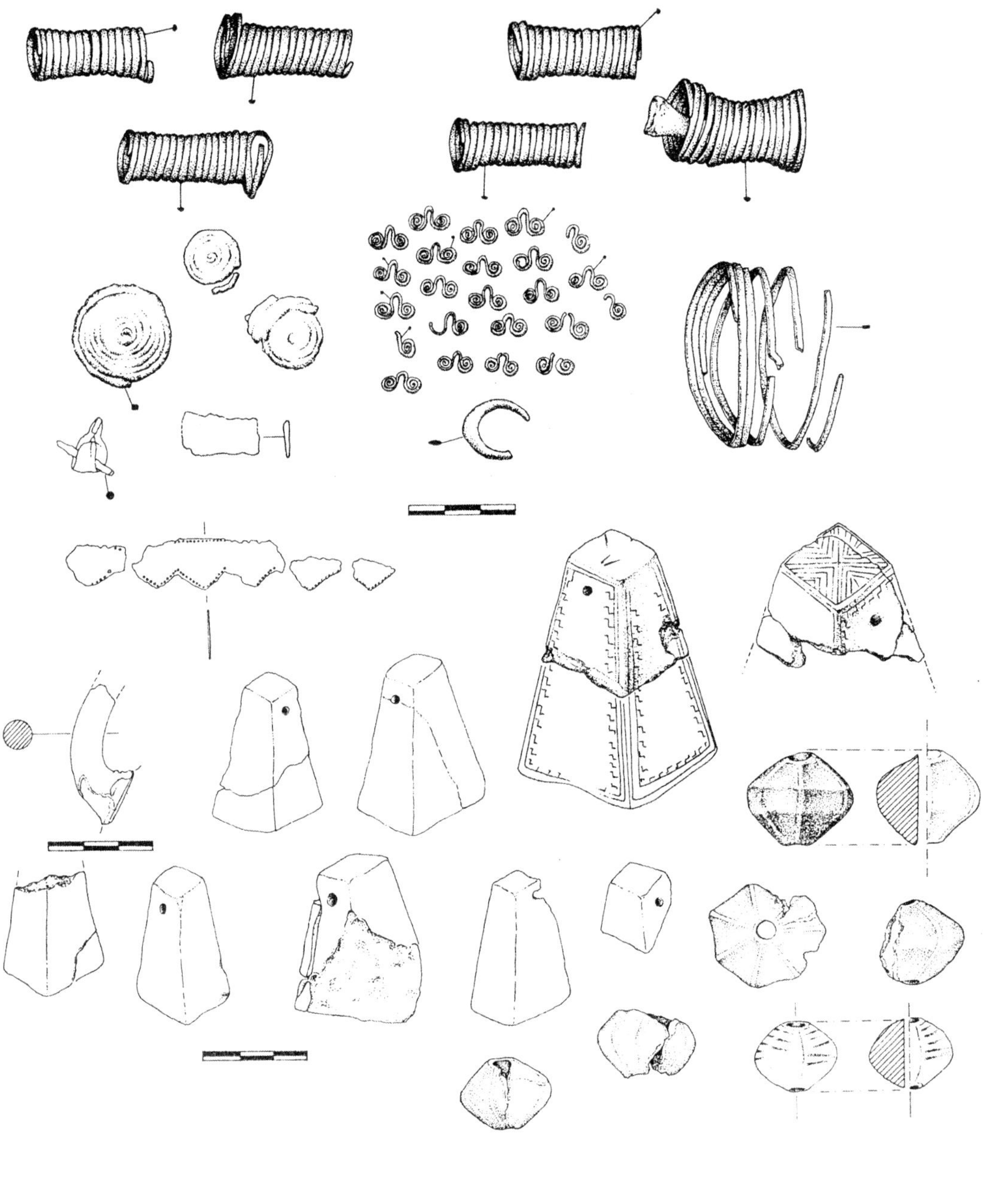

Figure 117. Tomb 462, Incoronata, 8th century BCE, with nine loom weights and five spindle whorls (After Chiartano 1994, 81 pl. 1).

the reconstruction of the organisation of society. The presence of textile implements in burials thus provides insights into the questions of gender, social role and status, skill or level of specialisation, and, possibly, age of the recipient of these burial gifts. It can be postulated that yarn spinning and textile weaving were activities carried out by and strongly associated with women in ancient Italy.

Gender

One of the primary aspects of an individual that is indicated by burial rites is gender.[2] Textile implements in general, and spindle whorls in particular, traditionally have been regarded as indicators of female burials (Bartoloni 1989). Sex determinations made on osteological grounds are rare, often because the bones are not sufficiently well preserved, a situation especially common in the case of cremation burials.[3] Furthermore, in many old excavations, bones were not collected, or were discarded, and sex-typing has frequently been based on the types of burial goods that are found with the remains of the deceased (Toms 1998, 172). Although the orthodox view that textile production was an exclusively female activity has been occasionally questioned,[4] it is supported by the archaeological and iconographic evidence in Italy. In the instances when osteological remains are available, they seem to support the gender attribution made on the basis of burial goods (Toms 1998, 171): the absolute majority of burials containing spinning and weaving implements are female.[5]

For instance, at the Iron Age cemetery of Osteria dell'Osa in Latium, where the bones were analysed in order to determine the sex and age of the deceased, "a single faceted spindle-whorl occurred in the majority of female graves, especially in age-classes 4 to 8 [adults]" (Bieti Sestieri 1992a, 108). Judith Toms (1998, 166) has argued convincingly that, for the Villanovan necropoleis of Tarquinia and Veio, spinning/weaving equipment and weapons/armour form distinct groups, which are independent of chronology and wealth.

There are, however, occasional instances when weapons, accepted indicators of the male sphere of activity, and tools associated with textile production are found in the same burial. For example, in Poggio Aguzzo Tomb 6 at Murlo, dated to the Orientalising period, three spindle whorls were associated with a spear, an iron knife and a group of arrow heads (Tuck 1996, 105). Iron Age Tomb 4 at Tor de'Cenci in Rome had a spindle whorl as well as a dagger blade (Bedini 1992, 224, 256). In a similar fashion, tomb 43 at Narce contained two bronze helmets, a cuirass, two razors, and two spindle whorls (Dohan 1942, 7–9). Warrior Tombs 16 and 24 at San Onofrio in Rocella Jonica contained loom weights among the burial goods (Chiartano 1981). Setting aside the cases where intrusion of the material may be argued due to looting or mixing of contexts during excavation, these finds have been interpreted as offerings deposited by a female relative of the deceased man (Nielsen 1998b, 70; Bartoloni 2000, 273). The association of textile implements with women is, thus, further reinforced: not only were they designating the gender of the deceased but also women's role in society and the family.

Only one textile tool appears regularly in male burials. Shears are found in Gallo-Roman male graves in North Italy, indicating that, at least among the Celts, the shearing of wool was a man's job. In Bologna's Benacci and De Luca necropoleis, for example, burials of men without arms but with shears have been excavated, suggesting that a certain class of men in Celtic society earned their prestige not as warriors but as shepherds (Grassi 1991, 85).

Iconographic sources also confirm the strong association of women with the textile craft (Rallo 1989a; Bietti Sestieri 1992a; Nielsen 1998b). One of these is the wooden throne found in Tomb 89 at Verucchio (Eles 2002). The intricately carved scenes, shown on the inner side of the back depict women in the process of spinning and weaving. The structure identified as loom is quite high and operated by women. Another object, bearing one of the most important representations of textile production in the ancient world, is the late 7th century BCE bronze *tintinnabulum*. Here, women are represented at all stages of textile production, save collection of wool and finishing. What these scenes suggest is a high status of the women depicted and of the craft they are involved in. Textile tools in burials seem to indicate the same. Thus, we can assume that Iron Age people saw textile production as expressly associated with women. The scenes also emphasise the collective aspect of this female task and the importance of communication between women during textile production.

Status

Burial and iconographic data thus indicate that, in Italy, starting in the Final Bronze Age, textile craft became a symbol of the female sphere of life, and women's contribution to the community as textile workers was expressed by the deposition of textile tools in their burials.

Implements used for spinning and weaving were generally made of fired clay, wood, or bone, and had little intrinsic value of their own. They were meaningful only to the person who used them and therefore carried symbolic value when placed in tombs (Barber 1991, 299). It is this symbolic value associated with spinning and weaving implements that led to their manufacture in precious materials. Finds of spinning tools in precious materials, such as bronze, silver, amber, and bone, support the notion that spindle and distaff were not only role indicators but also important status symbols for the ladies in the highest social classes.

Another object connected to spinning that may have expressed the high status of its owner may be a container – *cista* or *situla* – to keep raw wool (Bartoloni 2000, 274–275). Bronze cists and *situlae*, frequently found in rich female burials may have served for this purpose, as demonstrated by the situla from Casa di Ricovero Tomb 23, which contained a spindle, a distaff and two wool combs (Chieco Bianchi 1987, 212 fig. 32 fig. 58, 213 no. 13).

There is a clear contrast between the majority of relevant tombs that contain a single spindle whorl and the significantly smaller number of tombs with spools and/or distaff in addition to whorls. The latter combination is frequently found in the richest burials. Distaffs made in valuable materials, such as bronze, silver, amber and bone, were particularly important markers of elite female status during the 9th–7th centuries BCE (Bartoloni 1989; Nielsen 1998b, 74; Jannot 2004). Being impractical and probably too precious to be used, they are more likely to have been made purely for display purposes. The tools themselves symbolised the prowess of the deceased woman in textile craft and, hence, her social role in the community (Bietti Sestieri 1979, 142), while the precious materials of which they were made expressed her status as a wealthy and high-ranking member of that society (Baldoni 1994, 264). Although distaffs largely disappear from grave good assemblages after the 7th century BCE, their symbolism continues well into the Roman period, when brides carried a spindle and a distaff during wedding processions (Plin. *NH* 8.194; cf. Torelli 1984, 131, 133) and spinning tools were depicted on funerary monuments or included among the grave goods (Cottica 2007). Varro records that the distaff and wool of Tanaquil, wife of Tarquinius Priscus, the first Etruscan king of Rome, was displayed for centuries in a temple of Sancus erected in Rome, according to tradition, in 466 BCE (Plut. *Q.R.* 30; Plin. *NH* 8.194).

Another status indicator may be the presence of spools, implements probably used in tablet weaving.[6] The vast majority of the burials containing spools are quite rich, thereby implying that in particular elite women possessed the skill and/or the right to use these tools to make special ceremonial garments with tablet-woven borders. The weaving of the borders on the mantles of Verucchio, probably added to the garments when the base textile was already finished, was extremely laborious and time consuming, resulting in a highly prised product, and thus indicating that such mantles were a status symbol not only for those who wore them but also for those who made them (Stauffer 2002, 208 and 212).[7] Moreover, these textiles must have served as indicators of social rank or 'ceremonial' clothes, with the border functioning as the distinguishing element characterised by technique, pattern and colour (Stauffer 2002, 194, 208). Such borders are well illustrated in the later representational material, such as the Etruscan tomb paintings of Tarquinia, the terracotta statues of Murlo and Veio and bronze figurines (Bonfante 2003). Production of these textiles not only required highly specialised materials and skills, available only to the members of the elite classes of society, but also was in itself exclusive elite female occupation marked at death by the deposition of associated tools in the grave.

Status is most certainly indicated by the deposition of precious textile implements, such as the bronze distaffs, in infant burials. The symbolic character of these objects is especially clear when they are miniaturised, and thereby lose their function (Bergonzi 1981, 289; Bietti Sestieri, De Santis and La Regina 1991). Miniaturised versions of tools are known from Latial infant burials, such as Tomb 5 at Le Caprine (Guidonia), which contained a miniature type B distaff, a spindle, a spindle whorl and four spools (Figure 118) (Guidi and Zaraffini 1993, 191; Damiani *et al.* 1998, 206). Objects of normal size may have been offerings from the mother.

Agneta Strömberg found that the spindle whorl is not an item of high frequency in Athenian Geometric graves, possibly being a status symbol for the 'ladies of the rich families'. She concluded that, "it is not these women's status as individuals that is being emphasized. ...The grave offerings are exposed in the common fashion as prestige goods and what is being accentuated is rather the wealth and power of the families to which they belong" (Strömberg 1993, 95). I believe that, in Italy, on the contrary, the mere numbers of the spinning and weaving implements in burials as well as their rendering in precious materials indicate a very concrete association with a specific activity. Indeed, gender was not the main theme to be emphasised through the burial goods; rather, it was status and the degree of wealth. Yet, it would be hard to believe that status was measured only through material means. Activities – spinning and weaving in women's case – and the skill of a person involved in these activities may have been important as well.

Skill

Gender and status, in fact, are not the only aspects of the deceased demonstrated by the textile implements. It is also theoretically possible to differentiate the degree of skill or specialisation in spinning and/or weaving. Based on the material from the Early Iron Age Osteria dell'Osa, Anna Maria Bietti Sestieri has proposed a differentiation between 'spinners', defined by a single spindle whorl that originally must have

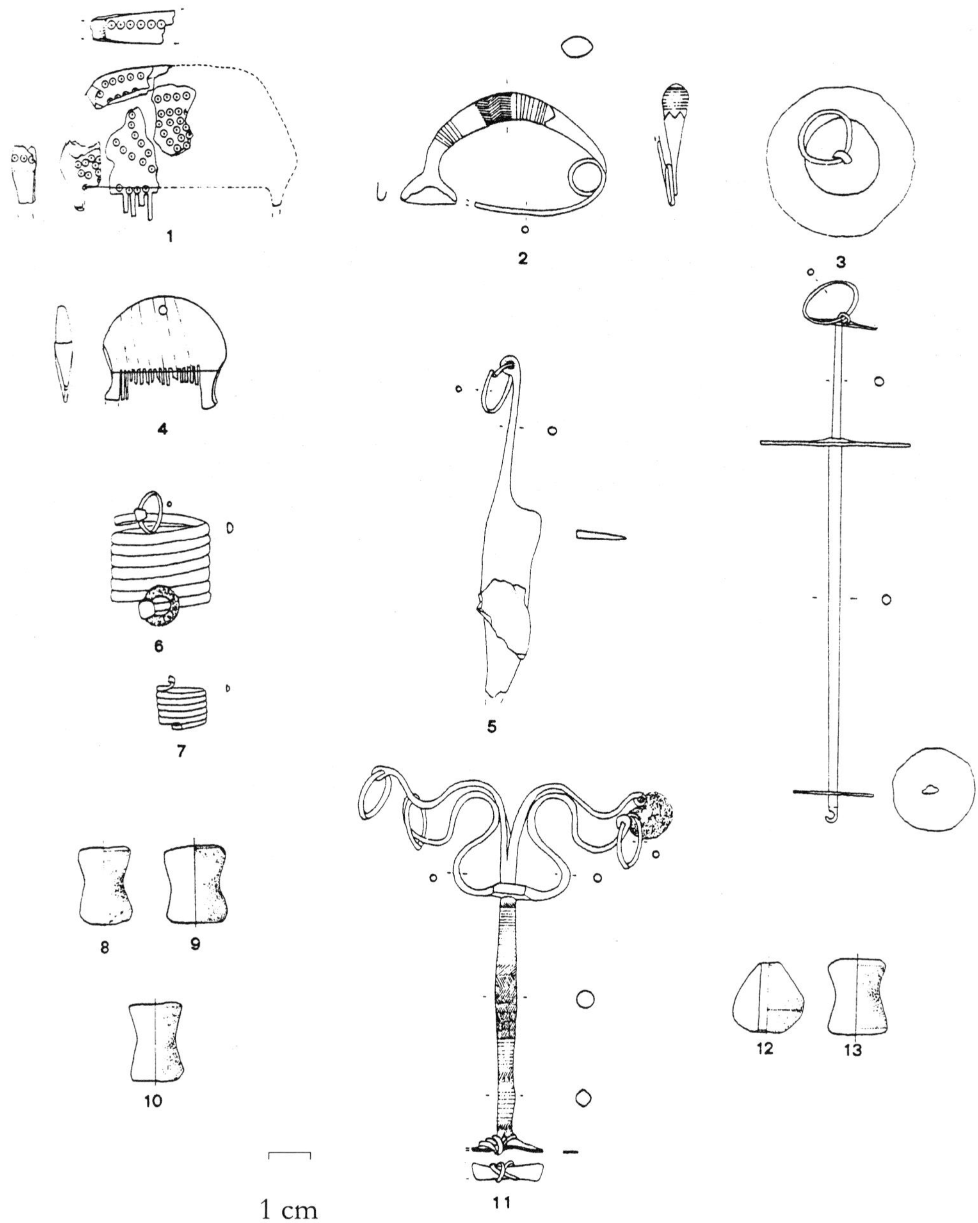

Figure 118. Tomb at Le Caprine (Guidonia), with miniature textile tools (After Giudi and Zaraffini 1993).

been associated with a wooden spindle and, sometimes, a distaff, and 'weavers', whose burial assemblages included several spindle whorls and numerous spools, the latter being more 'specialised' than the former (Bietti Sestieri 1979, 110, 143; 1992a, 102; Bartoloni 2000, 273). Such a division necessitates the separation of spinning and weaving in the textile production process and singles out weaving as its more complicated operation. Yet, although these tasks were technologically separate and required different tools, the production of textiles in Italy was considered a single process, as documented by both archaeological and representational material. For instance, the various textile production implements found in primary, settlement contexts usually occur together, indicating that spinning and weaving took place in the same areas. The scenes on the *tintinnabulum* and the Verucchio throne also illustrate continuity of the production process.

Cloth consumed by a small household was produced by women of that household. The 'spinners' did weave, and the 'weavers' spun so much that not one but several spindle whorls were often included in their burials in addition to the spools. Spinning rather than weaving became symbolic of the female sphere for two reasons: first, because, effectively, more time was spent on this task than on weaving. And second, because unlike a large, heavy and stationary loom, a spindle and a distaff were small and portable, thus enabling women to perform the task practically anywhere and making it more 'visible'. This daily visibility made distaff and spindle universally recognisable objects, which could easily be reproduced in precious materials and transformed into tangible symbols of female social role. I believe it is more likely that the whorl simply defined the deceased as female, based on her socio-economic role in the community, while the presence of spools and other implements signifies that this particular individual was a more skilled or specialist textile worker.

Iron Age burials provide indirect evidence for such differentiation. The tombs containing spools and multiple spindle whorls are few in comparison to those with just one spindle whorl. Assuming that the greater number of such implements deposited with the deceased symbolises that individual's skill in textile craft, we can see that such skill was far from common. The tools in these burials, then, express the specialisation of their owner in textile craft.

The 7th century BCE Tomb 3 of Caselle di San Lazzaro at Villanova di Castenaso contained 19 spindle whorls (Forte 1994, 243–245 pl. VI nos. 27–45). Seventeen spindle whorls, ranging in shape from globular to truncated conical, were among the burial goods of Quattro Fontanili Tomb Z 11–12 at Veii (Falconi Amorelli 1967, 213 no. 6, 214 fig. 75). Also included were 13 spools and a bronze distaff. Tomb R 3–4 in the same necropolis contained an astounding 96 biconical whorls, some of which were elaborately decorated, and no other textile equipment (Fabricotti *et al.* 1972, 310 fig. 80, 313 nos. 1–3). Seven spindle whorls were found in Tomb 328 and eight in Tomb 9 at Osteria dell'Osa (Bietti Sestieri 1992b, 707 nos. 9–15, fig. 3a.310; 746 nos. 5–11, fig. 4a.387). Tomb XVIII at Cuma, dated to the end of the 8th century BCE, had at least 27 impasto spindle whorls (Tocco Sciarelli 1985, 95 fig. 17.10, 96). Fornaci Tomb 363 at Capua had 14 spindle whorls (Johannowsky 1983, 136–137 nos. 53–66, pl. XXXI). Although rare, the practice of depositing large numbers of spindle whorls in burials is not limited to Italy. For example, the princely Tomb 200 at El Cigarralejo in south-east Spain, dated *c.* 400 BCE, had 59 whorls (Cuadrado 1968, 162, Fig. 10).

Multiple spindle whorls, usually present in the 'specialist' burials, often vary in size and shape, two of the parameters that affect the quality of yarn; this suggests a production of various types of yarn for different kinds of textiles.

Multiple spindle whorls are, in most cases, associated with spools, as in the case of 8th–7th century BCE Tomb 1 at San Cerbone, Populonia (Torelli 2000,

582 no. 129). It had 16 spindle whorls and 24 spools. Frequently, in addition to spindle whorls, several other textile tools are associated with spools in the same burial, further evidence to reinforce the specialist identification. Thus, at the Casa di Ricovero necropolis of Este, spools are often associated with numerous spindle whorls and loom weights, as in the case of Tomb 143, which contained 12 spools, five spindle whorls and seven loom weights (Alfonsi 1900, 540 no. 19; Chieco Bianchi and Calzavara Capuis 1985, 61–62 a–n, pl. 13; Chieco Bianchi 1988a, 33). Benacci Tomb 490 at Bologna had six spools, three spindle whorls and two distaffs (Amann 2000, 221). Quattro Fontanili Tomb HH 11–12 at Veii had 34 spools, a bronze distaff and a clasp (Cavallotti Batchvarova 1965, 128 n, 135 fig. 55). In Pontecagnano, Pagliarone Tomb 2066 had 16 spools, eight spindle whorls and a bronze distaff (Gastaldi 1992, 140, nos. 14–29, pl. 121).

Age

Finally, it is possible that certain textile implements or/and their combination may indicate that some activities were associated with women of a specific age group. The analysis of the Osteria dell'Osa Iron Age necropolis has provided an indication that spinning was a marker of young girls in that Latial community (Bietti Sestieri 1992a). There is, however, no evidence for such differentiation elsewhere. In fact, it is generally accepted that the crafts of spinning and, especially, weaving were the prerogatives of matrons, confirmed by the Roman practice of giving a bride a spindle and a distaff on her wedding day (Torelli 1997; Bartoloni 2000, 274; Cottica 2007).

Votive and sacred context

Spinning and weaving implements from votive deposits and sanctuaries constitute the smallest, chronologically most recent, and least explored category of contexts. These finds not only add another dimension to our view of ancient religious practices but are also important for our understanding of the development of social values and traditions associated with textile production. Unfortunately, contexts are not always clear for votive material and it is not always easy to differentiate textile implements that served as ex-votos from those used in the actual textile production process (cf. Bouma 1996, 24–30).[8] In the following discussion, only examples with relatively secure contexts are presented and they have been divided into three subgroups: actual votive deposits, foundation deposits and sanctuary workshops.

Votive deposits

Spindle whorls, spools and, especially, loom weights are common ex-votos in Etrusco-Italic sanctuaries, as well as in South Italy and in Sicily, where the practice was probably also influenced by Greek votive religion. For the most part, textile implements are present in deposits that are related to cult sites with a wide chronological range, as opposed to those which were created on a particular occasion (cf. Pascucci 1991, 469–471). The following list is far from exhaustive but it illustrates the importance of textile implements in votive ritual.

Votive deposits in North Italy

In North Italy, the large deposit of Baratella at Este yielded several spindle whorls, four needles, eight distaff elements, about 30 spools and over 300 loom weights (Figure 119) (Ghirardini 1898; Mastrocinque 1987, 97; Maioli and Mastrocinque 1992, 21). The loom weights offered as votives were for the most part decorated or inscribed. Smaller votive deposits in the area also yielded some textile implements (Maioli and Mastrocinque 1992, 32, 34–35). In Padova, the large votive deposit of San Pietro Montagnon (Morlungo) produced only two spindle whorls, two spools and seven small loom weights (Dämmer 1986, 76; Maioli and Mastrocinque 1992, 37).

Votive deposits in Central Adriatic Italy

Although not as large in number and scale as votive deposits elsewhere in Italy, those in the Central Adriatic region still contain small quantities of textile implements. Thus, one of the earliest and most important Picene sanctuaries at Cupramarittima contained spindle whorls and miniature ring loom weights (Baldelli 1997). The site of Monte Giove, frequented from the 6th century BCE through the Roman period, had all types of terracotta textile implements (d'Ercole, Cosentino and Mieli 2001, 339, 341–342 nos. 104–105, 109–111). A small Archaic-Hellenistic water sanctuary at San Vittore di Cingoli yielded three large trapezoidal loom weights (Landolfi and Baldelli 1997, nos. 3–5).

Votive deposits in Central Tyrrhenian Italy

Etruria

In Etruria, numerous large and small votive deposits have yielded textile implements. In Bologna, a relatively early deposit, dated to the second half of the 7th century BCE, included loom weights and spools (Miari 2000, 154–155). The necropolis sanctuary of Cannicella in Orvieto yielded two spindle whorls and a spool from the vicinity of altar 4 (Andrén 1967, 72 nos. 50–51, 73 no. 55). A Hellenistic deposit at the Santuario Campestre of Tessennano included five loom weights (Costantini 1995, 107–108 G, pl. 47 c and d). Similar Hellenistic votive deposits at the sanctuary of Carraccio della Osteria, Vulci, and at the temple at Castelsecco, near Arezzo, also produced some textile implements (Hackworth 1993, 56). The 3rd century BCE deposit of the Ara della Regina at Tarquinia yielded some loom weights as well (Comella 1982, 185, C7I–C7II, Pl. 101c). The Archaic Vigna Parocchiale deposit in Cerveteri,

Figure 119. Textile tools from the votive deposit of Baratella, Este, 6th–1st centuries BCE (After Tosi 1992, 96 fig. 95).

included 77 trapezoidal loom weights, of which 36 were decorated on top with simple motifs, as well as 15 spools and two spindle whorls (Moscati 1992). Various strata in the North sanctuary at Santa Severa (Pyrgi) also produced implements associated with textile production: seven loom weights and a fragmentary spool turned up in Temple A, two loom weights were excavated inside Temple B and another two in the entrance zone; finally, one spool and one spindle whorl came from the *piazza* of the sanctuary (*Pyrgi* 1970, 247, 263, 439, 542–543, 647). Material of the 6th through 2nd centuries BCE from the large Campetti deposit at Veio included three loom weights, three spindle whorls, and one spool (Vagnetti 1971, 103 no. 5, 153 nos. 2–5; Comella and Stefani 1990, 113, pl. 35a). Portonaccio area at the same site yielded three spindle whorls, three loom weights and one spool (Colonna 2002, 194–197). In the Faliscan territory, the Archaic–Hellenistic sanctuaries of Celle, Sassi Caduti and Vignale of Civita Castellana (ancient Falerii) had significant amounts of loom weights and spools (Comella 1986, 93–97 I and L, 118 Q6, pls. 49–50, 59e; Carlucci and De Lucia 1998, 49–67).

Latium

Weaving and spinning tools are even more abundant in the votive deposits of Latium. Loom weights were recovered from the votive deposit in the sanctuary of Sant' Omobono in Rome, dating to the Archaic period (*Enea nel Lazio* 1981, 145 C64, 147). The votive deposits, dated to the 6th–1st centuries BCE, of the Comitium, Clivo Capitolino and Colle Oppio, also in Rome, produced a number of spindle whorls, spools and loom weights (Gjerstad 1960, 240–242 fig. 150; Cristofani 1990, 68 nos. 70–72; Cordischi 1993, 42; Mangani 2004). A large number of miniature spools of cylindrical shape with concave ends were recovered from the deposit at Acquoria near Tivoli dated from the 7th to 2nd centuries BCE (Antonielli 1927, 230c, 235 no. 12; *Enea nel Lazio* 1981, 41 A79).

Pratica di Mare, ancient Lavinium, yielded large numbers of spindle whorls and loom weights, reportedly in votive contexts (*Enea nel Lazio* 1981, 218–219 D188 and D189; Fenelli 1991, 494 and 500). Numerous early votive deposits containing textile implements were excavated at Borgo Le Ferriere, ancient Satricum (Bouma 1996, 390). A large number of spindle whorls, approximately 200 objects, came from the votive deposit formed during the 6th–4th centuries BCE, at the Santa Cecilia sanctuary at Anagni, near Frosinone (Buddittu and Bruni 1985, 106–108; Gatti 1996, 135–138). Another sanctuary at Anagni, Osteria della Fontana, probably dedicated to Diana, also yielded textile tools: spindle whorls and dozens of loom weights some of which had incised decoration or letters (Gatti 2004, 86).

Sixteen loom weights were found in two *favissae* of the Hellenistic sanctuary of Juno in Gabii. The sanctuaries of Juno at Segni and Norba also had textile implements among votive dedications (Aldana Nàcher 1982). Finally, a Hellenistic deposit in Piazza Ungheria at Palestrina included loom weights (Quilici 1983).

Votive deposits in South Italy

In South Italy, loom weights are predominant textile implements in votive deposits. The 8th century BCE *Stipe dei Cavalli* in Pithekoussai had three loom weights (d'Agostino 1996b, 61 nos. 88–90, pl. XLI). At Metaponto, a large votive deposit was excavated next to the external face of an Archaic fortification wall adjacent to the sanctuary of Demeter and Kore; among the objects, dated to the 6th–3rd centuries, were five trapezoidal loom weights and numerous fragments of discoid weights with two holes (Adamesteanu, Mertens and D'Andria 1975, 282, 285 fig. 298 h, 287, 288 fig. 300; Lo Porto 1981, 316). Other votive deposits with textile implements were found in Timmari, and Cozzo Presepe (Lo Porto 1991, 169, pl. LXXIX; Morel 1970, 105–107, figs. 29–31). At Tiati, many loom weights were found, including two with Oscan inscriptions (Antonacci Sanpaolo 1995). Two golden needles found in the votive deposit of the sanctuary of Saturo in Taranto, were probably dedicated to Aphrodite or the local nymph Satyria (De Juliis 1984, 331, 366 nos. 337–338).

Votive deposits in Sicily

In Sicily, votive deposits containing textile implements have been noted only in Gela (Orlandini 1962, 371 no. 9, fig. 44a). The current state of Sicilian archaeological reports is confusing due to exigencies of early excavations. Furthermore, there is currently no information on the indigenous material since, until recently, the focus has been on the Hellenic sites of Sicily. At present, it is thus not possible to have a balanced assessment of Sicilian votive practices.

The examples above illustrate that, while frequently present in votive deposits, textile tools are never very numerous. More substantial quantities are noted in large sanctuaries, which were frequented for many centuries or in contexts for which other explanations can be suggested, as will be discussed below.

The meaning of tools in votive contexts

Most scholars have assumed that the presence of textile implements in votive deposits indicates a connection of the cult to divinities regarded as protectors of women and domestic activities (Mastrocinque 1987, 111). Athena/Minerva and Hera/Juno were, of course, the two most important recipients of women's gratitude: the first as the goddess of the loom, the second as the patroness of marriage and domestic activities. Textile implements are also common in sanctuaries of Demeter/Ceres and Kore/Proserpina, Aphrodite/Venus, Artemis/Diana and Herakles/Hercules.

Herakles was venerated in connection with transhumance and pastoral activities, and hence wool production (Santillo Frizell 2004b, 82). For this reason he is occasionally depicted with a spindle and a distaff (Di Giuseppe 1995, 141). Gem impressions representing Herakles appear on some unprovenanced loom weights (Ferrandini Troisi 1986, 97). The importance of sheep husbandry and transhumance between Samnium and Tiati is suggested by votives in the form of loom weights, found in the 3rd–1st century BCE sanctuary at Coppa Mengoni (Antonacci Sanpaolo 1995, 87–88).

Occasionally, the recipient of the votive gift is named by an inscription. The inscriptions in most cases were made before firing, so it is evident that the object was created specifically for the purpose of dedication. Thus, a loom weight fragment inscribed *VEI* was found by the western wall of Roselle; it has been dated *c.* 300 BCE (van der Meer 1987, 113). The excavator suggested that the object was created for the goddess Vea (Ceres/Demeter). Another inscribed loom weight of unknown provenance is currently on display in the Civic Museum at Viterbo (Emiliozzi 1974, 243 no. 527). It reads in Etruscan *ATI* and, according to Adriana Emiliozzi, may have been dedicated to Mater Matuta. Recently, loom weights with letters possibly signifying their votive purpose were excavated at Campo della Fiera, an area of Orvieto believed to have been the site of Fanum Voltumnae, a major Etruscan sanctuary and gathering place (pers. com. Simonetta Stopponi, 2004). Generally, however, the practice of inscribing loom weights with votive dedications is atypical in Etruria, yet common in South Italy, where it may have been brought by the Greek settlers.

In South Italy, we find loom weights inscribed with the names of Greek gods, such as Athena, Hera, Demeter and Herakles (Wuilleumier 1932, 38–39, no. 20 a–e; Orlandini 1953, 442; Ferrandini Troisi 1986, 97; Di Giuseppe 1995, 141). An unprovenanced loom weight from Sicily bears an inscription ΧΑΡΙΤΕΣ, interpreted as a dedication to the Graces, who were also regarded as patrons of textile crafts (Pace 1946, 482). Together with the Seasons (Horai), the Graces were credited with weaving Aphrodite's robe (Hom. *Il.* 5.338; Scheid and Svenbro 1996, 57).

Different views have been expressed as to the meaning of textile implements in votive and sacred contexts: they may have been attached to (matrimonial) garments given to the deity; used to close small bags, which contained food to be eaten at the sacrificial meal; given to ask protection over weaving; or to symbolize the work of a wedded woman (Bouma 1996, 392). It has also been suggested that this category of votive objects is connected to the rite of passage into adulthood or marriage (Chieco Bianchi 1988a, 67; Maioli and Mastrocinque 1992, 27).

Mingazzini argued that weights themselves were not the votive gifts but instead, were attached to garments or other textiles (Mingazzini 1974, 204–206). This view, however, does not explain their presence in small votive deposits and it also does not explain the presence of other textile implements, such as spindle whorls, spools and needles in votive contexts. It is also unlikely that anyone wanting to dedicate a garment would buy it on the market: more likely, these were woven at home with particular care, either by the dedicant herself if she was a woman, or, if a man – by his female relative. Mingazzini himself, in fact, mentioned a good proof for this (1974, 207): the list of the dedications to the temple of Artemis Brauronia contains a mention of a garment, in which the name of the dedicant was woven in, showing that it was made to be dedicated, in which case there would be no reason for any sort of 'tag'.

The low intrinsic value of a loom weight or a spindle whorl is not a sufficient reason for it to be unsuitable as a votive gift, since, as already shown for the textile implements found in funerary contexts, it is the symbolic significance of the object that made it an ex-voto *par excellence*. This interpretation is confirmed by numerous dedicatory epigrams in the Palatine Anthology, which record women dedicating their spinning and weaving tools to deities (Book VI, 39, 47, 48, 160, 174, 247, 284, 285, 288). Objects, all dedicated to Athena, include wool combs, spindles, distaffs, shuttles, wool baskets and yarn, representing all stages of textile production.

Foundation deposits

Loom weights have also been found in what have been identified as foundation deposits, which, so far, have only been found in Sicily. Initially thought to be associated with sacred buildings and areas, and hence with some kind of consecration ritual, loom weights have since turned up in foundation deposits of non-sacred structures. The use of loom weights in the foundation deposits of sacred buildings starting in the Archaic period likely developed from the symbolic significance these objects already acquired in votive contexts (Orlandini 1953, 443). The foundation ritual then spread to include non-sacred constructions as well.

Foundation deposits under sanctuaries

Along the south and north walls of an Archaic building in Vassallaggi, central Sicily, six trapezoidal weights were found under the pavement (Adamesteanu 1958, 309 no. 1). Three of these weights were deposited together with Late Corinthian pyxides, each systematically placed in a small *fossa* carved into the bedrock. A similar series of small rock-carved *fossae*, dated to the late 7th–early 6th century BCE, each containing Ionic cups and a single large trapezoidal loom weight, were uncovered within the eastern room of a sacred building at Lentini (Adamesteanu 1956, 406, 413). In the sanctuary of Demeter Thesmophoros in Bitalemi (Gela), 30 weights inscribed ΘΕΟΤΙΜΟΣ, and dated to the 5th century BCE, were recovered on the south side of room G2 (Orlandini 1966, 20, pl. X fig. 3). Orlandini (1966, 20) interprets the inscription as meaning "worthy of god", although it may simply be a personal name. In Siracusa, loom weights were deposited along with pottery in small *fossae* underneath the Altar of Hieron, dated to the 3rd century BCE (Adamesteanu 1956, 413). These small foundation deposits have been referred to as θυσίαι, defined by Hackens as "separate depositions of the remains of each single sacrifice" (Hackens 1963, 88–89).

Foundation deposits under city walls

A foundation deposit found under the Hellenistic walls of the acropolis in Morgantina contained a discoid weight (Sjöqvist 1960, 126 pl. 21, fig. 3). A group of truncated pyramidal weights without perforation have been found in the fortification walls at Terravecchia di Cuti (Militello 1960, 57). Given that these objects are without perforations, they may have had a function other than loom weights.

Foundation deposits under houses

In Gela, under the foundations of the north wall of the 4th century BCE building A, 25 loom weights were found deposited systematically with a small *lekythos* (Orlandini 1953, 442; Orlandini 1962, 352, fig. 8). Orlandini initially hypothesised that the building was a cultic area, because of the numerous terracotta figurines of Demeter found inside, but he later identified it as a private house. Not far from the building A, under the east foundation wall (E) of another late 4th century BCE house, 102 loom weights had been carefully deposited in direct contact with the foundation blocks (Orlandini 1962, 362 fig. 24, 365). At the nearby site of Manfria, underneath the northern foundation wall of the 4th century BCE farm-house, originally identified as an altar, three trapezoidal and three discoid weights were found under a large kylix (Adamesteanu 1958, 297 and fig. 8.3, 308).

Mingazzini argues that loom weights in these foundation contexts either were utilised as reinforcement material, or that they originally had some function other than loom weights (Mingazzini 1974, 212–213). While the first function is unlikely, it remains possible that the weights may have had some role in the foundation ritual that had nothing to do with textile production. Alternatively, the association of the loom with household activities and consequently with the house may have led to the use of loom weights as a token of the loom to symbolically strengthen the physical foundation of the house.

Sanctuary textile production

It was noted earlier, that frequently it is difficult to identify the exact nature of the contexts of the material classified as votive. Just because the function of a structure from which the material was recovered had been religious does not automatically endow all objects found inside and in its immediate vicinity with a sacred significance. As noted before, the number of textile implements in votive contexts is relatively small, excepting a few cases of exceptionally large deposits, which were used for considerable amount of time (Greco 1997, 193). When large concentrations of loom weights are found in sanctuaries, a different use for them can be postulated: they may have been utilised in the actual production of sacred garments within the sanctuary.

Literary evidence for textile production in sanctuaries

The weaving of garments for cult statues was a well-documented phenomenon in Archaic and Classical Greece. That weaving activities took place in sanctuaries is known from the literary sources (Paus. 3.16.2; 6.24.10). Moreover, "the essential feature of this type of ritual is the weaving and dedication of the garment" (Mansfield 1985, 443). Although dedication of garments to gods was a regular practice, the ritual weaving of textiles for the statue or temple was not common in every sanctuary.

The best known to us is the ritual of the peplos for the Athena Polias in Athens, woven every year by young girls called Arrhephoroi, who were assisted by Ergastinai. Together with the priestesses of Athena, the Arrhephoroi warped the loom on the day of Chalkeia, a festival of Athena Ergane and Hephaistos; then they wove the peplos (Mansfield 1985, 260, 262, 283). The Ergastinai spun the yarn for the peplos and may have assisted in its weaving (Mansfield 1985, 279, 285; Lefkowitz 1996, 79). It is possible that the two terms are synonymous, rather than titles of different groups of workers. On the other hand, because spinning takes a longer period of time, it is easy to imagine a more numerous group of women (such as the group of over a hundred Ergastinai mentioned in the late 2nd century BCE decrees IG II/III^2 1034 and 1036b?) who performed the task. The weaving, on the other hand, could easily be accomplished by two or four Arrhephoroi under the supervision of the priestess(es). The peplos took nine months to complete (Mansfield 1985, 283, but cf. n. 91). During the time when the peplos was woven, the Arrhephoroi lived on the Akropolis but it is not known whether the weaving also took place there. On the Athenian Akropolis, a small square building constructed at the end of the 5th–beginning of the 4th century BCE has been identified as the House of the Arrhephoroi (Stevens 1936, 490 fig. 40; Mansfield 1985, 275–276; Lefkowitz 1996, 79). It is possible that the loom was set up in the Temple of Athena Polias. The peplos was dedicated and placed on the statue of Athena Polias during the Panathenaic festival. The robe, supposedly being folded by an Arrhephoros and the chief priest during the Panathenaic procession, is probably shown on the Parthenon Frieze (Barber 1992a, 113 fig. 72; Neils 1996, 185–186 and fig. 8.6).

At the temple of Hera at Olympia a robe was woven for the goddess every fourth year by the Sixteen Women representing the cities of Elis (Paus. 5.16.2–3; Scheid and Svenbro 1996, 10–13). A special house located in the agora of Elis was assigned for the task (Paus. 6.24.10; Scheid and Svenbro 1996, 11).

A similar tradition may have existed in Argos where a *patos* robe was woven for Hera at the Argive Heraion (Mansfield 1985, 465–466 no. 8; Scheid and Svenbro 1996, 23). Besides the literary sources, here we also have archaeological evidence in the shape of textile implements: 8 spindle whorls, 54 loom weights, and 227 spools were found in the sanctuary (Waldstein 1905, 15, 43–44 nos. 283–292).

Finally, in Sparta, a chiton was made for the Apollo at Amyklai every year. Pausanias claims that the building in which it was made was located in the northern part of the city and was also called Chiton (Paus. 3.16.2).

It is notable that, in all cases, special structures existed for the production of sacred cloth and in all cases such a structure was within the sacred precinct of the temple. Thus we should re-examine large accumulations of textile tools (especially loom weights) excavated in votive deposits, as they may be evidence of production rather than dedication.

Archaeological evidence for textile production in sanctuaries

Several textile workshops have been identified archaeologically in sanctuaries. In the so-called Weaving Hut phase of *edificio* V at Francavilla Marittima (Figure 120), dated to the second half of the 9th and the 8th centuries BCE, a row of meander-decorated loom weights found *in situ* provides evidence for textile production in cult buildings already in the Early Iron Age (Maaskant-Kleibrink 1993, 4; Attema *et al.* 1998, 337; Kleibrink 2001, 49). The excavator of the site, Marianne Kleibrink, believes that "the looms associated with this house indicate special weaving activities" (Kleibrink 2001, 49). Because subsequent building phases of this building are demonstrably a succession of temples, it is usually assumed that the hut, too, was a temple, dedicated, according to Kleibrink, to the 'Lady of the Loom' (Kleibrink 2001, 48–52). The 7th century BCE terracotta pinakes from the site, identified as 'Athena seated in her house', have helped to identify the cult in the colonial period (Zancani Montuoro 1972; Maaskant-Kleibrink 1993, 8). In these depictions, Athena is shown seated and is holding a folded peplos in her lap. Another, contemporary terracotta image from the site, often called 'The Lady of Sybaris' (Figure 3) and identified as Athena, shows the goddess standing, wearing a skirt decorated with figural bands (Maaskant-Kleibrink 1993, 9 fig. 7a–b, 12). Terracotta figurines produced at the sanctuary during the 6th century BCE depict dedicants holding cloth (Kleibrink 2001, 51, 53 fig. 10). Other *pinakes* with the 'Ladies procession', believed to stem from Francavilla Marittima, may indicate a ritual procession connected with the dedication of the peplos to the goddess, not unlike the Panathenaic procession in Athens. The finds of spindle whorls and loom weights throughout the sanctuary confirm its association with the craft of textile production, and it seems likely that a special robe may have been woven for the patron divinity of the sanctuary starting in the Early Iron Age.

Later periods provide even more evocative examples. Large numbers of loom weights were found in and around the *Oikos* and the Rectangular Hall in the sanctuary of Santa Venera at Paestum. This shrine was dedicated to Aphrodite-Hera and was active during the late 5th and throughout the 4th century BCE. The Rectangular Hall yielded 93 weights, 47 of which were found in a group (Pedley 1990, 140, 148; 1993, *passim*). They were identified as votive offerings on the basis of the assumed sacred function of the buildings and the fact that they were not systematically arranged, as they would have been if in use on a loom. The context is, however, not primary, as far as one can judge, and none of the weights was found in accumulations consistent with a votive or foundation deposit.[9] It remains a possibility that loom weights may have been utilised in the production of sacred garments within the sanctuary (cf. Greco 1997, 196).

Another sanctuary in the area of Paestum, the Heraion alla Foce del Sele,

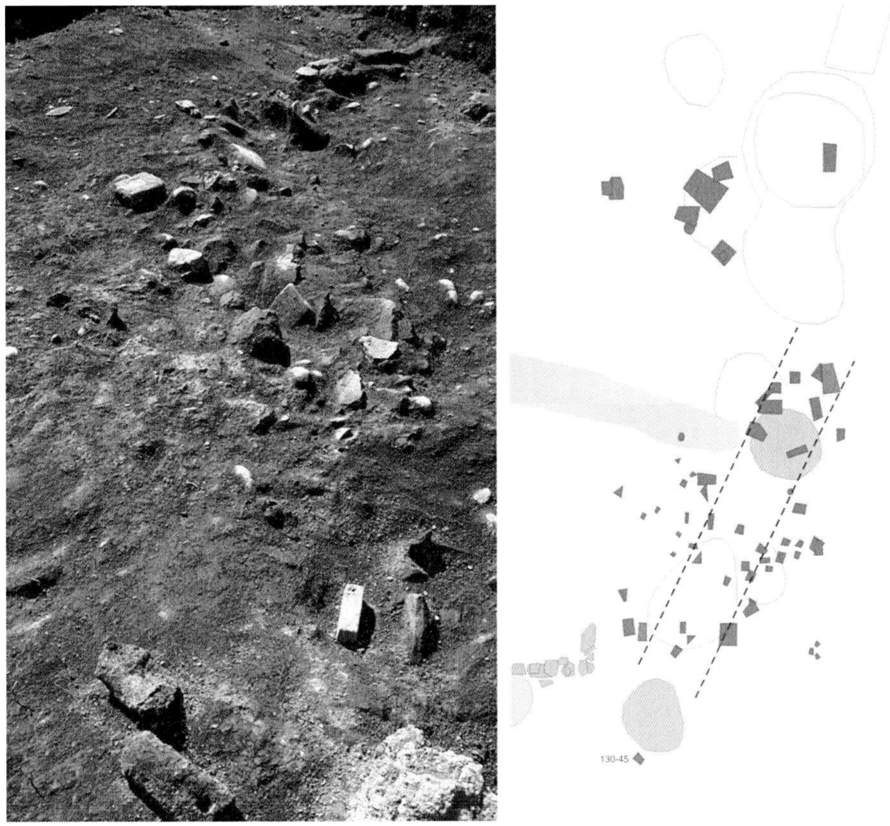

a b

Figure 120. The row of loom weights in the "Weaving Hut", Francavilla Marittima, 8th century BCE (Courtesy of Marianne Kleibrink).

also produced a significant number of loom weights. Specifically, 300 weights were found in the so-called Square Building, which has been dated to the late 4th–early 3rd centuries BCE (Zancani Montuoro 1966, 77; Greco and de la Genière 1996, 231–232 nos. 172.22–174.49). These objects, although initially published in an exemplary fashion, were then erroneously interpreted as instruments for measuring weight (Zancani Montuoro 1966, 61) – an unlikely interpretation since, usually, such instruments were made of stone or metal, materials that gave a more precise weight. A recent reassessment of the material established that, in fact, the Square Building might have functioned as a sanctuary textile workshop, which produced sacred garments, possibly for some sort of peplophoric ritual (Greco and de la Genière 1996, 225; Greco 1997). The number of loom weights indicates that at least three large looms could have been set up inside the workshop.

At Halae, also in South Italy, a large number of discoid loom weights were found inside room F' of the 4th century BCE West Building at the North Gate. The room may have been a textile workshop of the sanctuary of Athena (Goldman 1940, 479, 509–513; Mingazzini 1974, 208 note 24). One of the inscriptions found at the site mentions πεταμνυφάντεραι, interpreted as the weavers of the cloth, who may have been involved in official capacity in weaving a robe for the Athena of Halae (Goldman 1915, 448; Goldman 1940, 401).

Finally, Pratica di Mare (Lavinium) yielded more than 350 loom weights from the 5th–3rd centuries BCE, of which more than 200 came from the Archaic Building to the NE of the altar complex although a cultic function for this structure is not unequivocally accepted (*Enea nel Lazio* 1981, D189; Bouma 1996, 391). The number of loom weights strongly suggests a weaving activity within the sanctuary, and, given this sanctuary's association with matrimonial rites such a state of affairs should not be surprising.

So far, Pratica di Mare is the only non-Hellenic site providing evidence of textile activities in its sanctuary. However, written sources indicate that in Rome, the offering of a cloak to the statue of a goddess was performed in times of crisis (Scheid and Svenbro 1996, 40). In one recorded example, in the year 125 BCE, an oracle prescribed the offering of richly decorated garments to Persephone in order for the misfortunes to end (Paus. 5.16.5–6; Scheid and Svenbro 1996, 36). A century later, during the Second Punic War, Roman matrons offered a *palla* made with golden threads to Juno (Sil. Ital. *Pun.* 7.77–83).

Literary and archaeological sources, thus, provide ample evidence for the existence of temple- or sanctuary-based textile workshops in Italy. Specifically, such workshops are present in sanctuaries of Athena/Minerva and Hera/Juno. It may be argued that ritual weaving was a practice brought to Italy by Hellenic colonists. However, the Weaving Hut at Francavilla Marittima suggests a much older tradition in South Italy and evidence from Lavinium shows that similar practices existed in Archaic Latium. It is probable that future review of material excavated in Etruscan and other Italic sanctuaries will show that textile workshops existed in non-Hellenic temples as well. Textile implements, though common in votive deposits, are rarely present in very large quantities, suggesting that when significant accumulations of loom weights and other tools are found in votive and sanctuary contexts, especially concentrated in small areas, they may indicate production as an integrated part of religious activity.

Shipwrecks

One last unusual context should be considered. A number of ancient shipwrecks have a small number of terracotta loom weights, often called sinkers or ritual objects. Thus, at Gela, the 6th century BCE ship included two truncated pyramidal weights, while one such weight was found on the 5th century BCE ship (Panvini 2001, 62, 95). While it is unlikely that these objects were used on a loom, Richard Ellis suggested that they could have been used for various braided cords used on a ship, such as sennits, gaskets, robands *etc.* (pers. com. Richard Ellis, 2006). Shipwreck context is, thus, an example of interpretational problems that arise if context is not considered carefully.

Conclusions

In Italy, textile tools occur in three principal contexts: settlement, funerary and votive. A study of the number, distribution and morphology of textile tools from settlements can provide important information about technology, place and scale of textile production, as well as the raw materials and the final products. The data necessary for statistical analyses include presence, distribution, and physical parameters, such as preservation, size, weight, use wear, and decoration of specific artefacts at a given site. An investigation of textile tools from burials, on the other hand, can furnish interesting insights into gender, status, skill and age of textile workers. The occurrence of weaving implements in votive and foundation deposits, and the identification of textile production areas in sanctuaries, emphasise the important social and religious implications of spinning and weaving activities. On a broader scale, implements can be used to track both chronological and technological developments in textile production. Through the diachronic approach, it is possible to examine the development of textile workshops from household production.

PART 6

Technology, Production and Trade

The period from the 10th through the 6th centuries BCE in the Apennine peninsula was the time of development from small villages of mostly egalitarian type, to large urban centres with social stratification and specialised crafts (Andersen 1997). Organised production intensified steadily during these centuries, as did commercial exchange throughout and beyond the Italian peninsular sphere. Archaeological evidence points to the development of new or more effective production processes, standardisation and the manufacture of objects for specific purposes, which occurred mainly under the patronage of elites who required "the production of status markers and prestige goods" (Nijboer 1997, 384). This chapter explores how technological changes and trade patterns affected textile production in Italy during the first half of the 1st millennium BCE.

Technology and its changes

Technology may be defined as a "corpus of artefacts, behaviours, and knowledge for creating and using products that is transmitted intergenerationally" (Schiffer and Skibo 1987, 595). Thus, information about technology can be deduced from every archaeological artifact. Artefacts, in turn, represent a part of technology which can be recovered archaeologically, and, through their variability, attest technological changes. Aspects of artifact variability may be divided according to style-function dichotomy as having "utilitarian function (techno-function), socio-function and ideo-function" (Schiffer and Skibo 1987, 596). Thus, in the case of a loom weight, its weight reflects its techno-function, while its shape and decoration are related to its socio- and ideo-functions. In a bronze spindle, the ideo-/socio-function supersedes the techno-function, since its material is not suitable for textile production but reflects its value and, hence, symbolic meaning. It is this combination of artefact functions that constitutes the main source of technological change, evidence for which comes from the study of the formal properties of artefacts and their variability through time (Schiffer and Skibo 1987, 598, 600). These changes in technology may be caused by experimentation, change in demand involving function of the product, or economic processes such as competition (Schiffer and Skibo 1987, 598). As Johannes Nijboer points out, all of these factors can be discerned in pre-Roman Italy, especially during the 7th and 6th centuries BCE, a period of important cultural transformations (Nijboer 1998, 20).

The physical parameters of textile tools also reflect the technical expertise of the spinner/weaver, which touches upon a cognitive aspect of archaeology (Renfrew 1982). The concept of *chaîne opératoire* or logic of production has been prevalent in the studies of stone tools and ceramics (Lemmonier 1990). However, the manufacture of textiles, like no other material, requires both 'conceptual, abstract knowledge' and 'practical or procedural know-how' (Schlager 1994). Through understanding of the techniques associated with the making of textiles, the tools can be related to human actions and thus present a means of examining the cognitive aspects of technology.

While the processes and artefacts involved in textile technology were described already, archaeological evidence for the changes that occurred on various stages of production during the first half of the 1st millennium BCE is considered below.

Changes in animal husbandry

Procurement of raw materials is the first step of any craft. While the data are insufficient to support any conclusions about linen, changes in wool production can be traced archaeologically through bone assemblages. Thus, analysis of the animal bone data from the settlements of Central Tyrrhenian and Adriatic Italy showed an increase in ovicaprids "from 23.1 per cent in the Middle Bronze Age, to 34.2 per cent in the Recent Bronze Age and 49.7 per cent in the Final Bronze Age" (de Grossi Mazzorin 2001, 325–326, tbl. 2). In Iron Age Ancona, for example, Villanovan layers yielded 34.77 per cent ovicaprids versus 19.01 per cent in the preceding layers (Percossi Serenelli 1998, 37). Similar situation is observed at Moie di Polenza (Wilkens 1992, 39). This increase in sheep numbers has been interpreted as part of wealth accumulation of the emerging aristocracy (de Grossi Mazzorin 2001, 325). More significant are the mortality data, which indicate that, starting in the Early Iron Age, sheep were being kept increasingly for wool and other secondary products, as shown by the large numbers of older animals present in the assemblages (Barker and Rasmussen 1998, 73; de Grossi Mazzorin 2001, 326–327, fig. 1).

More complex is the question of breeds. The existence of numerous sheep varieties by the Roman period presupposes their development throughout the 1st millennium BCE. It is likely that, already in the Early Iron Age, intensive selective breeding was aimed at producing fleeces with specific qualities that would permit manufacture of highly specialised and differentiated textiles. More extensive fibre analysis of the existing textile corpus is needed in order to understand the development of wool in pre-Roman Italy.

Increase in number of tools and their standardisation

The increase in sheep bone percentages correlates with the large numbers of textile implements found on settlement sites. Large concentrations of spindle whorls and loom weights indicate a greater intensity of spinning and weaving activities. The increase in the numbers of textile implements is accompanied by their standardisation in shape and size and particularly in a general decrease in size, indicating finer products and higher expertise. At Murlo, for example, the vast majority of spindle whorls are small and more than 90 per cent of them are of the same truncated conical shape. Loom weights found at the contemporary site of Acquarossa, similarly, have very uniform sizes and shapes. Since such uniformity is not essential for weaving itself, it is likely due to the general standardisation of loom weight production. Compared to the preceding periods, implements were probably increasingly produced by specialist craftsmen, such as potters and coroplasts in the case of clay tools, and smiths in the case of metal implements. Standardisation of tools thus indicates that they were produced on a much larger scale than before, corresponding to the intensification of textile making activities and a demand for the appropriate equipment. Such a demand would be especially high in urban areas.

New tools

In addition to the development of the old implements, certain new tools make their appearance during the period under discussion, indicating new techniques and, consequently, new or improved products. Thus, shears came into use in the Iron Age, revolutionising the way raw wool was collected and triggering development of new breeds of sheep with non-shedding fleece (Ryder 1992, 137). The introduction of shears likely resulted in the intensification of wool production. The appearance of spools in the Final Bronze Age corresponds to the advent of tablet weaving in Italy. The fact that they are found in earlier contexts in the north seems to indicate that the technique arrived from Europe via Alps. And, in fact, similar spools are common in the Swiss and Austrian Bronze Age contexts. The presence of spools in the Early Iron Age contexts of Calabria points to a quick spread of this new technology throughout the peninsula, likely to have occurred through intermarriage.

New textiles: twill and other 'standard' weaves

One of the major turning points in textile history is the appearance of twill. Twill comes into common use during the Urnfield period in Europe (9th–8th century BCE), starting what Bender Jørgensen has termed 'the twill horizon' (Bender Jørgensen 1992, 120). Different types of twills developed in different geographic areas. In Italy, spin-patterned twill in single yarns of medium or fine quality is present in its fully developed form at Sasso di Furbara and Verucchio, that is, by the 9th–8th century BCE. This type of twill is otherwise common in the Eastern Hallstatt area (Bender Jørgensen 1992, 122–123): the implication is that Italian textile technology may have been influenced by the Eastern Hallstatt traditions (cf. Bonfante 1976, 1981; *AttraVerso le Alpi* 2002).

Twill requires a more complicated loom with four or more sheds, marking the appearance of more complex technology. Multiple sheds may result in more than two rows of loom weights, which should be archaeologically recoverable in primary settlement contexts. Since fewer threads would be attached to each loom weight, the implements would decrease in size and increase in number. Being a denser kind of weave, twill requires more yarn, leading to more intensive spinning activities. Finally, twill is especially suited for wool due to its elasticity (Bender Jørgensen 1992, 120). Thus, the growth in percentage of ovicaprids in the bone assemblages and the increase in implement numbers, documented at the same time as twill makes its appearance throughout Europe, may indicate a cause-effect relationship between the new weaving technique and changes in textile production in early 1st millennium BCE Italy. Whether the development of breeds with finer wools in the course of the 2nd millennium BCE led to the appearance of more complex weaves remains to be seen.

In addition to a variety of twills, other types of weaves become 'standard' in various areas of Europe in the early 1st millennium BCE. One such weave, common in Central Europe, is a tabby with plied yarn in one or both systems (Bender Jørgensen 1992, 122). This type of weave is already present in Italy at

Sasso di Furbara. Plain tabby z/z fabrics found in Central Europe are often linen (Bender Jørgensen 1992, 125). Such z/z tabbies, frequently woven in rep, have been identified in Orvieto, Veio and Satricum.

Another technical innovation, as mentioned above, was a tablet-woven textile. The finds at Verucchio and Sasso di Furbara fit well within the group of tablet borders in Central Europe (Bender Jørgensen 1992, 123; Ræder Knudsen 2002, 232).

The presence of these 'standard' Central European weaves in Italy indicates a wide spread of textile techniques and far-reaching networks of material and information exchange. The connection with Central Europe is not surprising since, already in the Early Bronze Age, Italian textile technology demonstrated many similarities with the Neolithic and Bronze Age Swiss material (Barber 1991, 174; Rast-Eicher 1997). Italy appears to have shared so many of its textile traits with Central Europe that it seems most likely that it is from this region that it adopted its weaving technology. The transfer of textile technology probably occurred through long-distance marriages and slave trade, phenomena both of which require further investigation. Elite gift exchange is also likely to have played an important role in the circulation and spread of the new textile types. It is no coincidence, that some of the earliest and most sophisticated twills have been found at the site of Verucchio, which had extensive trading connections with Central Europe, as indicated by the astounding quantity of Baltic amber found on the site. Already during the Late Bronze Age, the area had one of the major European amber-trading centres, Frattesina di Fratta Polesine (Bellintani *et al.* 1986; Càssola Guida 1999, 63–64). It is likely that textiles arrived in Italy by the same routes which also brought this highly prised substance.

Production

The manufacture of the high-quality 'standard' weaves requires a more organised textile production process. Each stage of this process is "dependent both on the assured supply of raw materials or unfinished goods from the preceding operation, and on the steady demand for its own products from the one which followed" (Bender Jørgensen 1992, 126). Technology combined with social relations defines a particular mode of production (Sassaman 1992). Unlike many other specialised crafts that appeared in Italy during this period, textile production was not a new craft. Instead, part of the production shifted from making subsistence products to the manufacture of non-essential or luxury goods. Thus, in addition to the adoption of new weaving techniques, technological changes were also induced by an organisational shift in production (Schiffer and Skibo 1987, 601). In the period between the 10th and the 6th centuries BCE in Italy a change in purpose, intensity and scale of organisation of textile production defined changes in textile technology.

Modes of production

Different modes of production have been suggested for the ancient crafts: the majority of theories deal with ceramic manufacture. Thus, Peacock, in analysing Roman pottery production recognises the following eight modes: household production, household industry, individual workshops, nucleated workshops, the manufactory, the factory, estate production and military and other official production (Peacock 1982, 6–11). Arnold simplifies the list to include only four modes of production: household production, household industry, workshop industry and large-scale industry (Arnold 1985, 225–231). While more generalised, Arnold's system is more suitable to the pre-Roman societies under consideration and less restrictive in terms of its applicability to other crafts and, specifically, to textile production.

Household production is the basic mode in which each household produces what it requires for its own consumption. Household industry produces not only for its own use, but also for sale/trade. In a workshop industry, the product is manufactured for sale and requires increased efficiency and specialisation on the part of a craftsman. Finally, in a large-scale industry, production takes place outside the household on a full-time basis and exclusively for sale, and requires capital investment and extensive product distribution.

In contrast to the other crafts, specialised textile manufacture develops through intensification of production rather than through new technology. Furthermore, in comparison to most other crafts, household textile production in pre-Roman Italy was never supplanted or replaced by other, more advanced modes of production; rather they developed in addition to as opposed to instead of it.

Specialisation

The last three modes of production listed by Arnold require a lesser or greater degree of specialisation on the part of the craftsmen involved in the production process. Specialisation is a concentration on a specific type of production (Dark 1995, 135). Until the advent of urbanism, most communities were so small that full-time specialisation was unlikely.

Most discussions of craft specialisation in ancient societies involve the question of sedentariness (*e.g.* Nijboer 1997). Both Peacock and Arnold cite the degree of sedentariness as a major defining factor for each mode of production because of its direct relation to craft specialisation. However, while the dispute about itinerant versus sedentary craft specialisation has been relevant to other types of production, such as metal and ceramic, the specifics of textile manufacture practically require it to be sedentary, also due to its particular 'gendered' nature.

Specialists: craftswomen and men

Part 5 dealt extensively with the evidence for the seemingly virtually exclusive role women played in textile production in pre-Roman Italy. The written, iconographic and archaeological sources are unequivocal on this matter. By Roman times, although spinning was still done for the most part by women, Pliny (*NH* 19.18) states that, "spinning flax is a respectable occupation even for men". Weaving of linen was also done by both women and men during the Roman period, as confirmed by the epigraphic evidence (Vicari 2001). The appearance of specialised male professional labour, however, does not precede the 3rd century BCE, when a more 'industrialised' mode of textile production was developing (Larsson Lovén 1998, 76).[1]

Labour organisation

Unfortunately, we know next to nothing about the organisation of textile production in pre-Roman Italy. Anna Maria Bietti Sestieri postulates a division between spinners and weavers on the basis of burial evidence (Bietti Sestieri 1979, 110, 143; Bietti Sestieri 1992a, 102). Antonia Rallo, on the other hand, distinguishes between high status *dominae*, who worked wool produced by the household, and lower status women, who worked linen and who, in addition to garments, produced more specialised types of cloth for sails, tents and *libri lintei* (Rallo 1989c, 153). With the more complex mode of production, division of tasks was likely to occur, culminating in the Roman period with the establishment of specific textile professions, such as carder, comber, spinner, weaver, fuller *etc.*, all attested in literary and epigraphic sources (Larsson Lovén 1998).

Exchange

Most textile production remained local and "trade expansion… began only with specialised production of very high-value fabrics", accumulation of which was regarded as an indicator of status and wealth (Horden and Purcell 2000, 359). As Horden and Purcell put it: "It reflects control of labour and of the productive landscape, and access to many different sources of supply; it stores social credit in a long-lasting and relatively low-bulk form; it stock-piles materials which may sometimes be hard to obtain" (Horden and Purcell 2000, 357). The elaborate textiles of Verucchio, found folded (*i.e.* displayed) in a way that would emphasise the elaborate tablet borders, embodied their owners' wealth and were meant to reflect their high status (Stauffer 2002, 198, 203).

Trade was also generated by the intensification of production due to demands of the ever-growing urban centres where concentrations of people extended beyond the local resources. While some products were produced in the surrounding countryside, others had to be imported over long distances. The find of cashmere in Lattes, France, is an example of just how far certain textiles could travel (Moulherat and Vial 2000). Such highly prised luxury textiles are likely to have been disseminated in high-level gift exchange.

During the first half of the 1st millennium BCE, Italy was becoming an increasingly dynamic place, with constant movement of people and established, long-distance exchange routes, many of which formed a continuum since the Bronze Age or even longer. The discovery of the Similaun Man has demonstrated the previously unimagined mobility of the Neolithic populations across the Alps (Spindler *et al.* 1995). Qualitative changes in textile technology, documented by the appearance of new implement types or new tool shapes, often may reflect influences from other regions of Italy and Central Europe, and from the Eastern Mediterranean during the period of colonisation. In any investigation of the interactions between various indigenous cultures of Italy as well as their rapport with Greek and Phoenician visitors and with the rest of Europe, it is important to separate the various modes of exchange that existed at the nexus points, since the result of these interactions differed from place to place (Renfrew and Bahn 1996, 335–368).

Trade

The first mode of exchange is the actual trade, which may have involved finished textiles, as well as raw materials like wool and linen, dyes and mordants, and, to a lesser extent, certain tools, mainly those made of high-value materials, such as metals. Examples of this mode of exchange are known, for the historical period, from literary sources: Sibari was famous for its textiles, Taranto for its sheep and purple cloth, and Puglia for its wool.

As an item of consumption, textiles range between luxury and necessity and are susceptible to the creation of specialised products, the manufacture of which is narrowly localised. Such a situation creates demand and necessitates redistribution, resulting in textile trade.

Textiles themselves were probably

not traded extensively during the early Iron Age. Robes such as the Verucchio mantles must have been made for specific ceremonial purposes and would probably only have circulated through gift exchange. It is not until the appearance of workshops in the late 7th century BCE that cloth production for commercial purposes began. Textile trade in pre-Roman Italy has been tied to copper (Peroni 1971, 100), tin, salt (Wells 1980, 43–44), amber, slaves (Nash Briggs 2007) and other commodities. Trade seems to be indicated by the spread of fashion from what was now Etruscan Central Italy to the northern areas, as attested in the Situla art (Bonfante 1981, 21).

Trade does seem to be indicated in the case of linen. The extant corpus of archaeological textiles indicates that linen was used extensively in Central Italy, yet it is highly unlikely that it was produced there. I believe that the majority of linen was brought to Etruria either from the Po region or from Campania, through trade routes established in an earlier period. Given the long duration of processing needed for linen fibre, it is probable that, unlike wool, linen was sold by the producers in a half-finished or finished state.

As far as the wool is concerned, an important factor that needs to be taken into consideration is transhumance, "the seasonal movement of animals from lowland to adjacent upland and back", which often involves great distances, and "the routes taken by the herders and their animals become major thoroughfares that may remain in use for centuries" (Horden and Purcell 2000, 63). Changes due to increasing population and agricultural pressure may have affected these patterns, which, in turn, influenced how and where wool was shorn and prepared, creating a complex system of raw wool exchange that continued into Roman times (Jongman 2000, 189).

While the above cases were dependent on the Italian peninsular network of exchange, textiles certainly travelled much longer distances on the pan-Mediterranean trade routes. The purple cloth was probably the most notorious item involved in long-distance trade, exchanged since the Bronze Age, as indicated by the ancient Near Eastern texts. In the early 1st millennium BCE, purple cloth was brought to Italy by the Phoenicians and traded as a finished product, together with other typically Phoenician luxury items, such as glass beads, faience trinkets and metal vessels. By the end of the Early Iron Age, the Phoenicians brought the technology of purple production to the West Mediterranean and established production sites in Italy, which continued to supply the long-distance purple cloth networks.

Gift exchange

The second, reciprocal, mode of exchange was gift giving, linked to the traditions of hospitality. This mode of exchange was practiced among the elite members of communities and hence involved luxury materials that certainly included clothing and textiles, but also implements that held a symbolic value, such as bronze distaffs. The presence of single bronze distaffs in the Calabrian Iron Age cemeteries of Torre Galli and Pantano di Cleto – Serra Aiello may indicate such an exchange, as both sites have been shown to have extensive ties with many regions throughout Tyrrhenian Italy, but especially with proto-Etruscan areas. Meanwhile, until more work is done on the surviving textiles, one is left to hypothesise that the presence of non-local fibulae in burials might indicate that they came with imported garments.

Information exchange

The last and most complex mode of exchange existed on the level of information or textile-related know-how involving ideas, symbols, inventions, fashions, values and, consequently, the technology and tools associated with them. The role of migrant craftsmen in spreading new technologies has been highlighted for ceramics, metalwork, and other crafts. In the case of textile production, since women were the principal agents of the craft, this last mode of exchange was probably often carried out through intermarriage. The practice is documented in the presence of burials of indigenous type in allogenous necropoleis, such as those at Pithekoussai (Coldstream 1993) and Pontecagnano. Because women were associated with cloth manufacturing activities, textile equipment allows us to track their movements, as well as the diffusion of technology. Women took with them the know-how that included 'recipes for action' and tools, and transmitted it to the next generation.

For example, women's burials at some of the sites in Campania and Calabria often include spools, which also appear in Iron Age contexts of North and Central Italy. The Campanian spools obviously indicate yet another connection with Etruria. But the ones found in the necropoleis of Torre Galli, Canale, Calanna and Ianchina in southern Calabria may suggest that technological exchange was the consequence of involvement of these sites in a larger Tyrrhenian trade network. Even more telling is the presence of such indigenous spools in Greek contexts at Incoronata/Metaponto.[4]

Another example illustrates the diffusion of values and the symbolism associated with textile production. In contrast to Central and North Italy, the cultures of Puglia, especially the northern, Daunian area, did not differentiate their dead by the role they played in society, and did not include textile implements amongst female burial gifts. In later periods, however, burials containing textile implements do appear and their number increases towards the Hellenistic period. For example, at Oria, Tomb 20, dated to the 5th century BCE contained 22 pyramidal loom weights of different sizes, 19 of which were decorated with animal stamps. It is likely that the practice was transmitted to Puglia from other cultural spheres.

Conclusions

The overview of the archaeological evidence for textile production in settlements, presented in previous chapters and the theoretical considerations discussed above, allows me to draw some preliminary conclusions about the development of textile production in Italy over time.

Apart from few exceptional sites, there is no evidence for major craft centres during the Bronze Age and textile production seems to be concentrated in the individual households, indicated by small groups of loom weights and single spindle whorls found throughout the settlements. Towards the end of the 2nd millennium BCE, there is a small increase in the numbers of implements, but production appears to be still geared towards making basic subsistence products for the individual household. The Early Iron Age brings more variety in terms of tools that correspond to a wider variety of textiles produced. It is also the period when textile implements become common in burials, indicating a shift in social perception of the craft and, possibly, specialisation among some members of the community. Textile production has moved to a new level, becoming one of the main economic activities and sources of wealth.

It is during the Orientalising and Archaic periods that we see a significant increase in the scale of textile production, indicated by the large number and standardisation of tools, as for example at Poggio Civitate. Frequently, textile implements are concentrated in specific areas where other kinds of production, such as ceramic or metal, have been documented, providing evidence for a household or even workshop mode of manufacture and the existence of at least part-time specialist craftspeople. Specialised production certainly took place in the sanctuaries, as indicated by the large numbers of textile implements concentrated in specific areas of the sacred precincts, like st the the Santa Venera sanctuary at Paestum or at Lavinium. In the same period, textile implements start appearing in votive contexts. At some sites specialisation begins in the manufacture of a particular kind of cloth, which is even more evident in the post-Archaic period, when relatively large-scale textile production can be postulated for some sites on the basis of archaeological evidence.[2] Intensification of textile production to increase the output for the purpose of exchange can be achieved very quickly without economic or technological revolution (Horden and Purcell 2000, 352). In this way, demand stimulated the specialisation of certain sites in textile manufacturing. Thus, at Gravina di Puglia, sheep husbandry was practiced in the Early Iron Age but an economy based on wool did not develop there until later periods (Watson 1992, 97). Large numbers of loom weights excavated in the Hellenistic levels of the settlement indicate a large-scale textile production for trading purposes.

Other textile-producing centres are known from literary sources; among these are Tarquinia, Sibari and, especially, Taranto. Thus, while the archaeological material for Taranto is problematic, we know from ancient authors that sheep husbandry produced the celebrated wool that was worked and transformed on the spot,[3] while on the coast of Mar Piccolo there existed dyers' workshops that produced purple (Forbes 1956, 136, 162; Morel 1978, 104ff; De Juliis 2000, 81). There is a mention of an estalopia or a clothing market, which may have been located in the agora (De Juliis 1984, 331). During the Hellenistic period, Taranto also may have been the centre of gold thread production (De Juliis 1984, 331).

While there is no evidence that textile production in Italy ever reached an industrial scale of organisation before the Roman period, there is strong indication of a manufacture mode, which greatly exceeded in quantity the simple subsistence production.

PART 7

Coda: Textile Production in its Social Context

Among the various sources that provide evidence for the study of textile production in pre-Roman Italy, archaeological material is the most direct as it documents the productive activities through tools, installations, raw materials and finished products. With the help of other, secondary sources of information, namely the representational material, ancient literature, experimental archaeology and ethnography it is then possible to reconstruct the sequence of production processes involved in textile manufacture. As one of the most important and labor-intensive crafts of the ancient world, textile production had great social significance in pre-Roman Italy. This is expressed in funerary ritual through the inclusion of textile implements among the burial goods, as well as in religious activities through the deposition of textile tools in votive and foundation deposits. Textile production was also an integral part of local and regional economies and local, regional and international trade. To set textile production in its social and economic context it is vital to investigate how textile technology relates to such important concepts as urbanisation, craft specialisation, gender, exchange, agricultural and cultural development.

The Early Iron Age (1000–750/700) was a time of geopolitical and social restructuring in Italy, characterised by the formation of proto-urban communities (Pacciarelli 2000). A rapid and growing socio-economic transformation followed during the second half of the 9th and the first quarter of the 8th centuries BCE. As Pacciarelli notes, "one of the motors of this change was the flourishing of trade and exchange fed both by these enormous proto-urban sites, which were natural places for 'markets' and craft production, and by the dominant rank's demand for luxury goods" (Pacciarelli 2000, 284). These luxury goods included textiles. The development of this uppermost social group reached its acme at the end of the Early Iron Age and is documented by the appearance of extremely rich tombs.

Apart from arms and armor, one class of objects stands out within these wealthy burials, namely implements associated with textile production. Archaeological, representational and literary evidence indicates that, in ancient Italy, spinning and weaving were practiced primarily by women (Barber 1991, 283–298; Bartoloni 1989; Torelli 1997; Toms 1998; Nielsen 1998b; Jannot 2005). Already at the end of the Bronze Age, the presence of spindle whorls, distaffs and spools in female funerary assemblages testifies to women's contribution to communities as textile workers, and indicates that the textile craft became a symbol of the female sphere. During the Iron Age, finds of spindles and distaffs in precious materials, such as bronze, silver, amber, and bone, also support the notion that these implements were important markers of elite female status.

The iconographic evidence, which is of great help to our understanding and reconstruction of this ancient technology, shows that there is still more complexity inherent in the symbolism/tradition of cloth production. The Bologna *tintinnabulum* (Figure 8) features four scenes depicting various stages of textile manufacture. Another important visual document is the wooden throne found in Tomb 89 at Verucchio (Figure 7). Both objects, the *tintinnabulum* and the throne, are prestige objects in their own right, indicating not only that textiles constituted a source of wealth and/or status for their owners, but also that specialised textile production, specifically the manufacture of ceremonial garments, was the prerogative of the elite women represented.

Textiles found at Verucchio may actually be such ceremonial garments. The mantles are at least 260 cm wide and 70 cm long and it appears that the weaving of this garment type respected some standard dimensions (Stauffer 2002, 197; Eles 2001, 265–266). Annemarie Stauffer, who has analysed these textiles, has suggested that the warp is the width of each mantle and thus that the loom on which these mantles were produced must have been very high indeed. Thus, the tall looms, depicted on the *tintinnabulum* and the wooden throne, are true-to-life renditions of looms observed in life. The specialised technique and the equipment are proof that such mantles could be woven by one person, thereby maintaining the exclusive nature of such work.

Both mantles have elaborate borders featuring a triangle motif and three horizontal lines made by the tablet weaving technique, in this case utilising 36 tablets with four holes each along the perimeter (Ræder Knudsen 2002, 222–225). Similar borders are well illustrated in later representational material. In fact, the *tebenna*, the Roman descendant of the Verucchio mantles, retained the border as the status symbol, in this case dyed purple (Bonfante 1975, 15, 39, 45, 48–55, 102; Granger Taylor 1982). The scenes on the *tintinnabulum* and the Verucchio throne demonstrate that only privileged persons were allowed to weave these garments, and it seems that this exclusivity was applied from the initial stages of preparation all the way through to the finishing process.

If such indeed was the case, then the presence of textile implements in the Iron Age female burials, especially the inclusion of the spools, highlights a new and crucial aspect of this craft. I have argued that these spools were in fact used as small weights for the sets of threads passing through the tablets used to make the borders on the Verucchio textiles. Tombs containing spools and multiple spindle whorls are both few and rich, thereby implying that elite women possessed the exclusive right to use such tools to make special ceremonial garments with tablet-woven borders.

Although quite distant from each other geographically, the textile finds from Sasso di Furbara and Verucchio show many similarities, including the use of thread with varied spin direction to create a pattern and the tablet weaving technique. They demonstrate that, by the 8th century BCE, a sophisticated technology with the capacity to produce highly complex and labor-consuming luxury textiles existed in Italy. The production of these textiles not only required highly specialised materials and skills, available only to the members of the elite classes of society, but also was in itself exclusive elite female occupation marked at death by the funeral deposition of associated tools.

Despite the high degree of specialisation that characterises this type of textile production in the Early Iron Age, it remained confined to the household level, as indicated by the regular finds of small quantities of textile instruments in settlement sites. By the 7th century BCE, however, a new production mode seems to come into play, with the appearance of certain sites, such as Murlo and Acquarossa, where enormous quantities of tools were found concentrated in small areas or in specific structures. The size, shape, material and, often, decoration, of the tools themselves show increasing standardisation, and they were most likely produced by specialists. What we seem to have now is something more akin to a household or even a workshop industry mode of production. This change coincides with the specialisation and professionalisation of other crafts, most notably metallurgy and ceramic production (Nijboer 1998).

By the mid-7th century BCE, after a long period of urban development throughout the Early Iron Age, large settlements emerged as centers of production and trade. Socio-political power became more and more concentrated in the hands of wealthy families who controlled both trade and production. Changes in economic demands led to "a gradual shift from production of luxury goods to subsistence goods and intensification of local specialised production" during the 7th and 6th centuries BCE (Nijboer 1997, 400). The picture that emerges from the funerary context is one of a luxury textile production that is no longer controlled by the elites in the same way as it was in the Early Iron Age. The bronze distaffs that were common during the Villanovan period disappear by the 6th century BCE, as do the spools, which are found mainly in tombs dated from the 10th through 6th centuries BCE, with a few exceptional cases dated to the 5th–4th centuries BCE. They continue to be found on settlement sites, but gradually disappear by the Roman period. The weaving of ceremonial textiles is no longer the prerogative of elite ladies but is now handled by specialists or even slaves on a more 'industrialised' level.

No textiles have been found in later periods of as high quality as the Verucchio textiles, although this might be an accident of preservation. The purple border of the Roman *tebenna*, the descendant of the Verucchio mantles, was woven into the garment – a technique that was labor-intensive but nowhere near as extravagant as in the case of our Iron Age examples. It is tempting to posit an intermediate stage in garment evolution where the borders were tablet-woven separately and sewn onto the base textile, as in the case of the Hohmichele example (Banck-Burgess 1999).

It is hard to tell whether textile production was a household, a workshop, or a large-scale industry. Probably, depending on time and place, it could have been either one of these or a combination of all three. It is also not always clear when textiles were made exclusively for household consumption or for trade as well. What the existing archaeological evidence does show is that we are dealing with surplus production by the end of the 7th century BCE.

The textile craft becomes specialised in Italy with the transformation of the social structure from an egalitarian to an aristocratic community. As in the preceding periods, women continue to make textiles at home, but only the ladies of the highest rank possess the right (or the skill) to weave the ceremonial garments. As the power of the military aristocracy declined, social and political power was transferred to the more mercantile element of society, and large urban centers developed that were able to afford to have specialists to produce not only luxury but also subsistence goods. Textile production thus became an enterprise on a much larger scale, with archaeological evidence indicating that, with time, it moved from individual specialists in the early Iron Age to a specialised workshop-based manufacture during the Orientalising and Archaic periods. It should come as no surprise that the urbanisation of Italy was mirrored by a domestic environment in such an important craft as textile production.

Notes

Part 1

1 For a recent overview of the 1st millenium BCE Italian chronology, see Turfa 2006b.

2 The region of Emilia-Romagna, usually considered part of northern Italy, will be discussed in the section on Etruria because of its strong cultural associations with Etruscan culture.

3 See, for example, the catalogue of the major exhibition on Magna Graecia that took place in Venice in 1996 (Pugliese Carratelli 1996), where the native populations of South Italy and Sicily are defined exclusively in terms of the influence that Greek colonies supposedly exercised on them.

Part 2

1 Such, for example, is the case of Pliny the Elder's description of purple production. For analysis of the text, see Steigerwald 1986; Doumet 1999.

2 Bonfante 2003 with bibliography of author's works on ancient dress; see review by Turfa and Gleba 2005.

3 Depictions of elaborate carpets are known from Persia and Assyria (Canby 1971; Albenda 2005). The earliest such carpet was found in a 5th century BCE Scythian burial in Pazyryk, Siberia: see Rudenko 1970; Polosmak and Barkova 2005, 154ff.

4 A proliferation of conferences on archaeological textiles shows growing interest in and importance of the topic. Over the last few decades, regular meetings have been held by the North European Symposium for Archaeological Textiles (NESAT), Centre International d'Études des Textiles Anciens (CIETA), Ancient Mediterranean Textiles and Dyes Symposium, Textile Society of America (TSA) – all with regularly published proceedings. In addition, various topical conferences have been held, the most recent of which include 1999 Lattes colloquium *Archéologie des textiles des origines au Ve siecle* (Cardon and Feugere 2000); *Ancient Textiles: Production, Craft and Society* conference held in Lund/ Copenhagen in 2003 (Gillis and Nosch 2007a); the first symposium on the Hallstatt textiles held in 2004 in Hallstatt, Austria (Bichler *et al.* 2005). In addition, some of the major international conferences like European Archaeologists' Association (EAA) and Society of American Archaeologists (SAA) have organised several textile sessions since 1999, which have experienced great attention and showed that textile research has potential – also for scholars from neighbouring fields.

5 See, however, excellent early works by Davidson and Burr Thompson 1943; Davidson 1952.

Part 3

1 Two classifications of textiles exist at the moment: one based on finished structures (Emery 1966) and another on techniques (Seiler-Baldinger 1994). For comparison, see Speiser 1991. On weave recording systems, see Hodges 1976, 137ff.

2 Also see Stage 1985, 69–84; 1975, 11–14; Bonfante 1985; 2003, 213. One of the cases documented by Stage should be reconsidered. The 6th century BCE *ziro* burial from Chiusi published by Dohan (1935) reportedly contained three bronze objects bearing traces of the same fabric. These objects are a chair, a table, and a belt, all now housed in the University of Pennsylvania Museum, Philadelphia (Inv. Nos. MS 2310, 2362, 2363). I have examined the pieces and found no textile traces. Furthermore, the authenticity of these objects has been recently put to question (Warden 2003, 40).

3 Such, for example, is the case of the armour lining from the Tomba del Guerriero in Tarquinia (Cat. No. 50), which was never analysed or even photographed and is now lost (Stage 1985, 84).

4 Exact provenance of the find is unknown; reportedly, it came from a hoard found between Montemerano and Saturnia, but it may have come from a burial.

5 The provenance attribution is based on the epigraphic analysis of the text; see Roncalli 1981.

6 The urn, belonging to a woman as indicated by the inscription *PETRUI TELMUS* is without known provenance. However, a very similar vessel (MS 3428) with the name in the inscription indicating a relation to our urn (*CAES. V.V.TELMU*, the name of a freedman, probably her husband), has been attributed to Tuscania; see Turfa 2005, no. 103.

7 For the history of the fibre and the earliest finds of cotton in the Near East see J. Alvarez-Mon 2004: http://www.achemenet.com/ressources/souspresse/annonces/Cotton%20from%20Elam.pdf.

8 For a long time it was thought that silk was already used in Europe as early as the 6th century BCE, based on the finds of Hohmichele and Hochdorf in Germany, and Athenian Kerameikos in Greece (Hundt 1969; Barber 1991, 32). Recent re-examination of these finds casts doubt that the material is silk (Banck-Burgess 1999). However, Good (1995, 966) confirms silk identification of the Hohmichele and Kerameikos fibres based on amino acid composition analysis.

9 Samples initially identified as sea-silk among the materials recovered in the Vesuvius area (D'Orazio *et al.* 2000, 745), under more detailed examination turned out to be a sponge; see Maeder 2002, 10. The earliest sea-silk textile is considered to be the 4th century CE fragment from Aquincum (Budapest), Hungary. This mollusc fibre has been used until recently in Sardinia, Sicily and Puglia.

10 For example, in a publication that specifically deals with the Italian landscape, Barker and Rasmussen (1998, 186 and 193) explore sheep husbandry in conjunction with wool production on numerous occasions but do not even mention flax.

11 Seeds are usually preserved when burned; however, exposure to fire generally affects only those species that were cooked. Uncarbonised flax seeds require special conditions in order to be preserved. They have been found in Swiss Neolithic and Bronze Age lake dwellings (Jacomet, Brombacher and Dick 1991, 273).

12 Small seeds (2–3 mm), according to Renfrew (1973, 273), indicate flax cultivation for fibre, while larger seeds (4–5 mm) imply their use for oil extraction.

13 The surviving artistic representations of the garment leave little doubt as to its material. Linen cuirasses are represented in tomb paintings, as, for example, on the two Ajaxes in the Tomba François in Vulci (Buranelli 1986; Moretti Sgubini 2004a), on Greek soldiers on the Prest's sarcophagus found in Tarquinia (Blanck 1983) or on Geryon in Tomba del'Orco in Tarquinia (Steingräber 1985, 329–332, pl. 129). For more information on linen corselets see Törnkvist 1969 and Gleba forthcoming.

14 The Colosseum had giant sails attached to capstans, some of which are still visible outside the building, and which were worked as an honour by the naval veterans (Coarelli 2001, 197). There is also evidence from Pompeii where 'posters' on houses invite to the amphitheatre

by proclaiming '*vela erunt!*'; see Graefe 1979 for epigraphic and iconographic evidence.

15 In some areas of Roman provinces these also were made of leather (van Driel Murray 1990).

16 For a discussion of linen production in Spain, see Blazquez 1978, 425–427; Alfaro Giner 1984, 53–55.

17 The same cultivation technology is still used today in the main European flax-growing countries: Belgium, France, the Netherlands, the Baltic region and Russia.

18 The difficulties associated with growing flax were well known in antiquity. Columella (*Rust.* 2.10.17), Virgil (*G.* 1.91–94), and Pliny (*NH* 19.2.7), believed that flax seriously depleted soil. In reality, flax was not exhaustive, since the amount of other crops planted subsequently in the same soil is unaffected and the amount of elements taken from soil and air by flax is similar to that absorbed by other crops, such as wheat. Unlike other crops, however, flax is pulled and all parts of the plant are used since the stems are harvested for fibre. In this way, nothing of the plant is left in the soil to restore the nutrients taken up by plants. See Moore 1922, 83–84.

19 Wool was originally identified in much earlier Çatal Hüyük in Turkey (Burnham 1965, 170), but later proven to be flax (Ryder 1965, Vogelsang-Eastwood 1988).

20 Until recently, in Abruzzo and Molise for example, thousands of shepherds travelled twice each year down to Apulian plains and then back to the mountain pastures with millions of sheep.

21 The earliest example of cloth dyed by this method was found in a 4th century BCE burial at Kertch, on the Crimean peninsula of modern Ukraine; see Barber 1991, 226. Later, resist-dyeing was also used in Roman Egypt.

22 Recent tests found presence of dyestuff components in a fragment dated to the Bronze Age as well, dating back the dyeing technology of Europe (Grömer forthcoming).

23 The finds are on display at the Archaeological Museum of Padova. On the value and possible apotropaic function of shells in burials and their association with women, see Chieco Bianchi and Calzavara Capuis 1985, 75–76.

24 Vicari (1994, 246) suggests that two toponyms, Purpurano near Parma and Porporana near Ferrara, may recall existence of dyeing workshops in the area in Roman times.

25 A magnificent surviving example comes from the so-called Philip's Tomb in Vergina, where the cremated remains of a young woman were wrapped in a gold-and-purple tapestry (Andronikos 1984; Flury-Lemberg 1988, 483; Cardon 2007, 573).

26 Pliny the Elder (*NH* 19.57) demonstrates a misunderstanding of the process claiming that gold could be spun and woven like wool.

27 For an overview of ancient gold thread finds, see Gleba 2008.

28 In Tomb 6, for example, a bronze cauldron containing cremated remains was apparently wrapped in linen cloth; see Bloesch and Mühletaler 1967; Bérard 1970, 13; Kurtz and Boardman 1971, 53.

Part 4

1 In Scandinavia, flax was combed with tools that had long iron teeth. While no such tools have been recorded in ancient contexts of Italy, they are worth noting.

2 See Wild 1970, 27–29, for the description of the process in Roman times.

3 Only well-preserved examples from Ornavasso are listed in Table 2. Bianchetti (1895) states that at least 38 items (14 from S. Bernardo and 24 from Persona) were recovered.

4 Pollux (10.126), however, describes shears as a female tool.

5 Two of the tombs have been securely identified as male, and six as female; see Bechtold 1999, 232.

6 I do not consider here shears excavated in Roman contexts, such as a pair from Room 43 of the Casa del Menandro in

Pompeii; see D'Orazio and Martuscelli 1999, 92–94, 142.

7 Shears were found for example in Athenean Kerameikos, dated 360–350 BCE (Kovascovics 1990, 17 no. 9, pl. 30 no. 4, 87 no. 2, pl. 48 no. 11).

8 On carding and combing, see Wild 1970, 24–26.

9 Similar items in bronze have been found in Greece at Lindos (Blinkenberg 1931, 136 no. 343, pl. 13), Olynthos (Robinson 1941, 376–377, pl. 119 nos. 1884–1892), Corinth (Davidson 1952, 173 and 176 nos. 1223–1228, pl. 78 nos. 1223–1228) and Olympia (Curtius and Ader 1890, 61 no. 422, pl. 23), and at Ephesos in Turkey (Trinkl 2007, 84).

10 See general typology in Liu 1978, 88.

11 Interesting comparative data is offered by Parsons 1975, who studied the distribution of Aztec cotton and maguey whorls in the Mexico Valley and noted that while the cotton whorl types were virtually the same across the region, the maguey whorls could be divided into three distinct groups based on the decoration and each type correlated with a particular geographic area. The author concluded that while the cotton spinning industry reflected centralised tribute service, the maguey spinning was more of a local manifestation.

12 Some information exists, however, for sites in other areas of Europe and the Near East. The published results of the analysis of spindle whorls from the Late Chalcolithic site of Hacinebi Tepe, in South-East Turkey, show a range of 7.6–158.2 g (Keith 1998, 506–7). Whorls from the Neolithic site of Anau, in western Turkestan, range 12–40 g (Barber 1991, 392). The range of weights within Bronze Age Cypriote assemblages is 10–169 g, with only eight whorls above 100 g (Crewe 1998, 13). More specifically, at the site of Kition, they range from 14 to 120 g (Smith 2002, 301). Barber (1991, 392) also provides data for the Late Bronze Age whorls from Zürich area that weigh 15–57 g. In Early Iron Age Gordion, large percentage of spindle whorls weighed between 10 and 20 g (Burke 2007, 68). Last but not least, a collection of over 2000 Hallstatt period spindle whorls from Smolenice Molpír in Slovakia had an average weight of 16 g (Belanová 2007, 42).

13 Spindle whorls are often difficult to distinguish from beads. Most beads are less than 2 cm in diameter, so any larger object would most likely be a spindle whorl. At the same time, most whorls are more than 2 g in weight. See Liu 1978, 91.

14 Comparable items dated to the Roman times have been found in Austria (Grömer 2004).

15 Despite the Mediterranean influences assigned to this scene (cf. D'Ercole 2000, 337), the loom structure with the pit underneath has no comparisons in the Mediterranean world.

16 Another reason for the use of the warp-weighted loom to weave linen may be the fact that flax fibre does not stretch easily and the warp-weighted loom provided better tension while weaving than the vertical two-beam loom.

17 Apparently, similar pyramids made of wood were known to be used in such a fashion in some regions of Romania. However, this adds another step in the process of textile production when there is a way to avoid it. The shuttles shown resting between the warp threads in representations of looms like the Chiusi vase, look like sticks or spindle shafts with thread. So, instead of unwinding the thread from the spindle, the whorl can be removed from the shaft, which then can be used as a shuttle and would also be convenient to part the warp while weaving. See Barber 1991, 107.

18 For site typologies, see Accesa (Camporeale 1997, 198–203); Adelfia (Tunzi Sisto 1988, 25–28); Locri (Dotta 1989; Arboletti 1992); Broglio di Trebisacce (Peroni and Trucco 1994, 248); Cavallino (Pancrazi 1979, 190–191).

19 Discoid loom weights of different, unrelated types are found in the Bronze Age contexts of central-eastern Mediterranean, particularly on Minoan sites (*e.g.* Becks and Guzowska 2004).

20 For example, see reconstructions for looms in Serra di Vaglio (Greco 1991, 63 fig. 129 and 64 fig. 131); Kleinklein (Dobiat 1990, 83 fig. 5).

21 Weights stored away in a set have been found at the Bronze Age Akrotiri, on Thera (Tzachili 2007) and at Early Iron Age Gordion (Burke 2007, 68).

22 On reading of inscriptions on loom weights see Mingazzini 1974; Ferrandini Troisi 1986. On Etruscan sigla, see de Grummond, Bare and Meilleur 2000. Loom weights with inscribed or stamped names from various sites in Italy have been found, for example, in Taranto (Ferrandini Troisi 1986; Ferrandini Troisi 1992), Terravecchia di Cuti (Militello 1960, 56, pl. XII), S. Martino in Pensilis (Capini and Di Niro 1991, 182 d170, pl. 19d). Similar weights are found in Greece (Davidson 1952, 151 fig. 24, 157 fig. 27, 156–169).

23 For gem impressions in general, see Boardman 1970, 234–235 and 405; Rutschmann 1985. For gem impressions and other decorative motifs on loom weights from specific sites, see: Gravina di Puglia (Wilson 1977, 134 d; Tatton Brown 1992, 223, fig. 100), Armento (Di Giuseppe 1995, 145 fig. 112); Locri (Dotta 1989, pl. XL), Monte Sannace (Scarfi 1962, 161 fig. 153), Pomarico Vecchio (Sartoris 1997, 101 pl. 93), Oppido Lucano (Lissi Caronna 1980, 220 fig. 133; 1983, 301 fig. 90, 330 fig. 125, 343 fig. 140; Lissi Caronna, Armignacco Alidori and Panciera 1992, 236 fig. 60, 284 fig. 109, figs. 246–247). For gem impressions on weights found in Greece, see Athens (Davidson and Burr Thompson 1943, 83 fig. 34, 92 fig. 39), Corinth (Davidson 1952, 154 fig. 25).

24 For example, Fratte (Serritella 1990, 164 no. 128, fig. 273), Cozzo Presepe (Lo Porto 1992, 401 fig. 115 no. 5), Adelfia (Tunsi Sisto 1988, 30, fig. 4 nos. 22–27), Monte Sannace (Scarfi 1962, 162, fig. 152), Cavallino (Pancrazzi 1979, fig. 29), Gravina di Puglia (Wilson 1977, 133 fig. 56 nos. 5–7, 134 c; Tatton Brown 1992, 222–223, fig. 101), Oppido Lucano (Lissi Caronna 1980, 154 no. 7, 220 no. 2, fig. 134; 1983, 332 nos. 1 and 9, fig. 128; Lissi Caronna, Armignacco Alidori and Panciera 1992, 239, fig. 66, 284 nos. 32, 34, fig. 108, 465 no. 35, fig. 245).

25 *Astragalli* often appear as votive offerings in some sanctuaries and are commonly associated with the passage from childhood to puberty, which explains their deposition in children's burials, especially in Magna Graecia; see discussion in Baglione 1991, 661–662. It is possible that impressions of *astragalli* on loom weights destined for votive purposes have a similar significance.

26 See the catalogue nos. 1–4, 8–11, 13–17, and 22 in Ferrandini Troisi 1986 and 1992. For others, also from Taranto, see Wuilleumier 1932, 37–41. Granted that the interpretation of these inscriptions as makers/ owners marks is correct, it is not surprising to find such a large number of 'weavers' in Taranto, which was famous for its wool and textiles in antiquity.

27 In Slovakia, over 200 loom weights, found in two rows at the Hallstatt site of Burgstallkogel by Kleinklein, were marked with crosses, stars, circles and other signs on the top, which possibly indicated the order of attachment or the pattern (Belanová 2007, 43).

28 Wuilleumier 1932, 37ff, however, interpreted these inscriptions as abbreviations of weight values and, hence, unconnected with weaving activity. Meanwhile, Ferrandini Troisi 1986, 94–95, suggests that these inscriptions referred either to the various parts of a textile in the cases of large pieces of cloth woven by several workers, or to the placement of a weight on the loom (*e.g.* mid-way). Mingazzini's interpretation seems the most plausible. On Tarentine loom weights inscribed with abbreviation of the word ἩΜΙΩΔΕΛΙΟΝ, interpreted as markers attached by the customs collectors to the sacks of goods, see Gardner 1883; Wuilleumier 1932, 36, 49; Mingazzini 1974, 215; Ferrandini Troisi 1986, 95–97 and cat nos. 5–7; Ferrandini Troisi 1992, nos. 69–71.

29 Not surprisingly, pyramidal weights with an owl have also been found in Athens.

30 See Davidson 1952, 173–174, for Byzantine Greek examples, and Wild 1970, 65–67, for Iron Age British finds.

31 In fine textiles, the optimal tension should be about 10 g per thread; personal communication by hand weaver Anna Nørgaard, 23 March 2003.

32 Site typologies have been published *e.g.* for Osteria dell'Osa (Bietti Sestieri 1992, 314–316); Accesa (Camporeale 1997, 203–205); Gran Carro (Tamburini 1995); Tarquinia (Bonghi Jovino *et al.* 2001, 65); Cavallino (Pancrazi 1979, 190); Pontecagnano (d'Agostino and Gastaldi 1988, 38).

33 In Greece, a different type of coarse spool-shaped objects has been found in the Late Bronze Age levels (12th–11th centuries BCE) at Ayia Irini (Keos), Lerna, Phylakopi (Davis 1986, 97–98), Asine (Frizell 1986, 41, 62), Kastanas (Hochstetter 1987, 90), Tiryns (Rahmstorf 2003), as well as at Palaikastro on Crete (Sackett and Popham 1965, 305, nos. 45–50, fig. 19). Similar spools are common on a number of Late Cypriote sites (12th–11th centuries BCE), although no satisfactory explanation has been given as to their function (Karageorghis 2002, 93, 100). At Maa-Palaeokastro, they have been frequently found in groups of 5–13, in ashy deposits, which suggests association with a wooden structure and possible use as loom weights (Karageorghis and Demas 1988, 222ff, pls. LVI, LIV, LXV). Similar examples were found at Enkomi (Dikaios 1969, 291, pl. 169 nos. 16–18) and Kition. Greek and Cypriote spools are cylindrical in shape and generally larger than Italian examples; they have no perforations and are generally unfired. Spools similar to those found on Cyprus have been recorded in contemporary Levantine sites, Ashkelon and Tel Mikne-Ekron (Stager 1991, 36; Stager 1995, 346). They have been interpreted as loom weights.

34 The same area also yielded over half of the textile implements excavated at the site, as well as a set of cisterns with drain channels, which could have supplied water for washing or dyeing cloth; see Hackworth 1993, 40.

35 Finds of single shells, noted in Part 3, do not necessarily indicate production of the dye.

36 One of the problems in identifying dying installations lies in the ability of archaeologists to recognize them as such; see Kardara 1961; Sagona 1999.

37 For knives, see Celano, Palombra Sabina and Fontana di Papa types in Bianco Peroni 1976, 56–57, 61–63, pls. 31, 34, 35.

38 For other examples, see Amann 2000, 42–43.

Part 5

1 The following is a brief summary of a more extensive study, which has been undertaken as research for a thesis written by the author in fulfillment of the requirements for the degree of Master of Arts at Bryn Mawr College, Pennsylvania (Gleba 1999); see also Gleba 2000.

2 On gender studies in Italian archaeology, see Whitehouse 1998.

3 See Toms 1998, 170. The recent methodologies are discussed in von Eles 2002, 277–281. For the analysis of skeletal remains from Osteria dell'Osa and Borgo le Ferriere (Satricum) see Becker 1984, 186–95; 1996, 186–88. For the summary of bone analysis of some Veio material, see Toms 1998, 170–171. For the discussion of cremated osteological material from Verucchio in general and Tomba del Trono in particular, see von Eles 2002, 281–290.

4 See, for example, Zuffa 1975, 253–257, who attempted to challenge the identification of spindle whorls in Villanovan tombs.

5 Despite the strong evidence for gender differentiation in graves, a large portion of burials in most necropoleis remain difficult to interpret, since they do not seem to have gender indicators, at least the ones that are clear to us. One of the reasons for their absence may have been the fact that not all members of the community had the economic means to express their gender role or were

not allowed to do so fo reasons such as social hierarchy. On the other hand, it is possible that these burials had gender indicators in perishable materials. One such indicator, we can be sure, was dress. Further, the gender of the deceased could have been expressed through the burial ritual itself. Another reason, however, may have been that not all members of the community had a well-defined gender role (Luttikhnuizen 2000, 143).

6 Bartoloni (2000, 273) suggests that spools represent a loom.

7 Time spent on production however, does not necessarily translate into monetary value; see Ciszuk 2007.

8 On problems of votive terminology, see Hackens 1963; Bouma 1996, 43–51; Turfa 2006a.

9 Pedley (1993, 19), however, uses the same reasoning to claim that the weights were votive offerings.

Part 6

1 Compare with the situation in Classical Athens, where the manufacture of cloth for sale was done by professional male weavers; see discussion of literary evidence by Thompson 1982.

2 This is also true of other regions in the Mediterranean; see examples in Horden and Purcell 2000, 352–353.

3 Morel 1978 has collected the literary evidence and suggests the Tarentine wool was traded in raw state.

4 Franchi 1986, 178–179, 184 nos. 142–143; Castoldi 1995, 110 nos. 12–15, 112 fig. 124–126; Stea 1997, 87 no. 1, 90 fig. 97, 149 fig. 261.

Bibliography

Acri, A. (1983) Sulla produzione laniera di Canosa. *AnnBari* 25–26, 189–200.

Adam, R. *et al.* (1992) I rapporti transalpini. In *Gli Etruschi e l'Europa*, 180–195. Milano, Fabbri.

Adamesteanu, D. (1956) Lentini. – Scavo nell'Area sacra della città di Leontini. *NSc* 10, 402–414.

Adamesteanu, D. (1958) Manfria (Gela). – Scavo di una fattoria-officina. *NSc* 12, 290–334.

Adamesteanu, D., ed. (1999) *Storia della Basilicata. 1. L'Antichità*. Bari, Laterza.

Adamesteanu, D., Mertens, D. and D'Andria, F. (1975) Metaponto I. *NSc* 29 *Suppl.*

Adriani, A. *et al.* (1970) *Himera I: Campagne di Scavo 1963–1965*. Roma, L'Erma di Bretschneider.

Aigner Foresti, L. (1992) Relazioni protostoriche tra Italia ed Europa centrale. In *Gli Etruschi e l'Europa*, 158–167. Milano, Fabbri.

Albanese Procelli, R. M. (2003) *Sicani, Siculi, Elimi: Forme di identità, modi di contatto e processi di trasformazione*. Milano, Longanesi.

Albenda, P. (2005) *Ornamental wall painting in the art of the Assyrian Empire*. Leiden, Brill.

Alberti, M. E. (2007) Washing and Dyeing Installations of the Ancient Mediterranean: Towards a Definition from Roman Times back to Minoan Crete. In M.-L. Nosch and C. Gillis (eds), *Ancient Textiles: Production, Craft and Society*, 59–63. Oxford, Oxbow Books.

Albiach Descals, R., Gallego Úbeda, A. and García Prósper, E. (2006) El procesado del lino en la villa romana de Cornelius. Primera valoración. In J.-P. Morel, J. Tresseras Juan and J. C. Matamala (eds), *The Archaeology of Crop Fields and Gardens, Proceedings of the 1st Conference on Crop Fields and Gardens Archaeology, Barcelona (Spain), 1–3 June 2006*, 75–86. Bari, Edipuglia.

Albore Livadie, C. (1985) Cuma preellenica. In *Napoli Antica*, 62–75. Napoli, Gaetano Macchiaroli.

Aldana Nàcher, C. (1982) Pesas de telar. In M. Almagro-Gorbea (ed.), *El santuario de Juno en Gabii: Excavaciones 1956–1969*, 569–571. Bibliotheca Italica 17. Roma, Monografías de la Escuela Española de Historia y Arqueología en Rom.

Alfaro Giner, C. (1984) *Tejido y cestería en la Península Ibérica*. Bibliotheca Praeistorica Hispana XXI. Madrid.

Alfaro Giner, C. (1998) Lo spazio destinato al pascolo sulle coste del Mediterraneo: il caso delle "isole delle capre". In M. Khanoussi, P. Ruggeri and C. Vismara (eds), *L'Africa romana. Atti del XII convegno di studio, Olbia, 2–15 dicembre 1996*, 863–877. Sassari, Editrice Democratica Sarda.

Alfaro Giner, C. (2002) Ebusus y la producción de púrpura en el Imperio romano. In M. Khanoussi, P. Ruggeri and C. Vismara (eds), *L'Africa romana. Lo spazio Marittimo del Mediterraneo occidentale: geografia storica ed economia. Atti del XVI convegno di studio, Sassari, 7–10 dicembre 2000*, 681–696. Sassari, Carocci.

Alfaro, C., Wild, J. P. and Costa, B., eds (2004) *Purpurae vestes: Actas del I Symposium Internacional sobre Textiles y Tintes del Mediterráneo en época romana (Ibiza, 8 al 10 de noviembre, 2002)*. València, Universitat de València.

Alfonsi, A. (1900) Giornale degli scavi eseguiti nell'orto della Pia Casa di Ricovero in Este tra gli anni 1895 e 1898. *NSc*, 523–551.

Algaze, G. (1993) Expansionary Dynamics of Some Pristine Early States. *AmerAnthr* 95, 304–333.

Allegro, N. (1976) L'Abitato – Il Quartere Est. In *Himera II. Campagne do scavo 1966–1973*, 471–574. Roma, l'Erma di Bretschneider.

Allgrove-McDowell, J. (2003) Ancient Egypt, 5000–332 BC. In D. Jenkins (ed.), *The Cambridge History of Western Textiles*, Vol. I, 30–39. Cambridge, Cambridge University Press.

Almagro Basch, M., Ripoll, E. and Muñoz, A. M. (1957) Excavaciones en la Caverna dei Pipistrelli, Finale Ligure. *Cuadernos de Trabajos de la Escuela Española de Arqueología en Roma* 9, 169–222.

Almagro-Gorbea, M. (1992) Gli Etruschi e la penisola iberica. In *Gli Etruschi e l'Europa*, 174–179. Milano, Fabbri.

Amann, P. (2000) *Die Etruskerin. Geschlechtsverhältnis und Stellung der Frau im frühen Etruria (9–5 Jh. V. Chr.)*. Vienna, Verlag der Österreichischen Akademie der Wissenschaften.

Ambrosini, L. (2002) I pesi da telaio con iscrizioni Etrusche. *ScAnt* 10 (2000), 139–162.

Ampolo, C. (1997) L'interpretazione storica della piú antica iscrizione del Lazio (dalla necropoli di Osteria dell'Osa, Tomba 482). In G. Bartoloni (ed.), *Le Necropoli arcaiche di Veio. Giornata di studio in memoria di Massimo Pallottino*, 211–217. Roma, Università degli studi di Roma "La Sapienza".

Anati, E. (1959) *La Grande Roche de Naquane*. Paris, Masson.

Anati, E. (1961) *Camonica Valley*. New York, Alfred A. Knopf.

Andersen, H. D. (1997) The Archaeological Evidence for the Origin and Development of the Etruscan City in the 7th to 6th Centuries BC. *ActaHyp* 7, 343–82.

Andersson, E. (2003) *Tools for Textile Production from Birka and Hedeby*. Birka Studies 8. Excavations in the Black Earth 1990–1995. Stockholm, Birka Project for Riksantikvarieämbetet.

Andersson, E. and Nosch, M.-L. B. (eds) (forthcoming) *Tools, Textiles and Contexts. Investigations of textile production in the Bronze Age Eastern Mediterranean*. Oxford, Oxbow Books.

Andrén, A. (1967) Il santuario della necropoli di Cannicella ad Orvieto. *StEtr* 35, 41–85.

Andronikos, M. (1984) *Vergina: The Royal Graves*. Athens, Ekdotike Athenon.

Annibaldi, G. (1960) Grottazzolina (Ascoli Piceno). – Rinvenimento di tombe picene. *NSc* 14, 366–392.

Antico Polesine (1986) *Testimonianze archeologiche e paleoambientali*. Padova, Antoniana.

Antonacci Sanpaolo, E. (1995) Ricerche archeoambientali nella Daunia antica. Paesaggio vegetale e allevamento tra documentazione archeologico-letteraria ed analisi dei reperti naturalistici. In L. Quilici and S. Quilici Gigli (eds), *Agricoltura e commerci nell'Italia antica*, 73–102. Roma, L'Erma di Bretschneider.

Antonielli, U. (1927) Tivoli. – Fossa votiva di età romana, republicana e con materiali arcaici, scoperta in contrada "Acquoria". *NSc* 3, 216–249.

Appleyard, J. M. (1960) *Guide to the Identification of Animal Fibres*. Leeds, The Wool Industries Research Association.

Arancio, M. L. and D'Erme, L. (1992) Una tomba del Bronzo Finale da Cerveteri: nuovi dati per la conoscenza del territorio. *Origini* 15 (1990–1991), 303–319.

Aranguren, B. M., Ducci, S. and Perazzi, P. (1987) Insediamento del Neolitico Finale in località Podere Casanuova, Pontedera (Pisa). *Atti della XXVI Riunione dell'Istituto Italiano di Preistoria e Protostoria, II Neolitico in Italia, Firenze, 7–10 novembre 1985*, 621–629. Firenze.

Arboletti, M. (1992) IX. I pesi da telaio. In M. Barra Bagnasco *et al.* (eds), *Locri Epizefiri IV. Lo scavo di Marasà Sud. Il sacello tardo arcaico e la "casa dei leoni"*, 295–302. Firenze, Casa Editrice Le Lettere.

Arnold, D. E. (1985) *Ceramic Theory and Cultural Process*. Cambridge, Cambridge University Press.

Arslan, E., and Vitali, D. (1991) Catalogue. In S. Moscati (ed.), *The Celts*. Milano, Bompiani.

Aspes, A., Bellintani, G. F. and Fasani, L. (1984) I materiali della stazione protoveneta di Villamarzana. *Padusa* 20, 37–54.

Aspes, A., Rizzetto, G. and Salzani, L. (1976) Catalogo. In *3000 anni fa a Verona*, 123–195. Verona, Museo Civico.

Attema, P. A. J. *et al.* (1998) Case studies in indigenous developments in early Italian centralization and urbanization: a Dutch perspective. *EJA* 1 (no. 3), 327–80.

AttraVerso le Alpi (2002) *Uomini, vie, scambi nell'antichita*. Trento, Provincia Autonoma di Trento.

Aubet, M. E. (1993) *The Phoenicians and the West: Politics, Colonies and Trade*. Transl. M. Turton. Cambridge, Cambridge University Press.

Baglione, C. (1991) Considerazioni sui santuari di Pyrgi e di Veio-Portonaccio. In G. Bartoloni, G. Colonna and C. Grottanelli (eds), *Anathema: Regime delle offerte e vita dei santuari nel Mediterraneo antico. Atti del convegno internazionale, Roma, 15–18 Giugno 1989. ScAnt* 2–4 (1989–1990), 651–667.

Baglione, C. and De Lucia Brolli, M. A. (1997) Veio e i Falisci. In G. Bartoloni (ed.), *Le necropoli arcaiche di Veio*, 145–174. Roma, Università degli studi di Roma "La Sapienza".

Bagnasco Gianni, G. (1999) L'acquisizione della scrittura in Etruria: materiali a confronto per la ricostruzione del quadro storico e culturale. In G. Bagnasco Gianni and F. Cordano (eds), *Scritture mediterranee tra il IX e il VII secolo a.C. Atti del seminario, Università degli Studi di Milano, Istituto di Storia Antica, 23–24 febraio 1998*, 85–106. Milano, Edizioni Et.

Bailo Modesti, G. (1980) *Cairano nell'età arcaica. L'abitato e la necropoli. AIONQuad* 1. Napoli.

Bailo Modesti, G. (1982) Oliveto-Cairano: l'emergere di un potere politico. In G. Gnoli and J.-P. Vernant (eds), *La mort, les morts dans les sociétés anciennes*, 241–256. Cambridge, Cambridge University Press.

Bailo Modesti, G., d'Agostino, B. and Gastaldi, P., eds (1974) *Seconda mostra della Preistoria e della Protostoria nel Salernitano.* Salerno, Pietro Laveglia.

Bailo Modesti, G., and Gastaldi, P., eds (1999) *Prima di Pithecusa. I più antichi materiali greci del golfo di Salerno.* Napoli, Arte tipografica.

Baines, P. (2003) *Flax and Linen.* Pembrokeshire, Shire Publications Ltd.

Baioni, M. *et al.* (2003) I pesi reniformi e le fusaiole piatte decorate della Cultura di Lagozza. Cronologia, distribuzione geografica e sperimentazioni. In M. Bazzanella *et al.* (eds), *Textiles. Intrecci e tessuti dalla preistoria europea*, 99–109. Riva del Garda (Trento), Provincia Autonoma di Trento.

Baldelli, G. (1997) Deposito votivo da Cupra Marittima, Località Sant' Andrea. In M. Pacciarelli (ed.), *Acque, grotte e Dei. 3000 anni di culti preromani in Romagna, Marche e Abruzzo*, 161–171. Imola, Musei Civici.

Baldoni, D. (1994) La necropoli di Villanova – Caselle di San Lazzaro: la Tomba 25. In M. Forte and P. von Eles (eds), *La pianura Bolognese nel Villanoviano. Insediamenti della prima età del Ferro*, 261–285. Firenze, All'Insegna del Giglio.

Balista, C. *et al.* (1985) Santorso (VI): osservazioni stratigrafiche e interpretative sullo scavo 1982. *QdAV* 1, 69–99.

Balmuth, M., ed. (1987) *Studies in Sardinian Archaeology III.* Oxford, BAR 387.

Banck-Burgess, J. (1999) *Hochdorf IV, Die Textilfunde aus dem späthallstattischen Fürstengrab von Eberdingen-Hochdorf (Kreis Ludwigsburg) und weitere Grabtextilien aus Hallstatt- und Latenezeitlichen Kulturgruppen.* Stuttgart, Theiss.

Barber, E. J. W. (1991) *Prehistoric Textiles. The Development of Cloth in the Neolithic and Bronze Ages.* Princeton, Princeton University Press.

Barber, E. J. W. (1992a) The Peplos of Athena. In J. Neils (ed.), *Goddess and Polis. The Panathenaic Festival in Ancient Athens*, 103–118. Princeton, Princeton University Press.

Barber, E. J. W. (1992b) An Archaeologist Tracks the Origins of Spinning and Weaving. *Shuttle, Spindle and Dyepot* 23, 46–48.

Barber, E. W. (1994) *Women's Work: The First 20,000 Years. Women, Cloth, and Society in Early Times.* New York, Norton.

Barber, E. W. (1999) *The Mummies of Ürümchi.* New York, Norton.

Barbieri, G. (1999) Castel D'Asso (Viterbo). Recenti interventi di scavo nella necropoli. *NSc* 9–10 (2000), 159–176.

Barfield, L. (1971) *Northern Italy before Rome.* London/New York, Thames and Hudson.

Barker, G. W. W. (1975) Prehistoric territories and economies in central Italy. In E. S. Higgs (ed.), *Palaeoeconomy*, 111–176. Cambridge, Cambridge University Press.

Barker, G. (1976) Chapter V: Animal Husbandry at Narce. In T. W. Potter (ed.), *A Faliscan Town in South Etruria*, 295–307. London, British School at Rome.

Barker, G. (1988) Archaeology and the Etruscan countryside. *Antiquity* 62, 772–785.

Barker, G. and T. Rasmussen. (1998) *The Etruscans.* Oxford, Blackwell Publishers.

Barletta, B. (2001) *The origins of the Greek Architectural Orders.* Cambridge/New York, Cambridge University Press.

Barnabei, F. (1894) Dei fittili scoperti nella necropoli di Narce. *MonAnt* 4, 165–320.

Barnabei, F. and Pasqui, A. (1894) Degli ogetti di ornamento personale, delle armi e degli altri istrumenti del corredo funebre. *MonAnt* 4, 347–398.

Barra Bagnasco, M. (1996) Pomarico Vecchio (Matera) – Scavi in un abitato indigeno (1989–1991). *NSc* 1992–1993, 147–232.

Barra Bagnasco, M., ed. (1997) *Pomarico Vecchio I. Abitato mura necropoli materiali.* Galatina, Congedo Editore.

Bartoloni, G. (1989) Marriage, Sale and Gift: A proposito di alcuni corredi femminili dalle necropoli populoniesi della prima età del ferro. In A. Rallo (ed.), *Le donne in Etruria*, 35–54. Studia archeologica 52. Roma, l'Erma di Bretschneider.

Bartoloni, G. (1994) La cultura laziale e il Villanoviano salernitano. Considerazioni sui rapporti tra le comunità del Lazio protostorico e le genti esterne. In *La presenza etrusca in Campania meridionale, Atti Salerno-Pontecagnano 1990*, 199–226. Firenze, Olschki.

Bartoloni, G., ed. (1997) *Le necropoli arcaiche di Veio.* Roma, Università degli studi di Roma "La Sapienza".

Bartoloni, G., ed. (2000a) *Principi Etruschi tra Mediterraneo ed Europa.* Venezia, Marsilio.

Bartoloni, G. (2000b) La donna del principe. In G. Bartoloni (ed.), *Principi Etruschi tra Mediterraneo ed Europa*, 271–277. Venezia, Marsilio.

Bartoloni, G. (2001) Le necropoli. In A. M. Moretti Sgubini (ed.), *Veio, Cerveteri, Vulci. Città d'Etruria a confronto*, 89–120. Roma, l'Erma di Bretschneider.

Bartoloni, G. (2002) *La Cultura Villanoviana. All'inizio della storia Etrusca. Nuova edizione aggiornata.* Roma, Carocci.

Bartoloni, G. (2003) *Le società dell'Italia primitiva. Lo studio delle necropoli e la nascita delle aristocrazie.* Roma, Carocci.

Bartoloni, G. *et al.* (1975) Castel di Decima (Roma). La necropoli arcaica. *NSc* 29, 233–408.

Bartoloni, G. *et al.* (1994) Veio tra IX e VI sec. a.C.: primi risultati sull'analisi comparata delle necropoli veienti. *ArchCl* 56, 1–46.

Bartoloni, G., Cataldi Dini, M. and Zevi, F. (1982) Aspetti dell'ideologia funeraria nella necropoli di Castel di Decima. In G. Gnoli and J.-P. Vernant (eds), *La mort, les morts dans les sociétés anciennes*, 257–273. Cambridge, Cambridge University Press.

Bartoloni, G., Colonna, G. and Grottanelli, C., eds (2001) *Anathema. Regime delle offerte e vita dei santuari nel Mediterraneo antico. Atti del convegno internazionale, Roma, 15–18 Giugno 1989. ScAnt* 2–4 (1989–1990).

Bar-Yosef, O. (1985) *A Cave in the Desert: Nahal Hemar.* Jerusalem, The Israel Museum.

Baserga, G. (1922) Nuove tombe alla Cà Morta, palette sacrali e frammenti di tessuto. *RivArchCom* 82, 121–131.

Bassi, C. (1994) Pesi per telaio. In E. Cavada (ed.), *Archeologia a Mezzocorona. Documenti per la storia del popolamento rustico di età romana nell'area atestina*, 195–198. Bolzano, Servizio beni culturali provincia autonoma di Trento.

Basso, P., Bonetto, J. and Ghiotto, A. R. (2004) Produzione, lavorazione e commercio della lana nella *Venetia* romana: le testimonianze letterarie, epigrafiche e archeologiche. In G. L. Fontana and G. Gayot (eds), *La lana: prodotti e mercati (XIII–XX secolo), Atti dell'Euroconferenza (Schio, Valdagno, Follina,*

Biella, 24–27 ottobre 2001), 49–78. Padova, Cleup.

Bats, M. and d'Agostino, B., eds (1998) *Euboica. L'Eubea e la presenza euboica in Cicladica e in Occidente. AIONQuad* 12. Napoli.

Battaglia, R. (1934) S. Anna di Alfaedo – Resti di un santuario veneto-gallico sul Monte Loffa. *NSc* 12, 116–143.

Battaglia, R. (1943) La palafitta del lago di Ledro nel Trentino. *Memorie del Museo di Storia Naturale della Venezia Tridentina* 7, 1–63.

Bazzanella, M. (1999) *La Tessitura nel Neolitico*. Udine.

Bazzanella, M. *et al.*, eds (2003) *Textiles. Intrecci e tessuti dalla preistoria europea*. Riva del Garda (Trento), Provincia Autonoma di Trento.

Bazzanella, M. and Mayr, A. (1995a) La lavorazione delle fibre tessili nell'Alto Garda durante l'età del rame. In A. Pedrotti (ed.), *Le statue stele di Arco. La statuaria antropomorfa alpina nel III millennio a.C.: abbigliamento, fibre tessili e colore*, 107–124. Riva del Garda (Trento), Provincia Autonoma di Trento.

Bazzanella, M. and Mayr, A. (1995b) L'uso del colore durante l'età del rame. In A. Pedrotti (ed.), *Le statue stele di Arco. La statuaria antropomorfa alpina nel III millennio a.C.: abbigliamento, fibre tessili e colore*, 125–139. Riva del Garda (Trento), Provincia Autonoma di Trento.

Bazzanella, M., Mayr, A. and Rast-Eicher, A. (2003) I telai preistorici tra Neolitico ed età del Bronzo. In M. Bazzanella *et al.* (eds), *Textiles. Intrecci e tessuti dalla preistoria europea*, 87–97. Riva del Garda (Trento), Provincia Autonoma di Trento.

Bazzanella, M. *et al.* (2005) Iron Age textile artefacts from Riesenferner/Vedretta di Ries (Bolzano/Bozen, Italy), in P. Bichler, K. Grömer, R. Hofmann-de Keijzer, A. Kern, H. Reschreiter (eds), *Hallstatt Textiles: Technical Analysis, Scientific Investigation and Experiment on Iron Age Textiles*, 151–160. BAR-IS 1351. Oxford.

Bechtold, B., ed. (1999) *La necropoli di Lilybaeum*. Palermo, L'Erma di Bretschneider.

Becker, M. J. (1984) Osteria dell'Osa: A preliminary report on the analysis of the human skeletal remains. In A. M. Bietti Sestieri (ed.), *Preistoria e Protostoria nel territorio di Roma.*, 186–195. Lavori e Studi di Archeologia 3. Roma, De Luca Editore.

Becker, M. J. (1996) Appendix 2: Skeletal remains from the Satricum excavations of 1907–1910. In B. Ginge (ed.), *Northwest Necropolis, Southwest Sanctuary and Acropolis: Excavations at Satricum 1907–1910*, 186–88. Amsterdam, Thesis Publishers.

Becks, R. and Guzowska, M. (2004) On the Aegean-type weaving at Troia. *Studia Troica* 14, 101–115.

Bedello, M. and Fabricotti, E. (1975) Veio (Isola Farnese). – Continuazione degli scavi nella necropoli villanoviana in località "Quattro Fontanili". *NSc* 29, 63–184.

Bedini, A. (1976) Castel di Decima. In G. Colonna *et al.* (eds), *Civiltà del Lazio primitivo*, 252–290. Roma, Multigrafica.

Bedini, A. (1983) Due nuove tombe a camera presso l'abitato della Laurentina. Nota su alcuni tipi di sepolture nel VI e V secolo a.C. *ArchLaz* 5, 28–37.

Bedini, A. (1985) Corredi protostorici dal Torrino. Osservazioni sull'affermarsi e la funzione delle aristocrazie terriere nell'VIII secolo a.C. nel Lazio. *QArchEtr* 11, 44–64.

Bedini, A. (1992) Tor De' Cenci (Roma). Tombe protostoriche. *NSc* 42–43 (1988–1989), 221–279.

Bedini, A. *et al.* (1975) Alfedena (L'Aquila). Scavi del 1974 nella necropoli. *NSc* 29, 409–481.

Belanová, T. (2007) Archaeological Textile Finds from Slovakia and Moravia. In A. Rast-Eicher and R. Windler (eds), *Archäologische Textilfunde – Archaeological Textiles. NESAT IX, Braunwald, 18.–21. Mai 2005*, 41–48. Ennenda, Ragotti + Arioli Print.

Bellinger, L. (1962) Textiles from Gordion. *The Bulletin of the Needle and Bobbin Club* 46, No. 1 and 2.

Bellintani, G. F. (1992) Frattesina di Fratta Polesine: il materiale ceramico conservato presso il Museo Civico di Rovigo. Classificazione, suddivisione in fasi e alcune considerazioni sulla cronologia del Bronzo Finale nella Pianura Padana orientale. *Padusa* 28, 245–297.

Bellintani, G. F. *et al.*, eds (1986) Preistoria e protostoria nel Polesine. *Padusa* 20.

Bellintani, G. F., Peretto, C. and Peretto, R. (1984) La stazione preistorica di Frattesina in Fratta Polesine – Rovigo. Notizie preliminari sul materiale raccolto in superficie. *Padusa* 20, 13–28.

Bellucci, G. (1910) Recenti scoperte paletnologiche nell'antichissima necropoli di Terni. *BPI* 35, 78–104.

Bender Jørgensen, L. (1986) *Forhistoriske textiler i Skandinavien. Prehistoric Scandinavian Textiles.* Nordiske Fortidsminder, ser. B, 9. Copenhagen.

Bender Jørgensen, L. (1992) *North European Textiles until AD 1000.* Aarhus, Aarhus University Press.

Bender Jørgensen, L., Banck-Burgess, J. and Rast-Eicher, A., eds (2003) *Textilien aus Archäologie und Geschichte. Festschrift für Klaus Tidow.* Neumünster, Wachholtz Verlag.

Bender Jørgensen, L. and Rinaldo, C., eds (1998) *Textiles in European Archaeology. Report from the 6th NESAT Symposium, 7th–11th May 1996 in Borås.* GOTARC Series A, Vol. 1. Göteborg, Göteborg University.

Benelli, E. 2005. Alle origini dell'epigrafia cerite. In O. Paoletti and G. Camporeale (eds), *Dinamiche di sviluppo delle città nell'Etruria Meridionale: Veio, Caere, Tarquinia, Vulci. Atti del XXIII Convegno di Studi Etruschi ed Italici, Roma, Veio, Cerveteri/Pyrgi, Tarquinia, Tuscania, Vulci, Viterbo, 1–6 ottobre 2001*, 205–208. Pisa/Roma, Istituti Editoriali e Poligrafici Internazionali.

Bérard, C. (1970) *L'hérôon à la Porte de l'ouest. Eretria III.* Berne, Éditions Francke Beme.

Berardinetti, A. and Drago, L. (1997) La necropoli di Grotta Gramiccia. In G. Bartoloni (ed.), *Le necropoli arcaiche di Veio*, 39–62. Roma, Università degli studi di Roma "La Sapienza".

Berggren, E. and Berggren, K. (1972) The Necropoleis of Porzarago, Grotte Tufarina and Montevangone. In *San Giovenale*, Vol. I. Stockholm, The Swedish Institute in Rome.

Berggren, E. and Berggren, K. (1980) *San Giovenale. Vol. III.1. The Iron Age Test Square in the North-East Part of Area D.* Stockholm, The Swedish Institute in Rome.

Berggren, E. and Berggren, K. (1981) *San Giovenale. Vol. II.2. Excavations in Area B, 1957–1960.* Stockholm, The Swedish Institute in Rome.

Bergonzi, G. (1981) Riti funebri, forme ideali, strutture sociali. In R. Peroni (ed.), *Necropoli e usi funerari nell'età del ferro*, 285–291. Bari, De Donato.

Bergonzi, G. (1985) Southern Italy and the Aegean during the Late Bronze Age: Economic Strategies and Specialized Craft Products. In C. Malone and S. Stoddart (eds), *Papers in Italian Arcaheology IV. Part III. Patterns in Protohistory*, 355–387. BAR-IS 245. Oxford.

Berkin, J. (2003) *The Orientalizing Bucchero from the Lower Building at Poggio Civitate (Murlo).* Philadelphia, The University Museum.

Berlin, S. (1999) Kumihimo the Easy Way. *Handwoven* (May/June), 31–34.

Bermond Montanari, G. (1969) S. Martino in Gattara (Com. di Brisighella, Prov. Di Ravenna). Scavi 1963. *NSc* 23, 5–37.

Bermond Montanari, G., ed. (1987) *La formazione della città in Emilia-Romagna. II. Prime esperienze urbane attraverso le scoperte archeologiche.* Bologna, Nuova Alfa.

Bernabò Brea, M. (1997) Gruppo di pesi da telaio dal vilaggio piccolo di Poviglio. Gruppo di pesi da telaio dal vilaggio grande di Poviglio. In M. Bernabò Brea, A. Cardarelli, and M. Cremaschi (eds), *Le Terremare più antica civiltà padana*, 550. Modena, Electa.

Bernabò Brea, M., Cardarelli, A. and Cremaschi, M., eds (1997) *Le Terremare più antica civiltà padana* (Cat. Mostra). Modena, Electa.

Bernabò Brea, L. and Cavalier, M. (1980) *Maligunis Lipàra IV. L'acropoli di Lipari nella preistoria.* Palermo, S. F. Flaccovia Editore.

Bernabò Brea, M. and Mutti, A. (1994) *"...Le Terremare si scavono per concimare i prati..."* Parma, Silva Editore.

Bernabò Brea, M., Bianchi, P. and Lincetto, S. (2003) La produzione tessile nell'età del Bronzo. Fusaiole e pesi da telaio nelle Terremare emiliane: esempi di studio dai vilaggi di S. Rosa di Poviglio (RE) e Forno del Gallo a Beneceto (PR). In M. Bazzanella *et al.* eds), *Textiles. Intrecci e tessuti dalla preistoria europea*, 111–120. Riva del Garda (Trento), Provincia Autonoma di Trento.

Bernabò Brea, M. *et al.* (2003) Strutture e aspetti culturali dei siti Chassey-Lagozza nel piacentino. In A. Ferrari and P. Visentini (eds), *Il declino del mondo neolitico. Ricerche in Italia centro-settentrionale fra aspetti peninsulari, occidentali e nord-alpini* (Atti del Convegno Pordenone 5–7 aprile 2001). Pordenone, Bianchettin.

Bernardini, E. (1977) *La preistoria in Liguria*. Genova, Sagep.

Bertani, M. G. (1995) Il 'banchetto dei morti' in Etruria padana (IX–IV secolo a. C.), risorse del territorio e alimentazione nelle testimonianze funerari. In L. Quilici and S. Quilici Girli (eds), *Agricoltura e commerci nell'Italia antica*, 41–64. Roma, L'Erma di Bretschneider.

Bettelli, M. (1982) Note sulla produzione tessile ad Atene in età classica. *Opus* 1, 261–278.

Bettelli, M. (2002) *Italia Meridionale e Mondo Miceneo*. Firenze, all'Insegna del Giglio.

Bevan, L. (2006) *Worshippers and Warriors. Reconstructing gender and gender relations in the prehistoric rock art of Naquane national Park, Valcamonica, Brescia, northern Italy*. BAR-IS 1485. Oxford.

Biagi, P. (1986) Nuovi materiali neolitici da Castelnuovo di Teolo (Padova). *Natura Bresciana* 21, 187–193.

Bianchetti, E. (1895) *I sepolcreti di Ornavasso*. Torino, G. B. Paravia.

Bianchin Citton, E. (1995) Concordia Sagittaria in età preromana: lo stato della ricerca. In P. Croce Da Villa and A. Mastrocinque (eds), *Concordia e la X Regio*, 229–54. Padova, Zielo Editore.

Bianchin Citton, E. (1999) Il Veneto orientale tra Età del Bronzo Medio-Recente e Prima Età del Ferro. In *Protostoria e Storia del "Venetorum angulus". Atti del XX Convegno di Studi Etruschi ed Italici 1996*, 31–45. Pisa, Istituti editoriali e poligrafici internazionali.

Bianchin Citton, E., Gambacurta, G. and Ruta Serafini, A., eds (1998) *"Presso l'Adige ridente..." Recenti rinvenimenti archeologici da Este a Montagnana*. Padova, Adle Edizioni.

Bianco, S. (1996) L'età del Ferro tra Agri e Sinni. In S. Bianco *et al.* (eds), *Greci, Enotri e Lucani nella Basilicata meridionale*, 31–36. Napoli, Electa.

Bianco, S. *et al.*, eds (1996) *Greci, Enotri e Lucani nella Basilicata meridionale*. Napoli, Electa.

Bianco, S. (1998) La prima Età del Ferro nel Matapontino e nella Siritide. In *Siritide e Metapontino. Storie di due territori coloniali*, 15–30. Naples/Paestum, Cahiers du Centre Jean Bérard XX.

Bianco, S. (1999) La prima Età del Ferro. In D. Adamesteanu (ed.), *Storia della Basilicata. 1. L'Antichità*, 137–182. Bari, Laterza.

Bianco Peroni, V. (1976) *I coltelli nell'Italia continentale*. PBF. Abt. VII. Band 2. München, C. H. Becksche Verlagsbuchhandlung.

Bichler, P. *et al.*, eds (2005) *Hallstatt Textiles: Technical Analysis, Scientific Investigation and Experiment on Iron Age Textiles*. BAR-IS 1351. Oxford.

Bietti Sestieri, A. M. (1976) Rivoli Veronese. In *3000 anni fa a Verona*, 103–110. Verona, Museo Civico.

Bietti Sestieri, A. M., ed. (1979) *Ricerca su una communità del Lazio protostorico*. Roma, De Luca.

Bietti Sestieri, A. M. (1980) L'abitato di Frattesina. In *Este e la civiltà Paleovenete a cento anni dale prime scoperte. Atti dell'XI Convegno di Studi Etruschi ed Italici*, 23–38. Firenze, Olschki.

Bietti Sestieri, A. M. (1984) L'Abitato di Frattesina. *Padusa* 20, 413–427.

Bietti Sestieri, A. M. (1988) The "Mycenaean Connection" and its Impact on the Central Mediterranean Societies. *DialArch* III, 6, 23–51.

Bietti Sestieri, A. M. (1992a) *The Iron Age Community of Osteria dell'Osa*. Cambridge, Cambridge University Press.

Bietti Sestieri, A. M. (1992b) *La necropoli Laziale di Osteria dell'Osa*. Roma, Quasar.

Bietti Sestieri, A. M. (1999) La tarda età del bronzo nell'area adriatica centrale. In L. Franchi Dell'Orto (ed.), *Piceni Popolo d'Europa*, 36–42. Roma, De Luca.

Bietti Sestieri, A. M. and De Santis, A. (2000) *The Protohistory of the Latin Peoples. Museo Nazionale Romano Terme di Diocleziano*. Milano, Electa.

Bietti Sestieri, A. M., De Santis, A. and La Regina, A. (1991) Elementi di tipo cultuale e doni personali nella necropoli laziale di Osteria dell'Osa. In G. Bartoloni, G. Colonna and C. Grottanelli (eds), *Anathema. Regime delle offerte e vita dei santuari nel Mediterraneo antico. Atti del convegno internazionale, Roma, 15–18 Giugno 1989. ScAnt* 2–4 (1989–1990), 65–88.

Bisi, A. M. (1970) Lilibeo (Marsala). – Scavi nella necropoli di Cappuccini. *NSc* 24, 524–559.

Blanck, H. (1983) Le pitture del 'Sarcofago del Sacerdote' nel Museo Nazionale di Tarquinia. *DialArch* (Terza Serie 1) 2, 79–84.

Blazquez, J. M. (1978) *Economia de la Hispania Romana.* Bilbao, Ediciones "Nájera".

Blinkenberg, C. (1931) *Lindos Fouilles de l'Acropole 1902–1914. I. Les petits objets.* Berlin, W. de Gruyter.

Bloesch, H. and Mühletaler, B.. (1967) Stoffreste aus Spätgeometrischen Gräbern südlich des Westtores von Eretria. *AntK* 10, 130–132.

Boardman, J. (1970) *Greek Gems and Finger Rings. Early Bronze Age to Late Classical.* London, Thames and Hudson.

Boardman, J. (1999) *The Greeks Overseas: their Early Colonies and Trade.* Fourth Edition. London, Thames and Hudson.

Bocci Pacini, P. (1985) Una casa nell'età orientalizzante a Roselle. In G. Colonna (ed.), *Santuari d'Etruria*, 53–57. Milano, Electa.

Bocquet, A. and Berretrot, F. (1989) Le travail des fibres textiles au Néolithique Récent à Charavines (Isère). In *Tissage, Corderie, Vannerie. IXe Rencontres Internationales d'Archéologie et d'Histoire, Antibes, Octobre 1988*, 113–128. Juan-les-Pins, Éditions Association pour la Promotion et la Diffusion des Connaissances Archéologiques.

Boiardi, A. (1991) La Tomba 2. In M. Forte (ed.), *Il dono delle Eliadi*, 137–154. Rimini, Arti Grafiche Ramberti.

Bonamici, M., Stopponi, S. and Tamburini, P. (1994) *Orvieto. La Necropoli di Cannicella.* Roma, L'Erma di Bretschneider.

Bonetto, J. (2004) Agricoltura e allevamento in Cisalpina: Alcuni spunti per una reflessione. In B. S. Frizell (ed.), *PECUS. Man and animal in antiquity. Proceedings of the conference at the Swedish Institute in Rome, September 9–12, 2002*, 57–66. Rome, The Swedish Institute in Rome.

Bonetto, J. and Ghiotto, A. R. (2004) Linee metodologiche ed esempi di approccio per lo studio dell'artigianato tessile laniero nella *Venetia et Histria.* In E. De Sena and H. Dessales (eds), *Metodi e approcci archeologici: l'industria e il commercio nell'Italia antica/Archaeological Methods and Approaches: Ancient Industry and Commerce in Italy*, 50–59. BAR-IS 1262. Oxford.

Bonfante, L. (1975) *Etruscan Dress.* Baltimore, The Johns Hopkins University Press.

Bonfante, L. (1976) Etruscan Influence in Northern Italy. *ArchNews* V, 4, 93–106.

Bonfante, L. (1981) *Out of Etruria. Etruscan Influence North and South.* BAR-IS 103. Oxford.

Bonfante, L. (1985) Etruscan Textiles. *AJA* 89, 325–326.

Bonfante, L. (1989) Aggiornamento: Il costume etrusco. In *Atti II Congresso Internazionale Etrusco, Firenze 1985*, 1373–1393. Roma, Giorgio Bretschneider.

Bonfante, L. (2000) Classical Nudity in Italy and Greece. In D. Ridgway *et al.* (eds), *Ancient Italy in Its Mediterranean Setting: Studies in Honour of Ellen Macnamara*, 271–293. Accordia Specialist Studies on the Mediterranean 4. London.

Bonfante, L. (2003) *Etruscan Dress. Updated Edition.* Baltimore, The Johns Hopkins University Press.

Bonghi Jovino, M. (1983) *La necropoli preromana di Vico Equense.* Cava dei Tirreni, Di Mauro Editore.

Bonghi Jovino, M. (ed.) (1990) *Artigiani e Botteghe nell'Italia Preromana.* Roma, L'Erma di Bretschneider.

Bonghi Jovino, M., ed. (2001) *Tarquinia. Scavi sistematici nell'abitato. Campagne 1982–1988. I materiali 2.* Tarchna III. Rome, Bretschneider.

Bonghi Jovino, M. and Chiaromonte Trerè, C., eds (1997) *Tarquinia. Testimonianze archeologiche e ricostruzione storica. Scavi sistematici nell'abitato. Campagne 1982–1983.* Tarchna I. Roma, Bretschneider.

Bonino, M. (1989) "Imbarcazioni archaiche in Italia: Il problema delle navi impiegate dagli Etruschi". In *Atti del Secondo Congresso Internazionale Etrusco, Firenze,*

1985. *StEtr* Suppl. III, 1517–36. Roma, Bretschneider.

Bonomi Ponzi, L. (1991) Le presenze umbre a oriente del Tevere. In F. Roncalli and L. Bonfante (eds), *Gens antiquissima Italiae. Antichità dall'Umbria a New York*, 141–170. Perugia, Electa Editori Umbri.

Borgard, P. and Puybaret, M. P. (2004) Le travail de la laine au début de l'Empire: l'apport du modèle pompéien. Quels artisans? Quels équipements? Quelles techniques? In C. Alfaro (ed.), *Purpurae vestes: Actas del I Symposium Internacional sobra Textiles y Tintes del Mediterráneo en época romana (Ibiza, 8al 10de noviembre, 2002)*, 47–59. València, Universitat de València.

Bottini, A. (1982) *Principi guerrieri della Daunia del VII secolo. Le tombe principesche di Lavello*. Bari, De Donato.

Bottini, A. (1996) The Impact of the Greek Colonies on the Indigenous Peoples of Lucania. In G. Pugliese Carratelli (ed.), *The Western Greeks*, 541–548. Milano, Bompiani.

Bottini, A. (2000) *Kestos himas poikilos. Ostraka* 9(2), 273–279.

Bottini, A., ed. (2003) *Moda Costume Bellezza nell'Antichità. Catalogo della Mostra. Firenze, Museo Archeologico Nazionale 25 Ottobre 2003 – 31 Marzo 2004*. Livorno, Sillabe.

Boulomié, B. (1992) Il commercio Marittimo nel Sud della Francia. In *Gli Etruschi e l'Europa*, 168–173. Milano, Fabbri.

Bouma, J. M. (1996) *Religio Votiva. The Archaeology of Latial Votive Religion*. Groningen, Institute of Archaeology.

Brandt, R. (1980) La vita quotidiana a Ficana. In *Ficana: Una pietra militare sulla strada per Roma*, 111–122. Roma, Libreria Editrice.

Brandt, J. R. (1997) *Scavi di Ficana. Vol. II.1. Il periodo protostorico e arcaico. Le zone di scavo 3b–c*. Roma, Libreria dello Stato.

Brandt, J. R. and Karlsson, L., eds (2001) *From Huts to Houses. Transformation of Ancient Societies*. Stockholm, The Swedish Institute in Rome.

Brem, H.-J., Bolliger, S. and Primas, M. (1987) *Eschenz, Insel Werd. III. Die römische and spätbronzezeitliche Besiedlung*. Züricher Studien zur Archäologie. Zürich.

Briquel, D. 2000. Le origini degli Etruschi: una questione dibattuta fin dall'antichità. In M. Torelli (ed.), *Gli Etruschi*, 43–51. Milano, Bompiani.

Brizio, E. (1889) Scavi dell'arcaica necropoli italica nel predio già Benacci, ora Caprara, presso Bologna negli anni 1887–88. *NSc* 6, 288–333.

Brizio, E. (1890) Bologna. – Sepolcri italici scoperti nell'arsenale militare a mezzodì di Bologna. *NSc* 7–8, 228–236.

Brizio, E. (1895) La necropoli di Novilara. *MonAnt* 5, 85–464.

Brizio, E. (1899) Il sepolcreto Gallico di Montefortino presso Arcevia. *MonAnt* 9.

Broncoli, M. (2001) Gli ultimi scavi nella necropoli di San Pietro in Campo – ex Poligrafico Alterocca di Terni: osservazioni preliminari. *AnnFaina* 8, 343–362.

Brusadin Laplace, D. (1984–1987) La necropoli protostorica del Sasso di Furbara. II. Montorgano ed altri sepolcreti protovillanoviani. *Origini* 13, 341–408.

Brusadin Laplace, D. and Patrizi-Montoro, G. and S. (1982) L'imbarcazione monossile della necropoli del Caolino al Sasso di Furbara. *Origini* 11, 355–379.

Brusadin Laplace, D. and Patrizi-Montoro, G. and S. (1992) Le necropoli protostoriche del Sasso di Furbara III: Il Caolino ed altri sepolcreti Villanoviani. *Origini* 16, 221–94.

Buchner, G. and Gialanella, C. (1994) *Museo Archeologico di Pithecusae, Isola d'Ischia*. Itinerari No. 22. Roma, Istituto Poligrafico e Zecca dello Stato.

Buchner, G. and Ridgway, D. (1993) *Pithekoussai I. La necropoli: tombe 1–723 scavate dal 1952 al 1961*. *MonAnt* 55. Roma, Bretschneider.

Buchner, G. and Rittmann, A. (1948) *Origine e passato dell'Isola d'Ischia*. Napoli, Gaetano Macchiaroli Editore.

Buckley, T. (1996) *Aspects of Greek History*. New York, Routledge.

Buddittu, I. and Bruni, L. (1985) Stipe votiva del VI–V secolo ad Anagni. *QArchEtr* 11, 106–108.

Buranelli, F. (1983) *La necropoli villanoviana "Le Rose" di Tarquinia*. Roma, Consiglio Nazionale delle Ricerche.

Buranelli, F., ed. (1987) *La Tomba François di Vulci*. Roma, Quasar.

Buranelli, F., Drago, L. and Paolini, L. (1997)

La necropoli di Casale del Fosso. In G. Bartoloni (ed.), *Le necropoli arcaiche di Veio*, 63–84. Roma, Università degli studi di Roma "La Sapienza".

Burgers, G. J. and Yntema, D. (2000) Town and countryside in pre-Roman Southern Italy. A regional perspective. In *Die Ägäis und das westliche Mittelmer. Beziehungen und Wechselwirkungen 8. bis 5. Jh. v. Chr. Akten des Symposions, Wien 24. bis 27. März 1999*, 95–104. Wien, Verlag der Österreichischen Akademie der Wissenschaften.

Burke, B. (2007) The Kingdom of Midas and Royal Cloth Production. In M.-L. Nosch and C. Gillis (eds), *Ancient Textiles: Production, Craft and Society*, 64–70. Oxford, Oxbow Books.

Burnham, H. B. (1965) Çatal Hüyük: the textiles and the twined fabrics. *AnatSt* 15, 169–174.

Calderone, A. (1982) Monte Saraceno. Scavo del abitato nel biennio 1978–79. *Kokalos* 26–27 (1980–1981), 601–612.

Callegari, A. (1940) Arquà Petrarca. – Scoperta di una necropoli di età gallo-romana. *NSc* 1, 145–163.

Calzavara Capuis, L. and Chieco Bianchi, A. M. (2006) *Este II. La Necropoli di Villa Benvenuti*. Monumenti Antichi VII. Roma, Bretschneider.

Calzavara Capuis, L. and Ruta Serafini, A. (1987) Per un aggiornamento della problematica del celtismo nel Veneto. In D. Vitali (ed.), *Celti ed Etruschi nell'Italia centro-settentrionale dal V sec. a.C. alla Romanizzazione*, 281–307. Bologna, University Press Bologna.

Camilli, A., De Laurenzi, A. and Setari, E. (2006) *Pisa. Un viaggio nel mare dell'antichità*. Milano, Electa.

Camporeale, G. (1967) *La Tomba del Duce a Vetulonia*. Firenze, Leo S. Olschki.

Camporeale, G. (1985) *L'Etruria mineraria*. Milano, Electa.

Camporeale, G. (1992a) Miniere e metalli alle origini dell'Etruria storica: la fase "villanoviana." In *Gli Etruschi e l'Europa*, 34–43. Milano, Fabbri.

Camporeale, G. (1992b) La vocazione marittima degli Etruschi. In *Gli Etruschi e l'Europa*, 44–52. Milano, Fabbri.

Camporeale, G. (1997) *L'abitato Etrusco dell'Acesa*. Roma, Bretschneider.

Camporeale, G. (2000) *Gli Etruschi*. Torino, UTET.

Camporeale, G. (2001) *Gli Etruschi fuori Etruria*. Verona, Arsenale Editrice.

Camporeale, G. and Giuntoli, S. (2000) *The Accesa Archaeological Park at Massa Marittima*. Follonica, Editrice Leopoldo II.

Canby, J. V. (1971) Decorated Garments in Ashurnasirpal's Sculpture. *Iraq* 33, 31–53.

Capini, S. and Di Niro, A. (1991) *Samnium. Archeologia del Molise*. Roma, Quasar.

Capini, S. and Nista, L. (2000) *Italia dei Sanniti*. Roma, Electa.

Capuis, L. and Chieco Bianchi, A. M. (2006) *Este II. La Necropoli di Villa Benvenuti. Monumenti Antichi VII*. Roma, Bretschneider.

Carancini, G. L., Peroni, P. and Peroni, R. (1996) L'Italia. *ActaArch* 67, 165–173.

Carancini, G. L. and Peroni, R. (1999) *L'età del bronzo in Italia: per una chronologia della produzione metallurgica*. Perugia, Aliano.

Carandini, A. (2007) *Roma Il Primo Giorno*. Roma, Laterza.

Cardarelli, A. (1992) L'età dei metalli nell'Italia settentrionale. In A. Guidi and M. Piperno (eds), *Italia preistorica*, 365–419. Bari, Laterza.

Cardarelli, A. and Cattani, M. (1989) La terramara di S. Ambrogio (Modena). In A. Cardarelli (ed.), *Modena dalle origini all'anno mille. Studi di archeologia e storia. Mostra*. Modena, Editore Panini.

Cardon, D. (2007) *Natural Dyes, Sources, Tradition, Technology and Science*. London, Archetype Publications.

Cardon, D. and Feugère, M., eds (2000) *Archéologie des textiles des origines au Ve siècle, Actes du colloque de Lattes, oct. 1999*. Monographies Instrumentum 14. Montpellier, Éditions Monique Mergoil.

Carlucci, C. and De Lucia, M. A. (1998) *Villa Giulia Museum. The Antiquities of the Faliscans*. Roma, L'Erma di Bretschneider.

Carroll, D. L. (1965) *Patterned Textiles in Greek Art*. Ph.D. diss., University of California at Los Angeles.

Carroll, D. L. (1973) An Etruscan Textile in Newark. *AJA* 77, 334–336.

Carroll, D. L. (1983) Warping the Greek Loom: A Second Method. *AJA* 87, 96–98.

Carter, J. C. (1998) *The Chora of Metaponto. The*

Necropoleis. Austin, University of Texas Press.

Casini, S. (1986) L'attività tessile. In *Como fra Etruschi e Celti*, 150–154. Como, Società Archeologica Comense.

Càssola Guida, P. (1995) Le prospettive della ricerca protostorica in Friuli. In P. Croce Da Villa and A. Mastrocinque (eds), *Concordia e la X Regio. Giornate di studio in onore di Dario Bertolini nel centenario della morte. Atti del convegno, Portogruaro, 22–23 ottobre 1994*, 197–205. Padova, Zielo.

Càssola Guida, P. (1999) Lineamenti delle culture altoadriatiche tra Bronzo Finale e Prima Età del Ferro. In *Protostoria e Storia del "Venetorum angulus". Atti del XX Convegno di Studi Etruschi ed Italici 1996*, 47–72. Pisa, Istituti editoriali e poligrafici internazionali.

Castelletti, L. (1978) Resti organici nelle tombe della Ca' Morta. In G. Frigerio and G. Luraschi (eds), *Età del Ferro a Como: Nuove scoperte alla Ca' Morta (Scavi 1975–76). Mostra in onore di F. Rittatore Vonwiller*, 99–102. Como, Societa Archeologica Comense.

Castiglioni, E. (1995) I macroresti vegetali. In M. A. Binaghi Leva (ed.), Somma Lombardo (VA)-Località Belcora. Tumuli dell'età del Bronzo finale. *Soprintendenza Archeologica della Lombardia, Notiziario* 1992–93, 27–28.

Castiglioni, E., Rettore E. And Rottoli, M. (in press) Olmo di Nogara. Resti Organici dalle tombe. In L. Salzani (ed.), *La necropoli dell'età del Bronzo di Olmo di Nogara*. Verona.

Castoldi, M. (1986) *I Greci sul Basento. Mostra degli scavi archeologici all'Incoronata di Metaponto 1971–1984*. Como, Edizioni New Press.

Castoldi, M. (1995) Oggetti diversi. In *Ricerche archeologiche all'Incoronata di Metaponto. 3. L'oikos Greco del saggio S. Lo scavo e reperti*, 109–111. Milano, Edizioni Et.

Catarsi Dall'Aglio, M. (1994) L'insediamento etrusco di Case Nuove di Siccomonte a Cabriolo di Fidanza (Parma). In F. Rebecchi (ed.), *Spina e il delta Padano. Riflessioni sul catalogo e sulla mostra ferrarese*, 247–252. Studia Archeologica 90. Roma, L'Erma di Bretschneider.

Cateni, G. (1998) *Volterra. La Tomba del Guerriero di Poggio alle Croci. Contributi allo studio del villanoviano a Volterra*. Firenze, Edizioni IFI.

Cavagnaro Vanoni, L. (1977) Tarquinia (Viterbo). – Sei tombe intatte nella necropoli dei Monterozzi in località Calvario. *NSc* 31, 157–204.

Cavallotti Batchvarova, A. (1965) Veio (Isola Farnese). – Continuazione degli scavi nella necropoli villanoviana in località "Quattro Fontanili". *NSc* 19, 65–225.

Cazzella A. *et al.* (2005) L'insediamento dell'età del Bronzo di Coppa Nevigata e la più antica attestazione della produzione della porpora in Italia. *Preistoria Alpina* Suppl. 1, 40 (2004), 177–182.

Ceccarini, T. *et al.* (1989) *Museo Civico di Velletri*. Roma, Quasar.

Cenciaioli, L. (ed.) (2004) *Il piacere della seduzione. Materiali dalle necropoli perugine*. Perugia.

Cerchiai, L. (1995) *I Campani*. Milano, Longanesi.

Chiaromonte Trerè, C., ed. (1999) *Tarquinia. Scavi sistematici nell'abitato. Campagne 1982–1988, I materiali 1*. Tarchna II. Roma, Bretschneider.

Chiartano, B. (1977) La necropoli dell'età del Ferro dell'Incoronata e di S. Teodoro (Scavi 1970–1974). *NSc* 31, 9–190.

Chiartano, B. (1981) Rocella Jonica (Reggio Calabria). Necropoli preellenica in contrada San Onofrio. *NSc* 35, 491–539.

Chiartano, B. (1994) *La necropoli dell'età del ferro dell'Incoronata e di San Teodoro. Scavi 1978–1985, 1*. Galatina, Congedo.

Chieco Bianchi, A. M. (1987) Dati preliminari su nuove tombe di III secolo da Este. In D. Vitali (ed.), *Celti ed Etruschi nell'Italia centro-settentrionale dal V sec. a.C. alla Romanizzazione*, 191–236. Bologna, University Press Bologna.

Chieco Bianchi, A. M. (1988a) I Veneti. In M. Pugliese Carratelli (ed.), *Italia, omnium terrarum alumna*, 3–90. Milano, Libri Scheiwiller.

Chieco Bianchi, A. M. (1988b) Este, Casa di Ricovero: La tomba 23/1984. In R. de Marinis (ed.), *Gli Etruschi al nord del Po*, Vol. I, 153–159. Mantova, Campanotto Editore.

Chieco Bianchi, A. M. (1999) *Museo Nazionale Atestino, Este*. Roma, Istituto Poligrafico e Zecca dello Stato.

Chieco Bianchi, A. M. and Calzavara Capuis, L. (1985) *Este*. I. Roma, Bretschneider.

Chieco Bianchi, A. M. and Tombolani, M., eds (1988) *I Paleoveneti*. Padova, Programma.

Chiesa, F. (1993) *Aspetti dell'Orientalizzante recente in Campania. La Tomba 1 di Cales*. Milano, Cisalpina.

Christiansen, C. A. 2004. A Reanalysis of Fleece Evolution Studies. In J. Maik (ed.), *Priceless Invention of Humanity – Textiles. NESAT VIII*, 11–18. Łódź, Łódzkie Towarzystwo Naukowe.

Ciancio, A., *et al.* (1989) *Monte Sannace. Gli scavi dell'acropoli (1978–1983)*. Galatina, Congedo.

Cianciani, F. and von Hase, F. (1979) *La tomba Bernardini di Palestrina*. Roma, Consiglio Nazionale delle Ricerche.

Ciattini, A. M. (1971) *Itinerari Etruschi*. Pistoia, Tellini.

Cimino, L. (1986) *La Collezione Mieli nel Museo Archeologico di Siena*. Roma, Bretschneider.

Cinquantaquattro, T. (2001) *Pontecagnano II.6. L'Agro Picentino e la Necropoli di località Casella. AIONQuad* 13. Napoli.

Cipolla, C. (1880) Povegliano. *NSc*, 236–241.

Cipriani, M. (1995) Eboli (Salerno). *StEtr* 60, 554–557.

Cipriani, M., and Longo, F. (1996) *Poseidonia e i Lucani*. Napoli, Electa.

Ciszuk, M. (2007) The Academic Craftsman – a Discussion on Knowledge of Craft in Textile Research. In M.-L. Nosch and C. Gillis (eds), *Ancient Textiles: Production, Craft and Society*, 13–16. Oxford, Oxbow Books.

Citton, E., Gambacurta, G. and Ruta Serafini A. (1998) *…"presso l'Adige ridente"… Recenti rinvenimenti archeologici da Este e Montagnana*. Padova, Adle Edizioni.

Clark, L. (1983) Notes on Small Textile Frames Pictured on Greek Vases. *AJA* 87, 91–96.

Close-Brooks, J. (1963) Veio (Isola Farnese). – Scavi in una necropoli villanoviana in località "Quattro Fontanili". *NSc* 17, 77–267.

Coarelli, F. (2001) *Roma*, 9th ed. Roma/Bari, Laterza.

Cocchi Genick, D., ed. (1996) *L'antica età del Bronzo. Atti del Congresso di Viareggio, 9–12 Gennaio 1995*. Firenze, Octavo.

Colburn, O. C. (1977) Torre del Mordillo (Cosenza). – Scavi negli anni 1963, 1966 e 1967. *NSc* 31, 423–526.

Coldstream, J. N. (1993) Mixed Marriages at the Frontiers of the Early Greek World. *OJA* 12, 89–107.

Coldstream, J. N. (1994) Prospectors and Pioneers: Pithekoussai, Kyme and Central Italy. In G. R. Tsetskhladze and F. de Angelis (eds), *The Archaeology of Greek Colonisation*, 47–59. Oxford, Oxford University Committee for Archaeology, Monograph 40.

Coldstream, J. N. (2003) *Geometric Greece 900–700 BC*. Second Edition. London and New York, Routledge.

Collingwood, P. (1982) *The Techniques of Tablet Weaving*. New York, Watson-Guptill.

Colonna, G. (1988) I Latini e gli altri popoli di Lazio. In G. Pugliese Carratelli (ed.), *Italia, omnium terrarum alumna*, 409–528. Milano, Libri Scheiwiller.

Colonna, G., ed. (1985) *Santuari d'Etruria*. Milano, Electa.

Colonna, G., ed. (2002) *Il santuario di Portonaccio a Veii. 1. Gli scavi di Massimo Pallottino nella zona dell'altare (1939–1940). MonAnt 58*. Roma, L'Erma di Bretschneider.

Colonna, G. (2005) Intervento. In G. Bartoloni and F. Delpino (eds), *Oriente e Occidente: metodi e discipline a confronto. Riflessioni sulla cronologia dell'età del ferro in Italia, Atti dell'Incontro di studi, Roma, 30–31 ottobre 2003. Mediterranea*, 1 (2004), 478–483. Pisa, Istituti Editoriali e Poligrafici Internazionali.

Colonna, G. *et al.* (1976) *Civiltà del Lazio primitivo*. Roma, Multigrafica.

Colucci Pescatori, G. (1971) Cairano (Avellino). – Tombe dell'età del Ferro. *NSc* 25, 481–537.

Comella, A. (1982) *Il deposito votivo presso l'Ara della Regina*. Roma, Bretschneider.

Comella, A. (1986) *I materiali votivi di Falerii*. Roma, Bretschneider.

Comella, A. and Stefani, G. (1990) *Materiali*

votvi del santuario di Campetti a Veio: Scavi 1947 e 1969. Roma, Bretschneider.

Cordischi, L. (1993) Nuove acquisizioni su un area di culto al Colle Oppio. *QArchEtr* 21, 39–44.

Cornaggia Castiglioni, O. (1955) Lo strumento tessile nella Cultura della Lagozza. Contributi alla conoscenza delle culture preistoriche della Valle del Po. *Rivista dell'Antica Provincia e Diocesi di Como* 136/7, 5–31.

Cornaggia Castiglioni, O. (1964) I "reniformi" della Lagozza. Origine e distribuzione eurasica dei pesi da telaio con fori apicali contrapposti. In *Comum. Miscellanea di Scritti in Onore di F. Frigerio*, 129–185. Como, A. Noseda.

Cornaggia Castiglioni, O. (1965) Aggiunte all'inventario eurasico dei ritrovamenti di pesi da telaio a fori apicali contrapposti. *Rivista dell'Antica Provincia e Diocesi di Como*, 321–323.

Cornell, T. J. (1995) *The Beginnings of Rome, Italy and Rome from the Bronze Age to the Punic Wars (c. 1000–264 BC)*. London and New York, Routledge.

Costantini, S. (1995) *Il deposito votivo del santuario campestre di Tessennano*. Archeologica 112. Roma, Bretschneider.

Cottica, D. (2007) Spinning in the Roman World: from Everyday Craft to Metaphor of Destiny. In C. Gillis and M.-L. B. Nosch (eds), *Ancient Textiles-Production, Crafts and Society*, 220–228. Oxford, Oxbow Books.

Cozza, A. (1895) Topografia di Narce e della sua necropoli. *MonAnt* 4, 105–164.

Cremonesi, G. (1976) *La Grotta dei Piccioni di Bolognano nel quadro delle Culture del Neolitico e dell'Età del Bronzo in Abruzzo*. Colana di Studi Paletnologici. Pisa, Giardini.

Cremonini, S. and Steffè, G. (1987) Elementi del Neolitico superiore dalle Fornace Grandi di Bondeno (Ferrara). In *Preistoria e protostoria nel bacino del basso Po*, 47–74. Ferrara, Belriguardo.

Crewe, L. (1998) *Spindle whorls: A study of form, function and decoration in prehistoric Bronze Age Cyprus*. Jonsered, Paul Åströms Förlag.

Cristofani, M. (1973) Volterra. Scavi 1969–1971. *NSc* 27 Suppl.

Cristofani, M. (1983) *Gli Etruschi del Mare*. Milano, Longanesi.

Cristofani, M. (1985) *I bronzi degli Etruschi*. Novara, Istituto Geografico De Agostini.

Cristofani, M., ed. (1990) *La grande Roma dei Tarquini* (Cat. Mostra). Roma, L'Erma di Bretschneider.

Croce Da Villa, P., *et al.* (1991) Concordia Sagittaria. Scavo nell'abitato protostorico. *QdAV* 7, 79–92.

Croce Da Villa, P. and Mastricinque, A. (1995) *Concordia e la X Regio*. Padova, Zielo Editore.

Crockett, C. (1977) *The Complete Spinning Book*. New York, Watsom-Guptill Publications.

Crowfoot, G. M. (1940) On the warp-weighted loom. *BSA* 37 (1936–37), 34–47.

CSE Denmark 1. (1981) H. Salskov Roberts. Copenhagen, Odense University Press.

Cuadrado, E. (1968) Tumbas principescas de El Cigarralejo. *MM* 9, 148–186.

Curtis, C. D. (1919) *The Bernardini Tomb*. Memoirs of the American Academy in Rome 3. Rome.

Curtius, E. and Ader, F. (1890) *Olympia IV. Die Bronzen*. Berlin.

d'Agostino, B. (1964) Oliveto Citra. Necropoli arcaica in Località Turni. *NSc* 18, 40–99.

d'Agostino, B. (1970) Tombe della prima età del ferro a S. Marzano sul Sarno. *MEFRA* 82, 571–619.

d'Agostino, B. (1977) *Tombe "Principesche" da Pontecagnano. Monumenti Antichi Serie Miscellanea* 2:1. Roma, Academia Nazionale dei Lincei.

d'Agostino, B. (1988) La gente della Campania antica. In G. Pugliese Carratelli (ed.), *Italia Omnium Terrarum Alumna*, 531–589. Milano, Libri Scheiwiller.

d'Agostino, B. (1996a) The Impact of the Greek Colonies on the Indigenous Peoples of Campania. In G. Pugliese Carratelli (ed.), *The Western Greeks*, 533–540. Venezia, Bompiani.

d'Agostino, B. (1996b) La "Stipe dei Cavalli" di Pitecusa. *AttiMGrecia* 1994–1995, 9–108.

d'Agostino, B. and Gastaldi, P. (1988) *Pontecagnano. II. La necropoli del Picentino. 1. Le tombe della prima Età del Ferro. AIONQuad* 5. Napoli.

d'Agostino, B. and Ridgway, D., eds (1994) *Apoikia. I più antichi insediamenti greci in*

Occidente: funzioni e modi dell'organizzazione politica e sociale. Scritti in onore di Giorgio Buchner. AION 16.

d'Agostino, B. and Soteriu, A. (1998) Campania in the Framework of the Earlier Greek Colonization in the West. In M. Bats and B. d'Agostino (eds), *Euboica. L'Eubea e la presenza euboica in Calcidica e in Occidente*, 355–368. *AIONQuad* 12. Napoli.

Dal Rì, L. (1996) I ritrovamenti presso il rifugio Vedretta di Ries/Rieserferner nelle Alpi Aurine (2850 m s.l.m.). Notizia preliminare. *Rivista di Scienze Preistoriche* 47, 367–388.

Dämmer, H.-W. (1986) *San Pietro Montagnon (Montegrotto): Ein Vorgeschichtliches Seeheiligtum in Venetien*. Mainz am Rhein, Philipp von Zabern.

Damiani, I. *et al.* (1998) Le Caprine. In N. Negroni Catacchio (ed.), *Protovillanoviani e/o Protoetruschi. Ricerche e scavi*, 203–214. Milano, Octavo.

D'Andria, F. (1988) Messapi e Peuceti. In G. Pugliese Carratelli (ed.), *Italia omnium terrarum alumna*, 651–715. Milano, Libri Scheiwiller.

D'Andria, F., ed. (1990) *Archeologia dei Messapi*. Bari, Edipuglia.

D'Andria, F. and Mannino, K. (1996) *Ricerche sulla casa in Magna Grecia e in Sicilia*. Galatina, Congedo.

Daquino, C. (1999) *I Messapi. Il Salento prima di Roma*. Lecce, Capone.

Daragan, M. N. (2004) Palaeoeconomics of the population of Zhabotin settlement. In *Paleoeconomics of the Early Iron Age on the Territory of Ukraine*, 32–65. Kyiv, Schliakh.

Dark, K. R. (1995) *Theoretical Archaeology*. London, Duckworth.

Davidson, G. R. (1952) *The Minor Objects. Corinth XII*. Princeton.

Davidson, G. R. and Burr Thompson, D. (1943) Small Objects from the Pnyx: I. *Hesperia Suppl.* 7. Athens.

Davis, J. L. (1986) *Ayia Irini: Period V*. Mainz am Rhein, Philipp von Zabern.

Davison, J. M. (1972) *Seven Italic Tomb Groups from Narce*. Firenze, Leo S. Olschki.

De Agostino, A. (1961) Populonia (Livorno. – Scoperte archeologiche nella necropoli, negli anni 1957–1960. *NSc* 15, 63–102.

De Angelis, F. (2001) Archaeology in Sicily 1996–2000. *AR* 2000–2001, 145–201.

De Benedittis, G. (1980) L'oppidum di Monte Vairano ovvero Aquilonia. In *Sannio. Pentri e frentani dal VI al I sec. a.C.*, 321–341. Roma, De Luca.

De Caro, S. and Gialanella, C. (1998) Novità pitecusane. L'insediamento di Punta Chiarito a Forio d'Ischia. In M. Bats and B. d'Agostino (eds), *Euboica. L'Eubea e la presenza euboica in Calcidica e in Occidente*, 337–353. *AIONQuad* 12. Napoli.

de Grossi Mazzorin, J. (1998) Alcune osservazioni sulla fauna dell'abitato protostorico di "Le Paludi" di Celano. In V. d'Ercole and R. Cairoli (eds), *Archeologia in Abruzzo. Storia di un metanodotto tra industria e cultura*, 209–220. Tarquinia, Arethusa.

de Grossi Mazzorin, J. (2001) Archaeozoology and habitation models: from a subsistence to a productive economy in Central Italy. In J. R. Brandt and L. Karlsson (eds), *From Huts to Houses Transformations of Ancient Societies. Proceedings of an International Seminar organized by the Norwegian and Swedish Institutes in Rome, 21–24 September 1997*, 323–330. Stockholm, Paul Åströms Förlag.

de Grossi Mazzorin, J. (2004) Some considerations about the evolution of the animal exploitation in Central Italy from the Bronze Age to the Classical period. In B. S. Frizell (ed.), *PECUS. Man and animal in antiquity. Proceedings of the conference at the Swedish Institute in Rome, September 9–12, 2002*, 38–49. Rome, The Swedish Institute in Rome.

de Grummond, N. T., ed. (2000) *Cetamura Antica: Traditions of Chianti*. Tallahassee, Florida State University.

de Grummond, N. T. (2001) Cult of the Kiln. *Archaeology* 54 (No. 1), 58–60.

de Grummond, N. T. (2005). Sestius at Cetamura and Lurius at Cosa? In J. Pollini (ed.), *Terra Marique: Studies in Honor of Anna Marguerite McCann*, 30–39. Oxford, Oxbow Books.

de Grummond, N. T., Bare, C. and Meilleur, A. (2000) Etruscan Sigla ("Graffiti"): Prolegomena and Some Case Studies." In R. R. Holloway (ed.), *Miscellanea*

Mediterranea, 25–38. Providence, Brown University.

Dejana, A. (1980) Una necropolis romana a Cardano al Campo (Varese). In *Studi in onore di Ferrante Rittatore Vonwiller, Parte II*, 129–135. Como, Società Archeologica Comense.

De Juliis, E. M. (1973) Ordona (Foggia). Scavi nella necropoli. *NSc* 27, 285–399.

De Juliis, E. (1984) *Gli ori di Taranto in Età Ellenistica*. Milano, Mondadori.

De Juliis, E. (1988) *Gli Iapigi. Storia e civiltà della Puglia preromana*. Milano, Longanesi.

De Juliis, E. (1996a) The Impact of the Greek Colonies on the Indigenous Peoples of Apulia. In G. Pugliese Carratelli (ed.), *The Western Greeks*, 549–554. Milano, Bompiani.

De Juliis, E. M. (1996b) *Magna Grecia. L'Italia meridionale dalle origini leggendarie alla conquista romana*. Bari, Edipuglia.

De Juliis, E. M. (1997) *Mille anni di ceramica in Puglia*. Bari, Edipuglia.

De Juliis, E. M. (2000) *Taranto*. Bari, Edipuglia.

de la Genière, J. (1968) *Recherches sur l'Âge du Fer en Italie méridionale*. Napoli, Publications du Centre Jean Bérard.

de la Genière, J. (1969) Scavi di Amendolara. *Klearchos* 11, 79–89.

de la Genière, J. (1977) Torano Castello (Cosenza). Scavi nella necropoli (1965) e saggi in contrada Cozzo la Torre (1967). *NSc* 31, 389–422.

de la Genière, J. (1992) Greci e indigeni in Calabria. *AttiMGrecia* 1992, 111–120.

De Lucia Brolli, M. A. (1998) Una tomba orientalizzante da *Falerii*. Contributo alla conoscenza della Necropoli dei Cappuccini. *ArchCl* 50, 181–211.

De Marinis, R. (1986) L'abitato protostorico di Como. In *Como fra Etruschi e Celti*, 25–38. Como, New Press.

De Marinis, R. (1988a) *Gli Etruschi al nord del Po*. Mantova, Campanotto Editore.

De Marinis, R. (1988b) Dall'età del Bronzo all'età del Ferro nella Lombardia orientale. In R. De Marinis (ed.), *Gli Etruschi a nord del Po,* Vol. 1, 21–39. Mantova, Campanotto Editore.

De Marinis, R. (1988c) L'abitato etrusco dal Forcello di Bagnolo S. Vito. In R. De Marinis (ed.), *Gli Etruschi a nord del Po*, Vol 2, 151–54. Mantova, Campanotto Editore.

De Marinis, R. (1988d) Liguri e Celti-Liguri. In G. Pugliese Carratelli (ed.), *Italia, omnium terrarum alumna*, 161–218. Milano, Libri Scheiwiller.

De Marinis, R. (1991) Golasecca Culture and Its Links with Celts beyond the Alps. In S. Moscati (ed.), *The Celts*, 93–102. Milano, Bompiani.

De Miro, E. (1989) Gli "indigeni" della Sicilia centro-meridionale. *Kokalos* 34 (I), 19–43.

De Miro, E. (1991) Eredità egeo-micenee e Alto Arcaismo in Sicilia. Nuove ricerche. In D. Musti (ed.), *La transizione del Miceneo all'alto arcaismo. Dal palazzo alla città*, 593–617. Roma, Centro Nazionale delle Ricerche.

De Natale, S. (1992) *Pontecagnano. II. La necropoli di S. Antonio: Propr. ECI. 2. Tombe della prima Età del Ferro. AIONQuad* 8. Napoli.

De Natale, S. (2000) La Campania nell'Età del Ferro. In M. Borriello and T. Giove (eds), *La collezione preistorica del museo archeologico nazionale di Napoli*, 14–20. Napoli, Electa.

Denti, A. (1982) Monte Saraceno di Ravanusa. Necropoli – Scavi nella necropoli occidentale (anni 1978 e 1979). *Kokalos* 26–27 (1980–1981), 620–641.

Deonna, W. (1938) *Le mobilier délien. Exploration archéologique de Délos XVIII*. Paris, Éditions de Boccard.

De Puma, R. D. and Small, J. P., eds (1994) *Murlo and the Etruscans: Art and Society in Ancient Etruria*. Madison, The University of Wisconsin Press.

d'Ercole, M. C. 2000. Immagini dall'Adriatico Arcaico. Su alcuni temi iconografici delle stele daunie. *Ostraka* 9, 327–349.

d'Ercole, V., Cosentino, S. and Mieli, G. (2001) Stipe votiva dal santuario d'altura di Monte Giove. In L. Franchi Dell'Orto (ed.), *Eroi e Regine. Piceni Popolo d'Europa*, 338–343. Roma, De Luca.

De Santis, A. (1997) Alcune considerazioni sul territorio Veiente in età orientalizzante e arcaica. In G. Bartoloni (ed.), *Le necropoli arcaiche di Veio*, 101–144. Roma, Università degli studi di Roma "La Sapienza".

De Santis, A., Merlo, R. and De Grossi Mazorin, J. (1998) *Fidene, una casa dell'età del ferro*. Roma, Mondadori Electa.

Descoeudres, J.-P. (1990) *Greek Colonists and Native Populations*. Oxford, Clarendon Press.

De Vido, S. (1997) *Gli Elimi: storie di contatti e di rappresentazioni*. Pisa, Scuola Normale Superiore.

Di Filippo Balestrazzi, E. *et al.* (1992) Concordia Sagittaria: Quartiere Nord Ovest. Relazione preliminare delle campagne 1990–1991. *QdAV* 8, 79–99.

Di Fraia, T. (1995) L'abitato dell'età del Bronzo Finale di Fonte Tasca (commune di Archi, Chieti). Studio Preliminare su alcune classi di manufatti. *Origini* 19, 447–478.

di Gennaro, F. (1982) Organizzazione del territorio nell'Etruria meridionale protostorica: applicazione di un modello graffico. *DialArch* 2, 102–112.

di Gennaro, F. (1988) Il popolamento dell'Etruria meridionale e le characteristiche degli insediamenti tra l'età del bronzo e l'età del ferro. In *Etruria meridionale: conoscenza, conservazione, fruizione. Atti del convegno Viterbo, 29/30 novembre – 1 dicembre 1985*, 59–82. Roma, Quasar.

Di Giuseppe, H. (1995) I pesi da telaio. In A. Russo Tagliente (ed.), *Armento. Archeologia di un centro indigeno. BolArch* 35–36, 141–149.

Dikaios, P. (1969) *Encomi. Excavations 1948–1958. Vol. I. The Architectural Remains. The Tombs*. Mainz am Rhein, Philipp von Zabern.

Di Sclafani, A. M. *et al.* (2005) Necropoli di Himera: Analisi di un frammento di tessuto. In *Intrecci vegetali e fibre tessili da ambiente umido. Analisi Conservazione e Restauro. Atti Trento 28–30 maggio 2003. Incontri di Restauro 4,* 352–355. Trento.

Di Stefano, C. A. (1993) *Lilibeo Punica*. Marsala, Centro socio-culturale "Luigi Sturzo".

Di Vita, A. (1956) Sui pesi da telaio: una nota. *ArchCl* 8, 40–44.

Di Vita, A. (1952–1954) Atena Ergane in una terracotta dalla Sicilia e il culto della dea in Atene. *ASAtene* 30–32, 141–154.

Dobiat, C. (1990) *Der Burgstallkogel bei Kleinklein I. Die Ausgrabungen der Jahre 1982 und 1984*. Marburger Studien zur Vor- und Frühgeschichte 13. Hitzeroth.

Dohan, E. H. (1935) A Ziro Burial from Chiusi. *AJA* 39, 198–209.

Dohan, E. H. (1942) *Italic Tombs in the University Museum*. Philadelphia, University of Pennsylvania Press.

Donati, L. (1994) *La Casa dell'Impluvium: Architettura etrusca a Roselle*. Roma, Bretschneider.

Dondero Bricchi, I. (1953) Roma (Palatino). – Saggi per uno studio topografico della casa di Livia. *NSc* 7, 126–150.

Donvito, A. (1982) *Monte Sannace. Archeologia e storia di un abitato peuceta*. Fasano, Schena Editore.

D'Orazio, L. *et al.* (2000) Nature, Origin and Technology of Natural Fibres of Textile Artefacts Recovered in the Ancient Cities around Vesuvius. *JAS* 27, 745–754.

D'Orazio, L. and Martuscelli, E. (1999) Il tessile a Pompeii: tecnologia, industria e commercio. In A. Ciarallo and E. De Carolis (eds), *Homo Faber. Natura, scienza e tecnica nell'antica Pompeii*, 92–94. Milano, Electa.

Dotta, P. (1989) I pesi da telaio. In M. Barra Bagnasco *et al.*, *Locri Epizefiri III. Cultura materiale e vita quotidiana,* 185–201. Firenze, Casa Editrice Le Lettere.

Doumet, J. (1999) De la teinture en pourpre des anciens par l'extraction du produit colorant des *Murex trunculus*, *brandaris* et des *Purpura haemastoma*. *National Museum News* 9, 10–18.

Ducati, P. (1911) Le pietre funerarie felsinee. *MonAnt* 20, 357–728.

Ducati, P. (1927) *Storia dell'arte etrusca*. Firenze, Rinascimento del Libro.

Duhn, F. K. von (1924) *Italische Gräberkunde*. Heidelberg, C. Winter.

Duhn, F. von and Messerschmidt, F. (1939) *Italische Gräberkunde*. Heidelberg, C. Winter.

Dunbabin, T. J. (1962) Monochrome Vases and Coarse Pottery. In T. J. Dunbabin (ed.), *Perachora II. The Sanctuaries of Hera Akraia and Limenia*, 315–333. Oxford, Clarendon Press.

Du Plat Taylor, J. *et al.* (1977) Gravina di Puglia III (Part two). *PBSR* 45, 132–136.

Egloff, M. (1991) Celtic Craftwork at la Tène. In S. Moscati (ed.), *The Celts*, 369–371. Milano, Bompiani.

Eles, P. von (1998) *Verucchio, Museo Civico Archeologico*. Rimini, Provincia di Rimini.

Eles, P. von (2002) *Guerriero e sacerdote. Autorità e comunità nell'età del ferro a Verucchio. La Tomba del Trono*. Firenze, All'Insegna del Giglio.

Eles, P. von (2006a) *Il potere e la morte. Aristocrazia guerrieri e simboli*. Verucchio, Pazzini.

Eles, P. von (2006b) Il rituale funerario nel Villanoviano dell'Emilia Romagna: considerazioni alla luce di nuovi scavi e studi. In P. von Eles (ed.), *La ritualità funeraria tra età del Ferro e Orientalizzante in Italia, Atti del convegno, Verucchio, 26–27 giugno 2002*, 67–78. Pisa and Roma, Fabrizio Serra Editore.

Emery, I. (1966) *The Primary Structures of Fabrics. An Illustrated Classification*. Washington D.C., The Textile Museum.

Emiliozzi, A. (1974) *La Collezione Rossi Danielli nel Museo Civico di Viterbo*. Roma, Consiglio Nazionale delle Ricerche.

Enea nel Lazio: archeologia e mito (1981) Roma, Fratelli Palombi.

Esposito, A. M. (1999) *I principi guerrieri. La necropolis etrusca di Casale Marittimo*. Milano, Electa.

Fabricotti, E., Buchanan, M. and Paton, S. (1972) Veio (Isola Farnese). – Continuazione degli scavi nella necropoli villanoviana in località "Quattro Fontanili". *NSc* 26, 195–384.

Falchi, I. (1887) Nuovi scavi della necropoli di Vetulonia. *NSc* 3, 472–530.

Falchi, I. (1891) *Vetulonia e la sua necropoli antichissima*. Firenze, Le Monnier.

Falconi Amorelli, M. T. (1967) Veio (Isola Farnese). – Continuazione degli scavi nella necropoli villanoviana in località "Quattro Fontanili". *NSc* 21, 87–280.

Falconi Amorelli, M. T. (1983) *Vulci Scavi Bendinelli (1919–1923)*. Roma, Paleani.

Falconi Amorelli, M. T. (1987) *Vulci Scavi Mengarelli (1925–1929)*. Roma, Borgia Editore.

Fasani, L. (1980) *I materiali preistorici della stazione palafitticcola di Cisano – Verona*. Roma, Bretschneider.

Feldtkeller, A. (2003) Nierenförmige Webgewichte – wie funktionieren sie? *ATN* 37, 16–19.

Fenelli, M. (1991) Culti a Lavinium: le evidenze archeologiche. In G. Bartoloni, G. Colonna and C. Grottanelli (eds), *Anathema. Regime delle offerte e vita dei santuari nel Mediterraneo antico. Atti del convegno internazionale, Roma, 15–18 Giugno 1989*. *ScAnt* 2–4 (1989–1990), 486–505.

Ferdière, A. (1989) Économie rurale et production textile en Gaule Romaine (Belgique, Lyonnaise, Aquitaine). In *Tissage, Corderie, Vannerie. IXe Rencontres Internationales d'Archéologie et d'Histoire, Antibes, Octobre 1988*, 181–191. Juan-les-Pins, Éditions Association pour la Promotion et la Diffusion des Connaissances Archéologiques.

Ferrandini Troisi, F. (1986) "Pesi da telaio" Segni e interpretazioni. *Decima Miscellanea Greca e Romana*, 91–114. Roma, Istituto Italiano per la Storia Antica.

Ferrandini Troisi, F. (1992) *Epigrafi "mobili" del Museo Archeologico di Bari*. Documenti e Studi Bari 12. Bari, Edipuglia.

Ficana. (1981) *Una pietra mileare sulla strada per Roma. Mostra degli scavi italo nordici a Ficana (Acilia) 1975–1980*. Roma, Viela.

Fioravanti, A. and Camerini, E. (1977) *L'abitato villanoviano del Gran Carro sommerso nel lago di Bolsena (1959–1977)*. Roma, Tipografia Fratelli di Gennaro.

Firmin, G. (1989) La teinture végétale, les plantes tinctoriales sauvages et cultivées en Europe occidentale et en France. In *Tissage, Corderie, Vannerie. IXe Rencontres Internationales d'Archéologie et d'Histoire, Antibes, Octobre 1988*, 149–158. Juan-les-Pins, Éditions Association pour la Promotion et la Diffusion des Connaissances Archéologiques.

Fiumi, E. (1976) *Volterra Etrusca e Romana*. Pisa, Pacini Editore.

Flury-Lemberg, M. (1986) Die Rekonstruktion des Liber Linteus zagrabiensis oder die Mumienbinden von Zagreb. *VAMZ* 3 s., 19, 73–82.

Flury-Lemberg, M. (1988) *Textile Research and Conservation*. Bern, Abegg-Stiftung.

Fogolari, G. (1975) La civiltà Paleoveneta. In

PCIA, vol. 4. Roma, Biblioteca di Storia Patria.

Fogolari, G. and Frey, O.-H. (1965) Considerazioni tipologiche e cronologiche sul II e il II periodo Atestino. *StEtr* 33, 237–293.

Fogolari, G. and Prosdocimi, A. (1988) *I Veneti antichi. Lingua e cultura*. Padova, Editoriale Programma.

Forbes, R. J. (1956) *Studies in Ancient Technology*. Vol. IV. Leiden, Brill.

Formigli, E., ed. (1993) *Antiche officine del bronzo. Materiali, strumenti, tecniche*. Siena, Nuova Immagine Editrice.

Forte, M., ed. (1994) *Il dono delle Eliadi: Ambre e oreficerie dei principi Etruschi*. Rimini, Arti Grafiche Ramberti.

Forte, M. and von Eles, P., eds (1994) *La pianura Bolognese nel Villanoviano. Insediamenti della prima età del Ferro.* (Cat. Mostra). Firenze, All'Insegna del Giglio.

Forti, L. and Stazio, A. (1986) Vita quotidiana dei Greci d'Italia. In G. Pugliese Carratelli (ed.), *Megale Hellas. Storia e Civiltà della Magna Grecia*, 643–713. Milano, Libri Scheiwiller.

Fortunati Zuccàla, M. (1979) Gropello Cairoli (Pavia). – La necropoli romana. *NSc* 33, 5–88.

Franchi, E. (1986) Pesi da telaio e rocchetti. In M. Castoldi (ed.), *I Greci sul Basento. Mostra degli scavi archeologici all'Incoronata di Metaponto 1971–1984*, 178–179. Como, Edizioni New Press.

Franchi Dell'Orto, L., ed. (1999) *Piceni popolo d'Europa*. Roma, De Luca.

Franchi Dell'Orto, L., ed. (2001) *Eroi e Regine. Piceni Popolo d'Europa*. Roma, De Luca.

Franco, M. C. *et al.* (1970) In Veio (Isola Farnese). – Continuazione degli scavi nella necropoli villanoviana in località "Quattro Fontanili". *NSc* 24, 178–329.

Frasca, M. (1982) La necropoli di Cugno Carrube in territorio di Carlentini. In *Scavi nelle Necropoli di Leontini (1977–1982). Cronache di Archeologia 21*, 11–35.

Frayn, J. M. (1984) *Sheep-Rearing and the Wool Trade in Italy during the Roman Period.* Liverpool, F. Cairns.

Frey, O.-H. (1988) I rapporti commerciali tra l'Italia settentrionale e l'Europa centrale dal VII al IV secolo a.C. In R. De Marinis (ed.), *Gli Etruschi a nord del Po*, 11–17. Mantova, Campanotto Editore.

Frey, O.-H. (1991) *Eine Nekropole der Frühen Eisenzeit bei Santa Maria d'Anglona.* Galatina, Congedo.

Frigerio, G. and Luraschi, G., eds (1978) *Età del Ferro a Como. Nuove scoperte alla Ca' Morta (Scavi 1975–76). Mostra in onore di F. Rittatore Vonwiller.* Como, Societa Archeologica Comense.

Frizell, B. S. (1986) *Asine II. Results of the Excavations East of the Acropolis 1970–1974. Fasc. 3. The Late and Final Mycenaean Period.* Stockholm, Swedish Institute in Rome.

Frost, H. (1981) Cordage. In H. Frost *et al.*, *Lilybaeum. NSc* 30 (*Suppl.*), 93–97.

Frontini, P. (1997) *Castellaro del Vhò. Campagna di scavo 1995.* Milano, Raccolta archeologica e numismatica.

Fugazzola Delpino, M. A. (1982) Saggi scavo sul Monte Bisenzio. In G. Bonucci Caporali and A. M. Sgubini Moretti (eds), *Archeologia nella Tuscia*, 153–157. Roma, Consiglio Nazionale delle Ricerche.

Fugazzola Delpino, M. A. (1984) *La cultura villanoviana. Guida ai materiali della prima età del Ferro nel museo di Villa Giulia.* Roma, Edizioni dell'Ateneo.

Gàbrici, E. (1913) La più antica civiltà dei calcidesi di Cuma. *MonAnt* 22, 214.

Gàbrici, E. (1941) Rinvenimenti nelle zone archeologiche di Panormo e di Lilibeo. *NSc* 2, 261–302.

Galli, E. (1930) Alla ricerca di Sibari. *AttiMGrecia* 1929, 7–147.

Gamba, M. (1987) Analisi preliminare della necropoli di Arquà Petrarca (Padova). In D. Vitali (ed.), *Celti ed Etruschi nell'Italia centro-settentrionale dal V sec. a.C. alla Romanizzazione*, 237–270. Bologna, University Press.

Gambari, F. M. (1998) Elementi di organizzazione sociale ed economica delle communità protostoriche piemontesi. In L. Mercando and M. Venturino Gambari (eds), *Archeologia in Piemonte. Vol. I. La Preistoria*, 247–260. Torino, Umberto Allemandi and Co.

Gandolfo, F. (1981) Analisi dei residui di tessuto sul pugnale del "Ripostiglio II" di Montemerano. In N. Negroni Catacchio (ed.), *Sorgenti della Nova. Una comunità protostorica e il suo territorio nell'Etruria Meridionale*, 362–365. Roma, Consiglio Nazionale di Ricerca.

Gardner, P. (1883) Clay Disks from Tarentum. *JHS* 4, 156–157.

Gastaldi, B. (1865) *Lake Habitations and Pre-historic Remains in the Turbaries and Marl-Beds of Northern and Central Italy*. London, Anthropological Society.

Gastaldi, P. (1979) Le necropoli protostoriche della valle del Sarno: proposta per una suddivisione in fasi. *AION* 1, 13–57.

Gastaldi, P. (1992) Pontecagnano. II.1. La necropoli del Picentino 1. Le tombe della Prina. Età del Ferro. *AIONQuad* 6. Napoli.

Gastaldi, P. (1994) Struttura sociale e rapporti di scambio nel IX sec. a.C. a Pontecagnano. In *La presenza etrusca in Campania meridionale, Atti Salerno-Pontecagnano 1990*, 49–61. Firenze, Olschki.

Gastaldi, P. (1998a) *Pontecagnano. II. 4. La necropoli del Pagliarone. AIONQuad* 10. Napoli.

Gastaldi, P. (1998b) La necropoli indigene della Valle del Sarno. In A. Pecorano (ed.), *Nuceria Alfaterna e il suo territorio. Dalla fondazione ai Longobardi*, 63–84. Nocera Inferiore, Aletheia Edizioni.

Gatti, S. (1996) Anagni (Frosinone): Localita S. Cecilia – Indagini nel santuario ernico: il deposito votivo arcaico. *NSc* 1994–1995, 5–153.

Gatti, S. (2004) Il Santuario Ernico di Osteria della Fontana di Anagni. In *Religio. Santuari ed ex voto nel Lazio Meridionale. Atti della giornata di studi*, 84–93. Terracina, Comune di Terracina.

Geijer, A. (1977) The Loom Representation on the Chiusi Vase. In V. Gervers (ed.), *Studies in Textile History*, 52–55. Toronto, Royal Ontario Museum.

Geijer, A. (1979) *History of Textile Arts*. London, Pasold Research Fund.

Gentili, G. V. (1985) *Il villanoviano verucchiese nella Romagna orientale ed il sepolcreto Moroni*. Studi e Documenti di Archeologia I. Bologna, Nuova Alfa.

Gentili, G. V. (1986) L'età del Ferro a Verucchio: chronologia degli scavi e scoperte ed evoluzione della letteratura archeologica, 1–41. Studi e Documenti di Archeologia II. Bologna, Nuova Alfa.

Gentili, G. V. (1988) Testimonianze dell'abitato villanoviano ed "Etruscoide" di Verucchio. In *La formazione della città preromana in Emilia Romagna*, 79–103. Studi e Documenti di Archeologia IV. Bologna, Nuova Alfa.

Gentili, G. V. (2003) *Verucchio Villanoviana. Il sepolcreto in località Le Pegge e la necropoli al piede della Rocca Malatestiana. MonAnt* 59. Roma, Bretschneider.

Ghedini, F. (1995) Cultura figurativa e trasmissione dei modelli: le stoffe. *RdA* 19, 129–141.

Ghedini, F. (1996) Le stoffe tessute e dipinte come fonte per la conoscenza della pittura antica. *RdA* 20, 101–118.

Ghirardini, G. (1898) La necropoli primitiva di Volterra. *MonAnt* 8, 101–216.

Gillard, R. D. *et al.* (1994) The mineralization of fibres in burial environments. *Studies in Conservation* 39, 132–140.

Gillis, C. and Nosch, M.-L. B., eds (2007a) *Ancient Textiles – Production, Craft and Society. Proceedings of the First International Conference on Ancient Textiles, held in Lund, Sweden and Copenahgen, Denmark, March 19–23, 2003*. Oxford, Oxbow Books.

Gillis, C., and Nosch, M.-L. B., eds (2007b) *First Aid for the Excavation of Archaeological Textiles*. Oxford, Oxbow Books.

Ginge, B. (1996a) Identifying Gender in the Archaeological Record: Revisiting our Stereotypes. *EtrSt* 3, 65–73.

Ginge, B. (1996b) *Excavations at Satricum (Borgo le Ferriere) 1907–1910: Northwest Necropolis, Southwest Sanctuary and Acropolis*. Amsterdam, Thesis Publishers.

Giontella, C. (2001) L'insediamento in loc. Maratta Bassa di Terni. *AnnFaina* 8, 363–385.

Gjerstad, E. (1956) *Early Rome II – The Tombs*. Lund, C. W. K. Gleerup.

Gjerstad, E. (1960) *Early Rome III: Fortifications, Domestic Architecture, Sanctuaries. Stratigraphic Excavations*. Lund, C. W. K. Gleerup.

Gjerstad, E. (1966) *Early Rome IV*. Lund, C. W. K. Gleerup.

Gleba, M. (1999) *Textile Production in Etruria: the Case of Poggio Civitate (Murlo)*. Masters thesis, Bryn Mawr College (unpub-lished).

Gleba, M. (2000) Textile Production at Poggio Civitate (Murlo) in the 7th c. BC. In D. Cardon and M. Feugère (eds), *Archéologie des textiles des origines au Ve siècle, Actes du colloque de Lattes, oct. 1999*, 77–81. Monographies Instrumentum 14. Montagnac, Éditions Monique Mergoil.

Gleba, M. (2003) Archaeology in Etruria 1995–2002. *AR* 49, 89–103.

Gleba, M. (2004a) Textile Production in Protohistoric Italy: The Use of Implements in Reconstructing the 'Missing' Craft. In E. De Sena and H. Dessales (eds), *Metodi e approcci archeologici: l'industria e il commercio nell'Italia antica/Archaeological Methods and Approaches: Ancient Industry and Commerce in Italy*, 1–11. BAR-IS 1262. Oxford.

Gleba, M. (2004b) Linen Production in Pre-Roman and Roman Italy, in C. Alfaro, J. P. Wild, B. Costa (eds), 2004, *Purpurae vestes: Actas del I Symposium Internacional sobre Textiles y Tintes del Mediterráneo en época romana (Ibiza, 8 al 10 de noviembre, 2002)*, 29–38. València, Universitat de València.

Gleba, M. (2008) Auratae vestes: Gold textiles in the ancient Mediterranean, in C. Alfaro and L. Karali (eds) *Purpurae Vestes II, Vestidos, Textiles y Tintes: Estudios sober la produccion de bienes de consumo en la antiguidad*, 63–80. València, Universitat de València.

Gleba, M. (Forthcoming) Linen Corslets in Etruscan Culture. In *Military and Textiles*. Oxford, Oxbow Books.

Gleba, M. and Turfa, J. M. (2007) Digging for archaeological textiles in museums: 'New' finds in the University of Pennsylvania Museum of Archaeology and Anthropology. In A. Rast-Eicher and R. Windler (eds), *Archäologische Textilfunde – Archaeological Textiles. NESAT IX, Braunwald, 18.–21. Mai 2005*, 35–40. Ennenda, Ragotti + Arioli Print.

Gli Etruschi da Genova ad Ampurias (2006) Atti del XXIV Convegno di Studi Etruschi ed Italici, Marseille-Lattes, 26 settembre – 1 ottobre 2002. Pisa/Roma, Istituti Editoriali e Poligrafici Internazionali.

Gli Etruschi e l'Europa (1992) Milano, Fabbri.

Gnade, M. (1992) *The Southwest Necropolis of Satricum*. Amsterdam, Thesis.

Gnoli, G. and Vernant, J.-P., eds (1982) *La mort, les morts dans les sociétés anciennes*. Cambridge, Cambridge University Press.

Godart, L. (1992) I "labirinti" di Francavilla ed il mondo egeo. *AttiMGrecia* 1, 195–201.

Goldman, H. (1915) Inscriptions from the Acropolis of Halae. *AJA* 19, 438–453.

Goldman, H. (1940) The Acropolis of Halae. *JHS* 9, 381–514.

Goldmann, A. (2007) Erste Versuche beim Aufrauen von Wollgeweben. In A. Rast-Eicher and R. Windler (eds), *Archäologische Textilfunde – Archaeological Textiles. NESAT IX, Braunwald, 18.–21. Mai 2005*, 224–229. Ennenda, Ragotti + Arioli Print.

Good, I. (1995) On the question of silk in pre-Han Eurasia. *Antiquity* 69, 959–968.

Good, I. (1998) Bronze Age cloth and clothing of the Tarim Basin: the Chärchän evidence. In V. Mair (ed.), *The Bronze Age and Early Iron Age People of Eastern Central Asia*, 656–668. Washington, DC, Institute for the Study of Man.

Good, I. (1999) *The Ecology of Exchange: Textiles from Shahr-I Sokhta, Eastern Iran*. Ph.D. diss., University of Pennsylvania.

Good, I. (2001) Archaeological Textiles: A Review of Current Research. *Annual Review of Anthropology* 30 (1), 209–226.

Gottschalk, R. von (1996) Ein Spätrömischer Spinnrocken aus Elfenbein. *ArchKorrBl* 26, 483–500.

Gould, D. G. (1973) *The Top: Universal Toy, Enduring Pastime*. New York, Clarkson N. Potter Inc.

Govi, E. (1994) I graffiti. In G. Sassatelli (ed.), *Iscrizioni e graffiti della città etrusca di Marzabotto*, 213–236. Bologna, University Press Bologna.

Graefe, R. (1979) *Vela erunt. Die Zeltdächer der römischen Theater und ähnlicher Anlagen*. Mainz, Philipp von Zabern.

Graham, A. J. (1982b) The Western Greeks. In J. Boardman and N. G. L. Hammond (eds), *CAH* III, pt. 3, 163–195. Cambridge, Cambridge University Press.

Granger-Taylor, H. (1982) Weaving Clothes to Shape in the Ancient World: The Tunic and Toga of the Arringatore. *Textile History* 13 (1), 3–25.

Granger-Taylor, H. (1987) The Emperor's

Clothes: The Fold Lines. *BClevMus* March 1987, 114–123.

Gras, M. (1985) *Trafics tyrrhéniens archaïques*. Roma, Bibliothèque des Écoles française d'Athènes et de Rome 258.

Grass, M. N. (1955) The Origins of the Art of Knitting. *Archaeology* 8 (no. 3), 184–190.

Grassi, M. T. (1991) *I Celti in Italia*. Milano, Longanesi.

Graue, J. (1974) *Die Gräberfelder von Ornavasso. Eine Studie zur Chronologie die späten Latene und frühen Kaiserzeit*. Hamburg, Buske.

Greco, G. (1991) *Serra di Vaglio. La "Casa dei* pithoi". Modena, Franco Cosimo Panini.

Greco, G. (1997) Des étoffes pour Héra. In *Héra. Images, espaces, cultes*, 185–199. Collection du Centre Jean Bérard 15. Naples, Centre Jean Bérard.

Greco, G. and de la Genière, J. (1996) L'Heraion alla foce del Sele: continuità e trasformazioni dall'età greca all'età lucana. In M. Cipriani and F. Longo (eds), *Poseidonia e i Lucani*, 223–233. Napoli, Electa.

Greco, G. and Pontrandolfo, A., eds (1990) *Fratte. Un insediamento etrusco-campano*. Modena, Franco Cosimo Panini.

Grelle, F. and Silvestrini, M. (2001) Lane apule e tessuti canosini. In *Epigrafia e territorio. Politica e società. Temi di antichità romane 6*, 91–136. Bari, Edipuglia.

Grömer, K. (2004) Aussagemöglichkeiten zur Tätigkeit des Spinnens aufgrund archäologischer Funde und Experimente. *ArchAustr* 88, 169–182.

Grömer, K. (2005a) The textiles from the prehistoric salt-mines at Hallstatt, in P. Bichler *et al.* (eds), *Hallstatt Textiles: Technical Analysis, Scientific Investigation and Experiment on Iron Age Textiles*, 17–40. BAR-IS 1351. Oxford.

Grömer, K. (2005b) Tablet-woven ribbons from the prehistoric salt-mines at Hallstatt, Austria – results of some experiments, in P. Bichler *et al.* (eds), *Hallstatt Textiles: Technical Analysis, Scientific Investigation and Experiment on Iron Age Textiles*, 81–90. BAR-IS 1351. Oxford.

Grömer, K. (forthcoming) Textile Craft in Iron Age Austria. In M. Gleba and U. Mannering (eds), *Textiles in Context*. Oxford, Oxbow Books.

Gualtieri, M and Fracchia, H., eds (1990) *Roccagloriosa I. L'abitato: scavo e ricognizione topografica (1976–1986)*. Naples, Centre Jean Bérard.

Guerreschi, G. (1966–1967) La Lagozza di Besnate e il Neolitico Superiore Padano. *Rivista dell'Antica Provincia e Diocesi di Como* 148–149, 5–352.

Guerreschi, G. (1976–1977) La stratigrafia dell'Isolino del Varese dedotta dall'analisi della ceramica (Scavi Bertolone 1955–1959). *Sibrium* 1, 29–258.

Guidi, A. (1985) An application of the rank size rule to protohistoric settlements in the middle Tyrrhenian area. In C. Malone and S. Stoddart (eds), *Papers in Italian Archaeology IV. Vol. 3.: Patterns in Protohistory*, 217–242. BAR-IS 245, Oxford.

Guidi, A. (1993) *La necropoli veiente dei Quattro Fontanili, nel quadro della fase recente della prima età del ferro Italiana*. Biblioteca di "Studi Etruschi" 26. Firenze.

Guidi, A. (1998) The Emergence of the State in Central and Northern Italy. *ActaArch* 69, 139–161.

Guidi, A. and Piperno, M., eds (1992) *Italia preistorica*. Bari, Laterza.

Guidi, A. and Zaraffini, A. (1993) Guidonia. Rinvenimenti d'età pre- e protostorica. *QArchEtr* 21, 183–194.

Guzzo, P. G. (1996) Cronache dall'impero. In S. Bianco *et al.* (eds), *Greci, Enotri e Lucani nella Basilicata meridionale*, 69–78. Napoli, Electa.

Hackens, T. (1963) Favisae. In *Études Étrusco-Italiques. Mélanges pour le 25e anniversaire de la Chaire d'Étruscologie à l'Université de Louvain*, 71–99. Leuven, Bureaux du recueil, bibliothèque de l'Université.

Hackworth, L. (1993) *Evidence of Textile Production at Cetamura del Chianti, Italy: A Catalog and a Content*. MA thesis. Tallahassee, The Florida State University (unpublished).

Hägg, I. (1994) Über quantitative Bearbeitung und Deutung eines archäologischen Textilmaterials. *ActaArch* 65, 197–202.

Hagy, J. W. (1986) 800 years of Etruscan ships. *IJNA* 15, 221–250.

Hald, M. (1980) *Ancient Danish Textiles from Bogs and Burials*. Copenhagen, National Museum of Denmark.

Hall, J. F., ed. (1996a) *Etruscan Italy. Etruscan Influences on the Civilizations of Italy from Antiquity to the Modern Era*. Provo, Brigham Young University.

Hall, J. F. (1996b) From Tarquins to Caesars: Etruscan Governance at Rome. In J. F. Hall (ed.), *Etruscan Italy. Etruscan Influences on the Civilizations of Italy from Antiquity to the Modern Era*, 149–189. Provo, Brigham Young University.

Hamlyn, P. *et al.* (2000). DNA analysis for the routine identification of speciality fibres. *Symposium on Biotechnology in the Textile Industry*. Póvoa de Varzim, Portugal.

Harding, A. F. (2000) *European Societies in the Bronze Age*. Cambridge, Cambridge University Press.

Haubrichs R. (2005) L'étude de la pourpre: histoire d'une couleur, chimie et expérimentations. *Preistoria Alpina*, Suppl. 1, 40 (2004), 133–160.

Hayes, J. W. (1977) Some Etruscan Textile Remains in the Royal Ontario Museum. In V. Gervers (ed.), 144–148. *Studies in Textile History*. Toronto, Royal Ontario Museum.

Haynes, S. (1977) The Isis-Tomb: Do its Contents form a Consistent Group? In *La civiltà arcaica di Vulci e la sua espansione* (Atti del X Convegno di Studi Etruschi e Italici), 17–30. Firenze, Leo S. Olschki.

Haynes, S. (2000) *Etruscan Civilization: A Cultural History*. London/Los Angeles, British Museum Press and the J. Paul Getty Trust.

Haynes, S. (2001) Nuovi e vecchi ritrovamenti nella necropoli di Casteluccio e Tolle. In G. Paolucci (ed.), *Antiche genti di Castelluccio la Foce e Tolle. Collezionismo antiquario e ricerche recenti*, 10–11. Siena, Protagon Editori Toscani.

Helbaek, H. (1959) Notes on the Evolution and History of Linum. *Kuml* 1959, 103–129.

Helbig, W. (1874) Oggetti trovati nella tomba Cornetana detta del Guerriero. *AnnInst* 46, 257–259.

Hellström, P. (1975) *Luni sul Mignone. The Zone of the Large Iron Age Building*. Stockholm, The Swedish Institute in Rome.

Hencken, H. (1968a) *Tarquinia, Villanovans and Early Etruscans*. Cambridge, American School of Prehistoric Research.

Hencken, H. (1968b) *Tarquinia and Etruscan Origins*. London, Thames and Hudson.

Herring, E. (1991) Socio-political change in the south Italian Iron Age and Classical periods: an application of the peer polity model. *Accordia Research Papers* 2, 32–54.

Hochstetter, A. (1987) *Kastanas. Ausgrabungen in einem Siedlungshügel der Bronze- und Eisenzeit Makedoniens, 1975–1979. Die Kleinfunde*. Berlin, Volker Spiess.

Hodder, I. and Orton, C. (1976) *Spatial Analysis in Archaeology*. Cambridge, Cambridge University Press.

Hodges, H. (1976) *Artifacts. An Introduction to Early Materials and Technology*. London, John Baker Bradford Row.

Hodos, T. (1999) Intermarriage in the Western Greek Colonies. *OJA* 18, 61–78.

Hoffmann, M. (1964) *The Warp-Weighted Loom. Studies in the History and Technology of an Ancient Implement*. Studia Norvegica 14. Oslo.

Hofmann-de Keijzer, R., M. R. van Bommel and Joosten, I. (2005) Dyestuff and element analysis on textiles from the prehistoric salt-mines of Hallstatt, in P. Bichler *et al.* (eds), *Hallstatt Textiles: Technical Analysis, Scientific Investigation and Experiment on Iron Age Textiles*, 55–72. BAR-IS 1351. Oxford.

Hogarth, D. (1908) *Excavations at Ephesus*. London, British Museum.

Holloway, R. R. (1965) Conventions of Etruscan Painting in the Tomb of Hunting and Fishing at Tarquinii. *AJA* 69, 341–347.

Holloway, R. R. (1970) *Satrianum. The Archaeological Investigations Conducted by Brown University in 1966 and 1967*. Providence, Brown University Press.

Holloway, R. R. (1991) *The Archaeology of Ancient Sicily*. London and New York, Routledge.

Holloway, R. R. (1994) *The Archaeology of Early Rome and Latium*. London and New York, Routledge.

Horden, P. and Purcell, N. (2000) *The Corrupting Sea. A Study of the Mediterranean History*. London, Blackwell Publishers.

Hundt, H.-J. (1967) Vorgeschichtliche Gewebe

aus dem Hallstätter Salzberg. *JRGZM* 14, 38–65.

Hundt, H.-J. (1969) Über vorgeschichtliche Seidenfunde. *JRGZM* 16, 59–71.

Iaia, C. (1999) *Simbolismo funerario e ideologia alle origini di una civiltà urbana. Forme rituali nelle sepolture "villanoviane" a Tarquinia e Vulci, e nel loro entroterra.* Grandi contesti e problemi della protostoria italiana, 3. Firenze, All'Insegna del Giglio.

Il Museo Archeologico dell'Antica Capua. (1995) Napoli, Electa.

Invernizzi, R. (1998) Tombe a cremazione di età tardo la Tenè e romana. *Soprintendenza Archeologica della Lombardia Notiziario.*

Iozzo, M. and Galli, F. (2003) *Museo Archeologico Nazionale Chiusi.* Chiusi, Edizioni Luì.

Isler, H. P. (1986) Monte Iato: la sedicesima campagna di scavo. *SicArch* 62, 29–48.

Jacomet, S., Brombacher, C. and Dick, M. (1991) Palaeoethnobotanical work on Swiss Neolithic and Bronze Age lake dwellings over the past ten years. In J. M. Renfrew (ed.), *New Light on Early Farming: Recent Developments in Palaeoethnobotany*, 257–276. Edinburgh, Edinburgh University Press.

Jakes, K. A. and Sibley, L. R. (1984) An Examination of the Phenomenon of Textile Fabric Pseudomorphism. In J. B. Lambert (ed.), *Archaeological Chemistry III*, 403–424. Washington D. C., American Chemical Society.

Jakes, K. A. and Howard, J. H. (1986) Replacement of protein and cellulose fibers by copper minerals and the formation of textile pseudomorphs. In H. L. Needles and S. Haig Zeronian (eds), *Historic Textile and Paper Materials: Conservation and Characterization*, 277–287. Washington, DC, American Chemical Association.

Janaway, R. C. (1987) The preservation of organic materials in association with metal artifacts deposited in inhumation graves. In A. Boddington, A. N. Garland, and R. C. Janaway (eds), *Death, decay and reconstruction. Approaches to archaeology and forensic science.* Manchester, Manchester University Press.

Jannot, J.-R. (2004) 'Assemblées de femmes: une survivance clusienne des valeurs familiales archaïques' in *RA* 2004, 1, 33–49.

Jenkins D., ed. (2003) *The Cambridge History of Western Textiles*, Vol. I. Cambridge, Cambridge University Press.

Jensen, L. B. (1963) Royal Purple of Tyre. *JNES* 22, 104–118.

Johannowsky, W. (1983) *Materiali di Età Arcaica dalla Campania.* Napoli, Macchiaroli.

Jones, A. H. M. (1960) The Cloth Industry under the Roman Empire. *Economic History Review* 13, 183–192.

Jongman, W. (2000) Wool and Textile Industry of Roman Italy: A Working Hypothesis. In E. Lo Cascio (ed.), *Mercati permanenti e mercati periodici nel mondo romano, Atti degli Incontri Capresi di Storia economica antica (Capri, 13–15 ottobre 1997)*, 187–197 Bari, Edipuglia.

Jucker, I. (1991) *Italy of the Etruscans.* Jerusalem/ Mainz, Philipp von Zabern.

Karageorghis, V. (2002) *Cipro. Crocevia del Mediterraneo orientale 1600–500 a.C.* Milano, Electa.

Karageorghis, V. and Demas, M. (1988) *Excavations at Maa-Palaeokastro 1979–1986.* Nicosia, Department of Antiquities.

Kardara, C. (1961) Dyeing and Weaving Works at Isthmia. *AJA* 65, 261–266.

Keith, K. (1998) Spindle Whorls, Gender, and Ethnicity at Late Chalcolithic Hacinebi Tepe. *JFA* 25, 497–515.

Keuls, E. C. (1983) Attic Vase-Painting and the Home Textile Industry. In W. G. Moon (ed.) *Ancient Greek Art and Iconography*, 209–230. Madison, Wisconsin, University of Wisconsin Press.

Kilian, K. (1970) *Früheisenzeitliche Funde aus der Südostnekropole von Sala Consilina (Provinz Salerno). Archäologische Forschungen in Lukanien III.* Heidelberg, Kerle Verlag.

Kilian, K. (1977) Das Kriegergrab von Tarquinia. *JdI* 92, 24–98.

Kinoshita, M. (1980) Kumihimo: an ancient Japanese bobbin braid. *Shuttle, Spindle and Dyepot* 11 (no. 3), 19–21, 64–65.

Kleibrink, M. (2001) The Search for Sybaris: an Evaluation of Historical and Archaeological Evidence. *BaBesch* 76, 33–59.

Klemm, S. (1992) *Die Malleiten bei Bad Fischau NÖ. Monographie zu Grab und Siedlungsfunden der urgeschichtlichen Höhensiedlung.* Diss. Univ. Wien.

Køie, M. (1943) Tøj fra yngre bronzealder fremstillet af nælde. *Årbøger for Nordisk Oldkyndighed og Historie.*

Körber-Grohne, U. (1991) The determination of fibre plants in textiles, cordage and wickerwork. In J. M. Renfrew (ed.), *New Light on Early Farming. Recent Developments in Palaeoethnobotany*, 93–104. Edinburgh, Edinburgh University Press.

Kossack, G. (1992) Lebensbilder, mythische Bilderzählung und Kultfestbilder. Bemerkungen zu Bildszenen auf einer Thronlehne von Verucchio. In *Festschrift zum 50jähringen Bestehen des Institutes für Ur- und Frühgeschichte der Leopold-Franzens-Universität Innsbruck*, 231–246. Universitätsforschungen zur Prähistorischen Archäologie 8. Bonn.

Kossack, G. (1999) *Religiöses Denken in dinglicher und bildlicher Überlieferung Alteuropas aus der Spätbronze- und frühen Eisenzeit (9.–6. Jahrhundert v. Chr. Geb.).* Bayerische Akademie der Wissenschaften. Phil.-Hist. Klasse. Abhandlungen N. F., Heft 116. München.

Kovascovics, W. K. (1990) *Die Eckterrasse an der Gräberrstrasse des Kerameikos. Kerameikos XIV.* Berlin/New York, Walter de Gruyter.

Krall, J. (1892) *Die etruskischen Mumienbinden des Agramer National-Museums.* Denkschriften der Kaiserlichen Akademie der Wissenschaften. Wien, Tempsky.

Kudlien, F. (2001) Drei antike textilverarbeitende Berufe und ihr Vertreter. In H. Bellen and H. Heinen (eds), *Fünfzig Jahre Forschungen zur Sklaverei an der Mainzer Akademie*, 269–280. Stuttgart, Franz Steiner.

Kurz, D. C. and Boardman, J. (1971) *Greek Burial Customs.* London/New York, Thames and Hudson.

La civiltà nuragica (1990) Milano, Electa.

Lambert, J. B. (1997) *Traces of the Past. Unraveling the Secrets of Archaeology through Chemistry.* Reading, Helix Books.

Landes, C., ed. (2003) *Les Étrusques en France. Archéologie et collections (Catalogue de l'Exposition).* Lattes, Imago-musée de Lattes.

Landolfi, M. and Baldelli, G. (1997) San Vittore di Cingoli. In M. Pacciarelli (ed.), *Acque, grotte e dei. 3000 anni di culti preromani in Romagna, Marche e Abruzzo*, 180–183. Imola, Musei Civici.

Lanzi, L. (1907) Terni. Scoperte nel suburbia. *NSc*, 646–650.

Lanzi, L. (1914) Terni. A. Scoperte di antichità in contrada S. Pietro in Campo presso la stazione ferroviaria. *NSc*, 62–63.

La Regina, A. (1989) I Sanniti. In G. Pugliese Carratelli (ed.), *Italia omnium terrarum parens*, 301–432. Milano, Libri Scheiwiller.

La Rosa, V. (1996) The Impact of the Greek Colonists on the non-Hellenic Inhabitants of Sicily. In G. Pugliese Carratelli (ed.), *The Western Greeks*, 523–532. Milano, Bompiani.

Larsson Lovén, L. (1998) Male and Female Professions in the Textile Production of Roman Italy. In L. Bender Jørgensen and C. Rinaldo (eds), *Textiles in European Archaeology. Report from the 6th NESAT Symposium, 7–11th May 1996 in Borås*, 73–78. GOTARC Series A, Vol. 1. Göteborg, Göteborg University.

Lassen, A. W. (2007) *Weaving with crescent shaped loom weights.* Lejre (unpublished).

Leadbeater, E. (1976) *Handspinning.* Bradford, Charles T. Branford Company.

Lefkowitz, M. R. (1996) Women in the Panathenaic and Other Festivals. In J. Neils (ed.), *Worshipping Athena. Panathenaia and Parthenon*, 78–91. Madison, The University of Wisconsin Press.

Leighton, R. (1993a) *The Protohistoric Settlement on the Cittadella. Morgantina Studies IV.* Princeton, Princeton University Press.

Leighton, R. (1993b) Sicily during the Centuries of Darkness. *CAJ* 3:2, 271–283.

Leighton, R., ed. (1996) *Early Societies in Sicily.* Accordia Specialist Studies on Italy 5. London, Accordia Research Centre.

Leighton, R. (1999) *Sicily Before History. An Archaeological Survey from the Palaeolithic to the Iron Age.* Ithaca, Cornell University Press.

Lemmonier, P., ed. (1993) *Technological Choices.*

Transformation in Material Cultures since the Neolithic. London, Routledge.

Lenticchia. A. (1921) Struttura e natura di un tessuto scoperto in una tomba preromana di Albate (Como). *RivArchCom* 78–91, 33–41.

Lentini, A. and Scala, G. (2002) Identification and technology of textile fibres and leather from Roman ships of Pisa San Rossore, Italy. *Revue d'Archéométrie* 26, 171–176.

Lentini, A. and Scala, G. (2005) Il cordame dalle navi di Pisa San Rossore: identificazione della tecnologia della lavorazione delle funi, dei tessuti, dei materiali da intreccio e delle fibre tessili. Caratterizzazione della tecnologia di concia del cuoio. In *Intrecci vegetali e fibre tessili da ambiente umido. Analisi Conservazione e Restauro. Atti Trento 28–30 maggio 2003. Incontri di Restauro 4*, 198–214. Riva del Garda (Trento), Provincia Autonoma di Trento.

Levi, D. (1935) *Il Museo Civico di Chiusi*. Roma, La Libreria dello Stato.

Levi, D. (1950) Le necropoli puniche di Olbia. *StSard* 9, 5–120.

Levi, D. (1965–1966) Bolli d'anfora e pesi fittili da Festos. *ASAtene* 27–28, 569–588.

Liber Linteus Zagabriensis. (1986) *VAMZ* 19. Zagreb.

Lilliu, G. (1982) *La civiltà nuragica*. Sassari, Delfino.

Lipponen, S. (2007) Textile-making tools from Ficana, Italy, Zone 4a. *ATN* 44, 2–5.

Lissi Caronna, E. (1972) Labirinti? *AttiMGrecia* 1970–71, 93–98.

Lissi Caronna, E. (1980) Oppido Lucano (Potenza) – Raporto preliminare sulla seconda campagna di scavo (1968). *NSc* 34, 119–297.

Lissi Caronna, E. (1983) Oppido Lucano (Potenza). Raporto preliminare sulla terza campagna di scavo (1969). *NSc* 37, 353–412.

Lissi Caronna, E. (1984) Oppido Lucano: Quattro case di IV–III sec. a.C. *AttiMGrecia* 24–25 (1983–1984), 193–212.

Lissi Caronna, E. (1986) Oppido Lucano (Potenza) – Raporto preliminare sulla terza campagna di scavo (1969). *NSc* 37 (1983), 215–352.

Lissi Caronna, E., Armignacco Alidori, V. and Panciera, S. (1992) Oppido Lucano (Potenza). Raporto preliminare sulla Quarta campagna di scavo (1970). Materiale archeologico rinvenuto nel territorio del Comune. *NSc* (1990–91), 185–488.

Liu, R. K. (1978) Spindle Whorls: Part I. Some Comments and Speculations. *The Bead Journal* 3, 87–103.

Lochner, M. (1986) Ein urnenfelderzeitliches Keramikdepot aus Oberravelsbach. *ArchAustr* 70, 295–316.

Lochner, M. (1991) *Studien zur Urnenfelderkultur im Waldviertel (Niederösterreich)*. Mitteilungen der Prähistorischen Kommission 25. Wien.

Lollini, D. G. (1956) L'abitato preistorico e protostorico di Ancona. *BPI* 65, 237–262.

Long, L., Pomey, P. and Sourisseau, J.-C. (2002) *Les Étrusques en mer. Épaves d'Antibes à Marseille*. Aix en Provence, Édisud.

Lo Porto, F. G. (1963) Leporano (Taranto) – La stazione protostorica di Porto Perone. *NSc* 17, 280–380.

Lo Porto, F. G. (1964) Satyrion (Taranto) – Scavi e ricerche nel luogo del più antico insediamento laconico in Puglia. *NSc* 18, 177–279.

Lo Porto, F. G. (1969) Metaponto. – Tombe a tumulo dell'età del Ferro scoperte nel suo entroterra. *NSc* 23, 121–170.

Lo Porto, F. G. (1977) Recenti scoperte in Puglia. In *Atti Taranto 1976*, 725–745.

Lo Porto, F. G. (1981) Metaponto (Matera). Nuovi scavi nella città e nella sua necropoli. *NSc* 35, 289–391.

Lo Porto, F. G. (1991) *Timmari. L'abitato, le necropoli, la stipe votiva*. Roma, Bretschneider.

Lo Porto, F. G. (1992) Metaponto (Matera) – Rinvenimenti nella città antica e nel suo retroterra ellenizzato. *NSc* 1988–89, 299–444.

Lo Schiavo, F. (1994) Bronzi nuragici nelle tombe della prima Età del Ferro di Pontecagnano. In *La presenza Etrusca nella Campania meridionale, Atti Salerno Pontecagnano 1990*, 61–82. Firenze, Olschki.

Lucentini, N. (1992) Nuove tombe Picene a Montedinove. In *La civiltà picena nelle Marche. Studi in onore di Giovanni Annibaldi.*

Atti del convegno. Ancona 1988, 464–505. Ripatransone, Maroni.

Lukesh S. S. (1999) Early Bronze Age Sicilian Geometric Decoration: Its Origin and Relationship to Vessel Form. In S. S. Lukesh (ed.), *Interpretatio Rerum*, 1–13. Archaeologica Transatlantica XVII. Providence, Brown University.

Lundgren, M.-B. and Wendt, L. (1982) *Acquarossa, Results of Excavations, Zone A*. Vol. III. Stockholm, Paul Åströms Förlag.

Luppino, S. (1982) Versante Nord-occidentale del fiume Savuto. In G. Maddoli (ed.), *Temesa e il suo territorio*, 75–78. Taranto, Istituto per la Storia e l'Archeologia della Magna Grecia.

Luttikhuizen, C. L. (2000) Differenze di *gender* nelle necropoli arcaiche della zona medio-adriatica italiana. *BaBesch* 75, 127–146.

Mårtensson, L. (2007) Investigating the Function of Mediterranean Bronze Age Textile Tool Using Wool and Flax Fibres. In *Experimentele Archäologie in Europa Bilanz. Heft* 6, 97–106. Oldenburg.

Maaskant-Kleibrink, M. (1987) *Settlement excavations at Borgo le Ferriere "Satricum". Vol. I. (The Campaigns 1979, 1980, 1981)*. Groningen, Egbert Forsten.

Maaskant-Kleibrink, M. (1992) *Settlement excavations at Borgo le Ferriere "Satricum". Vol. I. (The Campaigns 1983, 1985 and 1987)*. Groningen, Egbert Forsten.

Maaskant-Kleibrink, M. (1993) Religious activities on the 'Timpone della Motta', Francavilla Marittima–and the identification of Lagaría. *BaBesch* 68, 1–47.

Maaskant-Kleibrink, M. (1997) Dark Age or Ferro I? A tentative answer for the Sibarite and Metapontine Plains. *Caeculus III, Debating Dark Ages. Papers on Mediterranean Archaeology*, 63–89. Groningen, Groningen Archeological Institute.

Maaskant-Kleibrink, M. (2000) Early Cults in the *Athenaion* at Francavilla Marittima as evidence for Pre-Colonial Circulation of *nostoi* Stories. In *Die Ägäis und das Westliche Mittelmeer. Beziehungen und Wechselwirkungen 8. bis 5. Jh. V. Chr.*, 165–184. Wien, Verlag der Österreichischen Akademie der Wissenschaften.

Macellari, R. (1983) Pesi da telaio. In D. Vitali (ed.), *Monterenzio e la Valle dell' Idice*, 119–126. Bologna, Banca del Monte di Bologna e Ravenna.

Macheboeuf, C. (2004) Fabrication et commerce de la pourpre en Italie et en Sicile dans l'antiquité romaine. In E. De Sena and H. Dessales (eds), *Metodi e approcci archeologici: l'industria e il commercio nell'Italia antica/Archaeological Methods and Approaches: Ancient Industry and Commerce in Italy*, 25–37. BAR-IS 1262. Oxford.

McCorriston, J. (1997) The Fiber Revolution. *CurrAnthr* 38, 517–549.

MacKinnon, M. (2004) The role of caprines in Roman Italy: idealized and realistic reconstructions using ancient textual and zooarchaeological data. In B. S. Frizell (ed.), *PECUS. Man and animal in antiquity. Proceedings of the conference at the Swedish Institute in Rome, September 9–12, 2002*, 54–60. Rome, The Swedish Institute in Rome.

McLauchlin, B. K. (1981) New Evidence on the Mechanics of Loom Weights. *AJA* 85, 79–81.

Macnamara, E. (1976) Chapter III: The Small Finds. In T. W. Potter (ed.), *A Faliscan Town in South Etruria. Excavations at Narce 1966–71*, 127–178. London, British School at Rome.

Macnamara, E. (1992) The metalwork. In A. Small (ed.), *An Iron Age and Roman Republican Settlement on Botromagno, Gravina di Puglia. Excavations of 1965–1974. Volume II. The Artifacts*, 230–248. London, British School at Rome.

Maeder, F. (2002) The Project Sea-silk – Rediscovering an Ancient Textile Material. *ATN* 35, 9–11.

Maeder, F., Hänggi, A. and Wunderlin, D. (2004) *Bisso marino. Fili d'oro dal fondo del mare*. Basel, Naturhistoriches Museum and Museum der Kulturen.

Maier, F. (1991) The *Oppida* of the Second and First Centuries B.C. In S. Moscati (ed.), *The Celts*, 411–425. Milano, Bompiani.

Maioli, M. G. and Mastrocinque, A. (1992) *La stipe di Villa di Villa e i culti degli antichi veneti*. Roma, Bretschneider.

Malavolti, F. (1942) La stazione del Pescale,

Modena. *Atti della Società di Naturalisti e Matematici* 78, 151.

Malcus, B. (1984) Area D (ovest). In S. Forsberg and B. E. Thomasson (eds), *San Giovenale: Materiali e problemi*. Stockholm, The Swedish School in Rome.

Malkin, I. (1987) *Religion and Colonization in Ancient Greece*. Leiden, Brill.

Malnati, L. (1984) Tombe arcaiche di S. Maria d'Anglona (Scavi 1972–1973). In M. Castoldi and L. Malnati, *Studi e Ricerche in Basilicata*, 41–95. Milano, Cisalpino-La Goliardica.

Malnati, L. (1989) Lo scavo di una fattoria Etrusca a Baggiovara – Localita Case Vandetti. In A. Cardarelli (ed.), *Modena dalle origini all'anno mille. Studi di archeologia e storia. Mostra*, 262–271. Modena, Editore Panini.

Malnati, L. and Manfredi, V. (1991) *Gli Etruschi in Val Padana*. Milano, Il Saggiatore.

Malone, C. and Stoddart, S., eds (1985) *Papers in Italian Archaeology IV. Vol. 3.: Patterns in Protohistory*. BAR-IS 245. Oxford.

Mamez, L. and Masurel, H. (1992) Étude complémentaire des vestiges textiles trouvés dans l'embarcation de la nécropole du Caolino à Sasso di Furbara. *Origini* 16, 295–310.

Mancassola, N. *et al.* (1999) Lavagno. Progetto San Briccio: notizie preliminari. *QdAV* 15, 49–57.

Mandolesi, A. (1999) *La "prima" Tarquinia*. Firenze, All'Insegna del Giglio.

Mandolesi, A. (2000) Viterbo. Località Norchia. – Scavo di livelli abitativi protostorici presso il "grande" fossato (1973). *NSc* 1998–99, 59–156.

Manfredi, V. M. and Malnati, L. (2003) *Gli Etruschi in Val Padana*. 2nd ed. Milano, Mondadori.

Mangani, E. (2004) Le stipi votive di Roma e del Lazio Meridionale nel Museo Pigorini. In *Religio. Santuari ed ex voto nel Lazio meridionale. Atti della giornata di studi*, 59 83. Terracina, Comune di Terracina.

Mangani, E. and Pacciani, E. (1992) Castelnuovo Berardenga (Siena) – L'orientalizzante recente in Etruria settentrionale: tomba A della necropoli principesca del Poggione (1980). *NSc* 1988–89, 5–84.

Mannering, U. and Gleba, M. (forthcoming) *Designed for Life and Death*. Copenhagen.

Manning, S. W. *et al.* (2001) Anatolian Tree Rings and a New Chronology for the East Mediterranean Bronze-Iron Ages. *Science* 294 (5551), 2532–2535.

Mannoni, T. and Giannichedda, E. (1996) *Archeologia della produzzione*. Torino, Einaudi.

Mansfield, J. M. (1985) *The Robe of Athena and the Panathenaic* Peplos. Ph.D. diss., University of California, Berkeley.

Marchi, A. R. (1983) La filatura. In D. Vitali (ed.), *Monterenzio e la Valle dell' Idice*. Bologna, Banca del Monte di Bologna e Ravenna.

Martelli, M. (1994) Sulla produzione di vetri orientalizzanti. In M. Martelli (ed.), *Tyrrhenoi Philotechnoi*, 75–97. Roma, Gruppo Editoriale Internazionale.

Martin, C. (1984) Kumihimo: the Japanese art of braid-making. *Crafts* 69 (July/August), 30–35.

Marzatico, F. (1997) *I materiali preromani della Valle dell'Adige nel Castello del Buonconsiglio*. Trento, Servizio Beni Culturali della Provincia Autonoma di Trento.

Maspero, A. (1998) I resti di tessuto. In E. Bianchin Citton, G. Gambacurta, and A. Ruta Serafini (eds), *"Presso l'Adige ridente…" Recenti rinvenimenti archeologici da Este a Montagnana*, 62–67. Padova, Adle Edizioni.

Mastrocinque, A. (1987) *Santuari e divinità dei paleoveneti*. Padova, La Linea.

Masurel, H. (1982a) Les vestiges textiles retrouvés dans l'embarcation. *Origini* 11, 381–414.

Masurel, H. (1989) Méthodologie pour l'étude des tissus protohistoriques. In *Tissage, Corderie, Vannerie. IXe Rencontres Internationales d'Archéologie et d'Histoire, Antibes, Octobre 1988*, 165–180. Juan-les-Pins, Éditions Association pour la Promotion et la Diffusion des Connaissances Archéologiques.

Mautendorfer, H. (2005) Genähtes aus dem prähistorischen Hallstatt. In P. Bichler *et al.* (eds), *Hallstatt Textiles: Technical Analysis, Scientific Investigation and Experiment on Iron Age Textiles*, 41–54. BAR-IS 1351. Oxford.

Médard, F. (2000) *L'artisanat textile au Néolithique. L'exemple de Delley-Portalban II (Suisse) 3272–2462 avant J.-C.* Montagnac, Éditions Monique Mergoil.

Meola, E. (1996) *Necropoli di Selinunte I. Buffa.* Palermo, Academia Nazionale di Scienze Lettere Arti.

Mercati, C. (2003) *Epinetron: Storia di una forma ceramica fra archeologia e cultura.* Perugia, Petruzzi Editore.

Mertens-Horn, M. (1994) In *Sibari e la Sibaritide, Atti del trentaduesimo Convegno di studi sulla Magna Grecia : Taranto-Sibari, 7–12 ottobre 1992*, 489–492. Taranto.

Metaponto II (1977) *NSc* 31 Suppl I.

Miari, M. (2000) *Stipi votive dell'Etruria padana. CSVI 11, Reggio III, 3.* Roma, Bretschneider.

Militello, E. (1960) *Terravecchia di Cuti.* Palermo, Fondazione 'Ignazio Mormino'.

Militello, E. and La Piana, S. (1969) Mineo (Catania). – La necropoli detta Molino della Badia: nuove tombe in contrada Madonna del Piano. *NSc* 23, 210–276.

Miller, N. (1991) The Near East. In W. van Zeist, K. Wasylikowa and K.-E. Behre (eds), *Progress in the Old World Palaeoethnobotany. A Retrospective View on the Occasion of 20 Years of the International Work Group for Palaeoethnobotany*, 133–160. Rotterdam, Balkema.

Miller, S. G. (1993) *The Tomb of Lyson and Kallikles: A Painted Macedonian Tomb.* Mainz am Rhein, Philipp von Zabern.

Mingazzini, P. (1974) Sull'uso e sullo scopo dei pesi da telaio. *RendAcadNazLinc* 29, 201–220.

Minto, A. (1917) Populonia. Relazione intorno agli scavi governativi eseguiti nel 1915. *NSc* 1917, 69–93.

Minto, A. (1921a) *Marsiliana d'Albegna.* Firenze, Istituto di Edizioni Artistiche.

Minto, A. (1921b) Populonia. Scavi governativi eseguiti nell'autunno del 1920 nella zona di Porto Baratti. *NSc* 1921, 301–336.

Minto, A. (1925) Populonia. Scavi e scoperte fortuite nella località Porto Baratti durante il 1924–25. *NSc* 1, 346–373.

Minto, A. (1934) II. Populonia. – Scoperte archeologiche fortuite dal 1931 al 1934. *NSc* 12, 351–428.

Minto, A. (1936) Orvieto. – Trovamenti archeologici nella zona di S. Domenico. *NSc* 14, 251–269.

Minto, A. (1943) *Populonia.* Firenze, Istituto di Edizioni Artistiche.

Moeller, W. O. (1976) *The Wool Trade of Ancient Pompeii.* Leiden, Brill.

Möller-Wiering, S. (2002) *Segeltuch und Emballage. Textilien im mittelalterlichen Warentransport auf Nord- und Ostsee.* Internationale Archäologie 70. Rahden, Verlag Marie Leidorf.

Montelius, O. (1895–1910) *La civilisation primitive en Italie depuis l'introduction des métaux.* Stockholm, Imprimerie royale.

Moore, S. (1922) *Linen.* New York, The Macmillan Company.

Morel, J.-P. (1970) Fouilles à Cozzo Presepe, près de Métaponte. *MEFRA* 82, 73–116.

Morel, J.-P. (1978) La laine de Tarente (De l'usage des textes anciens en histoire économique). *Ktèma* 3, 93–110.

Moretti Sgubini, A. M., ed. (2001a) *Veio, Cerveteri, Vulci. Città d'Etruria a confronto.* Roma, l'Erma di Bretschneider.

Moretti Sgubini, A. M., ed. (2001b) *Tarquinia etrusca una nuova storia.* Roma, l'Erma di Bretschneider.

Moretti Sgubini, A. M., ed. (2004a) *Eroi Etruschi e Miti Greci: gli affreschi della Tomba François tornano a Vulci.* Calenzano, Soprintendenza per i Beni Archeologici dell'Etruria Meridionale.

Moretti Sgubini, A. M. (2004b) La necropoli di Ponte Rotto. In A. M. Moretti Sgubini (ed.), *Eroi Etruschi e Miti Greci: gli affreschi della Tomba François tornano a Vulci*, 17–28. Calenzano, Soprintendenza per i Beni Archeologici dell'Etruria Meridionale.

Morigi Govi, C. (1971) Il tintinnabulo della 'Tomba degli Ori' dell'Arsenale Militare di Bologna. *ArchCl* 23, 211–35.

Morigi Govi, C., ed. (1982) *Il Museo Archeologico di Bologna.* Bologna, Bononia University Press.

Morigi Govi, C. and Vitali, D. (1979) Zona della Fiera. *StEtr* 47, 467–68.

Morigi Govi, C. and Vitali, D. (1988) *Il Museo Civico Archeologico di Bologna.* Bologna, Bononia University Press.

Moscati, P. (1992) Pesi da telaio, rocchetti,

fuseruole. In M. Cristofani (ed.), *Lo scarico arcaico della Vigna Parocchiale*, 467–476. Roma, Consiglio Nazionale delle Ricerche.

Moscati, S., ed. (1988) *The Phoenicians*. Milano, Bompiani.

Moscati, S., ed. (1991) *The Celts*. Milano, Bompiani.

Moser, M. E. (1996) The Origins of the Etruscans: New Evidence for an Old Question. In J. F. Hall (ed.), *Etruscan Italy. Etruscan Influences on the Civilizations of Italy from Antiquity to the Modern Era*, 29–43. Provo, Brigham Young University.

Moss, H. E. (1987) Kumihimo: the art of Japanese braiding. *Fiberarts Magazine* 14, 20–21.

Mosso, A. (1908) Stazione preistorica di Coppa Nevigata presso Manfredonia. *MonAnt* 19, 305–386.

Moulherat, C. and Vial, G. (2000) Première attestation d'un tissu en laine de chèvre cachemire en Gaule. In D. Cardon and M. Feugère (eds), *Archéologie des textiles des origines au Ve siècle, Actes du colloque de Lattes, oct. 1999*, 107–113. Monographies Instrumentum 14. Montpellier, Éditions Monique Mergoil.

Mozia. (1989) *Itinerari IV*. Roma.

Müller-Karpe, H. (1959) *Beiträge zur Chronologie der Urnenfelderzeit nördlich und südlich der Alpen*. Berlin, De Gruyter.

Müller-Karpe, H. (1974) *Beiträge zu italienischen und griechischen Bronzefunde. PBF XX, 1.* München, Beck.

Muñoz, A. M. (1958) Prospecciones y excavaciones arqueológicas en la region de Torano: La Grotta dell'Olivo (Savona, Italia). *Cuadernos de Trabajos de la Escuela Española de Arqueología en Roma* X, 173–200.

Munro, R. (1890) *Lake Dwellings of Europe*. London, Methuen.

Murray Threipland, L. (1963) Excavations beside the North-West gate of Veii, 1957–58. Part II: Pottery. *PBSR* 31, 33–73.

Mutti, A. *et al.* (1988) *La terramara di Castione dei Marchesi*. Studi e Documenti di Archeologia V. Bologna, Nuova Alfa.

Nagy, G. (1999) *Ürschhausen-Horn. Keramik und Kleinfunde der spätestbronzezeitlichen Siedlung*. Forschungen im Seebachtal 2, Archäologie im Thurgau 6. Frauenfeld.

Nash Briggs, D. (2007) Invisible trade: piracy and land-based slave trade in Etruscan Italy and France from the ninth to the fifth centuries BC. *EtrStud 9*, 153–176.

Naso, A. (1996) *Architetture dipinte. Decorazioni parietali non figurate nelle tombe a camera dell'Etruria meridionale (VII–V sec. a. C.)*. Roma, l'Erma di Bretschneider.

Naso, A. (2000a) *I Piceni*. Milano, Longanesi.

Naso, A. (2000b) Etruscan and Italic artefacts from the Aegean. In D. Ridgway (ed.), *Ancient Italy in its Mediterranean Setting*, 193–207. Accordia Specialist Studies on the Mediterranean 4. London, Accordia Research Institute.

Naso, A., Giacummo, M. and Mandolesi, A. (2003) Ricerche di superficie a Botromagno (Gravina di Puglia). Rapporto preliminare. *Taras* 18, 2 (1998), 247–262.

Nava, M. L. (1988) *Stele Daunie I*. Firenze, Sansoni Editore.

Negroni Catacchio, N. (1975) Le fasi finale dell civiltà di Golasecca nell'ambito degli aspetti culturali della Valpadana fino alla romanizzazione. In *PCIA* 4, 329–345. Roma, Biblioteca di Storia Patria.

Negroni Catacchio, N., ed. (1982) *Sorgenti della Nova: una comunita protostorica ed il suo territorio nell'Etruria meridionale*. Roma, Vision.

Negroni Catacchio, N., ed. (1995) *Sorgenti della Nova. L'abitato del Bronzo Finale*. Firenze, Istituto Italiano di Preistoria e Protostoria.

Neils, J. (1992) *Goddess and Polis: The Panathenaic Festival in Ancient Athens*. Princeton, Princeton University Press.

Nenci, G. (1999) Siti e insediamenti nel territorio elimo. In M. Barra Bagnasco, E. De Miro and A. Pinzone (eds), *Magna Grecia e Sicilia. Stato degli studi e prospettive di ricerca. Atti dell'Incontro di Studi, Messina 2–4 dicembre 1996*. Messina, Di.Sc.A.M.

Nieling, J. (2007) Brettchengewebte Borten der späten Hallstattzeit in Prunkgrab von Hochdorf und auf den sog. Daunischen Stelen in Apulien. In S. Burmeister, H. Derks and J. von Richthofen (eds) *Zweiundvirzig. Festschrift für Michael Gebühr zum 65,* 159–166. *Geburtstag* Internationale Archäologie 25. Rahden/Westf., Leidorf.

Nielsen, E. (1998a) Bronze Production at Poggio Civitate (Murlo). *EtrStud* 5, 95–107.

Nielsen, M. (1998b) Etruscan Women: A Cross-Cultural Perspective. In L. Larsson Lovén and A. Strömberg (eds), *Aspects of Women in Antiquity*, 69–84. Jonsered, Paul Åströms Förlag.

Niemeyer, H. G. (1990) The Phoenicians in the Mediterranean: A Non-Greek Model for Expansion and Settlement in Antiquity. In J.-P. Descoeudres (ed.), *Greek Colonists and Native Populations*, 469–489. Oxford, Clarendon Press.

Nijboer. A. J. (1997) The Role of Craftsmen in the Urbanization Process of Central Italy (8th to 6th Centuries BC). *ActaHyp* 7, 383–406.

Nijboer, A. J. (1998) *From Household Production to Workshops.* Groningen, Rijksuniversiteit.

Noè, E. (1974) La produzione tessile nella Gallia Cisalpina in età romana. *RendIstLomb* 108, 918–932.

Nosch, M.-L. (2004) Red Coloured Textiles in the Linear B Inscriptions. In L. Cleland and K. Staers (eds), *Colour in the Ancient Mediterranean World*, 32–39. BAR-IS 1267. Oxford.

Nylander, C. and Pelagatti, P. (1986) *Architettura Etrusca nel Viterbese. Ricerche svedesi a San Giovenale e Acquarossa 1956–1986.* Roma, Leonardo Arte.

Olinder, B. and Pohl, I. (1981) *San Giovenale. Vol. II.4. The semi-subterranean building in Area B.* Stockholm, The Swedish Institute in Rome.

Orlandini, P. (1953) Scopo e significato dei cosidetti "pesi da telaio". *RendAcadNazLinc* 8, 441–444.

Orlandini, P. (1956) Gela. – Ritrovamenti vari. *NSc* 10, 203–401.

Orlandini, P. (1962) Gela. L'acropoli di Gela. *NSc* 16, 340–408.

Orlandini, P. (1966) Lo scavo del Thesmophorion di Bitalemi e il culto delle divinità ctonie a Gela. *Kokalos* 12, 8–39.

Orlandini, P. (1971) Vassallaggi. Scavi 1961. I. La necropoli meridionale. *NSc* 25 *Suppl.*

Orsi, P. (1905) Necropoli e stazioni sicule di transizione. Necropoli al Molino della Badia presso Grammichele (Catania). *BPI* 31, 96–133.

Orsi, P. (1906) Gela. Scavi del 1900–1905. *MonAnt* 17.

Orsi, P. (1912a) Locri Epizefirii. Necropoli sicula di Canale, Janchina, Patariti. *NSc* 1912 Suppl., 22–56.

Orsi, P. (1912b) Cotrone. Materiale inedito del Museo civico. *NSc* 1912 Suppl., 60–64.

Orsi, P. (1926) Le necropoli preelleniche calabresi di Torre Galli, e di Canale, Ianchina, Patariti. *MonAnt* 31, 5–376.

Osborne, R. (1998) Early Greek Colonization? The nature of Greek settlements in the West. In N. Fisher and H. van Wees (eds), *Archaic Greece. New Approaches and New Evidence*, 251–270. London, Duckworth.

Oscarson, U. (1994) *The Överhogdal Weaves.* Östersund, Jämtland County Museum.

Östenberg, C. E. (1967) *Luni sul Mignone e problemi della preistoria d'Italia.* Lund, The Swedish Institute in Rome.

Östenberg, C. E. (1975) *Case Etrusche di Acquarossa.* Roma, Multigrafica.

Östenberg, C. E. (1983) Acquarossa (Viterbo). – Rapporto preliminare. Cenni introduttivi, le necropoli e I periodi preistorici e protostorici. *NSc* 37, 25–96.

Pacciarelli, M. (1999) *Torre Galli. La necropoli della prima età del ferro (Scavi Paolo Orsi 1922–23).* Soveria Manelli (Catanzaro), Rubbettino Editore.

Pacciarelli, M. (2000) *Dal vilaggio alla città. La svolta protourbana del 1000 a.C. nell'Italia tirrenica.* Firenze, All'Insegna del Giglio.

Padden, A. N. and John, P. (2000) Indigo-reducing *Clostridium isatidis* Isolated from a Variety of Sources, Including a 10th-Century Viking Dye Vat. *JAS* 27, 953–956.

Padova Preromana (1976) Padova, Nuovo Museo Civico agli Eremitani.

Pallottino, M. (1939) Sulle facies culturali arcaiche dell'Etruria. *StEtr* 13, 85–129.

Pallottino, M. (1952) *Etruscan Painting.* Geneva, Skira.

Palm, J. (1952) Veiian tomb groups in the Museo Preistorico, Rome. *OpArch* 7, 50–86.

Pancrazzi, O. (1979) *Cavallino I. Scavi e Ricerche 1964–1967.* Galatina, Congedo Editore.

Panvini, R. (2001) *The Archaic Greek Ship at Gela (and Preliminary Exploration of a Second Greek Shipwreck)*. Palermo, Salvatore Sciascia Editore.

Paoletti, O. and Tamagno Perna, L. (2002) *Etruria e Sardegna centro-settentrionale tra l'età del Bronzo Finale e l'arcaismo. Atti del XXI Convegno di Studi Etruschi ed Italici. Sassari-Alghero-Oristano-Toorralba, 13–17 ottobre 1998*. Pisa/Roma, Istituti Editoriali Poligrafici Internazionali.

Paolucci, G. (1997) *Museo Civico Archeologico delle Acque di Chianciano Terme*. Siena, Protagon Editori Toscani.

Pare, C. F. E. (1996) Chronology in Central Europe at the End of the Bronze Age. *ActaArch* 67, 99–120.

Pareti, L. (1947) *La Tomba Regolini-Galassi del Museo Gregoriano Etrusco e la civiltà dell'Italia Centrale nel sec. VII A.C.* Città del Vaticano, Tip. Poliglotta Vaticana.

Paribeni, R. (1905) Civitella S. Paolo – Scavi nella necropoli capenate. *NSc* 2, 301–362.

Paribeni, R. (1906) Necropoli del territorio capenate. *MonAnt* 16, 277–490.

Paribeni, R. (1928) Capodimonte – Ritrovamento di Tombe Arcaiche. *NSc*, 434–467.

Parise Badoni, F. and Ruggeri Giove, M. (1980) *Alfedena. La necropoli di Campo Consolino. Scavi 1974–1979*. Chieti, Soprintendenza Archeologica dell'Abruzzo.

Parker, A. J. (1992) *Ancient Shipwrecks of the Mediterranean and the Roman Provinces*. BAR IS 580. Oxford, Tempus Reparatum.

Parsons, M. H. (1975) The distribution of Late Postclassic spindle whorls in the Valley of Mexico. *AmerAnt* 40, 207–215.

Pascucci, P. (1991) I depositi votivi paleoveneti: diversi livelli di "religiosità" in rapporto con il territorio e con le strutture sociali. In G. Bartoloni, G. Colonna and C. Grottanelli (eds), *Anathema. Regime delle offerte e vita dei santuari nel Mediterraneo antico. Atti del convegno internazionale, Roma, 15–18 Giugno 1989. ScAnt* 2–4 (1989–1990), 265–486.

Pasqui, A. (1894) Delle tombe di Narce e dei loro corredi. *MonAnt* 4, 399–548.

Pasqui, A. (1902) Mazzano Romano. – Scavi del principe Del Drago, nel territorio di questo commune. *NSc* 10, 321–355.

Pautasso, A. (2007) *Picturae in textili* on the shoulder busts in Hellenistic Sicily. In M.-L. Nosch and C. Gillis (eds), *Ancient Textiles: Production, Craft and Society*, 215–219. Oxford, Oxbow Books.

Peacock, D. P. S. (1982) *Pottery in the Roman World: An Ethnoarchaeological Approach*. London/New York, Longman.

Peacock, E. E. (2001) The contribution of experimental archaeology to the research of ancient textiles. In P. Walton Rogers, L. Bender Jorgensen and A. Rast-Eicher (eds), *The Roman Textile Industry and its Influence. A Birthday tribute to John Peter Wild*, 181–192. Exeter, Oxbow Books.

Peacock, E. E. (2005) The biodeterioration of textile fibres in wet archaeological contexts with implications for conservation choices. In *Intrecci vegetali e fibre tessili da ambiente umido. Incontri di restauro 4*, 29–61. Trento, Provincia Autonoma di Trento.

Pedley, J. G. (1990) *Paestum. Greeks and Romans in Southern Italy*. London, Thames and Hudson.

Pedley, J. G. (1993) *The Sanctuary of Santa Venera at Paestum. Vol. I.* Archeologia Perusina 11. Roma, Bretschneider.

Pedrotti, A., ed. (1995) *Le statue stele di Arco. La statuaria antropomorfa alpina nel III millennio a.C.: abbigliamento, fibre tessili e colore*. Riva del Garda (Trento), Provincia Autonoma di Trento.

Pellegrini, G. (1903) Tombe greche arcaiche e tomba greco-sannitica a tholos della necropoli di Cuma. *MonAnt* 13, 201–294.

Percossi Serenelli, E., ed. (1998) *Museo Nazionale delle Marche. Sezione protostorica. I Piceni*. Falconara, Soprintendenza Archeologica per le Marche.

Perini, R. (1970) Una decorazione su tessuto dalla palafitta di Ledro. *Natura Alpina* 21, 28–32.

Perini, R. (1971) I depositi preistorici di Romagnano-Loch (Trento). *Preistoria Alpina, Rendiconti* 7, 7–106.

Perini, R. (1978) *2000 anni di vita sui Montesei di Serso*. Trento, Provincia autonoma di Trento.

Perini, R. (1987) *Scavi archeologici nella zona palafitticola di Fiavè-Carera. Parte II: resti della cultura materiale: metallo, osso, litica, legno*. Trento, Servizio Beni Culturali della Provincia Autonoma di Trento.

Perini, R. (1994) *Scavi archeologici nella zona palafitticola di Fiavè-Carara. Parte III: resti della cultura materiale, ceramica*. Trento, Servizio Beni Culturali della Provincia Autonoma di Trento.

Pernier, L. (1907) Corneto Tarquinia. Nuove scoperte nel territorio tarquiniese. *NSc*, 43–82, 227–261, 321–352.

Peroni, R. (1971) *L'età del Bronzo nella penisola Italiana*. Firenze, Leo S. Olschki.

Peroni, R. (1987) La protostoria. In S. Settis (ed.), *Storia della Calabria Antica*, 67–136. Bari, Gangemi.

Peroni, R. (1989) *Protostoria dell'Italia continentale. La penisola italiana nelle età del Bronzo e del Ferro*. Roma, Spazio Tre.

Peroni, R. (1993) La Sibaritide prima di Sibari. In *Sibari e la Sibaritide. Atti del trentaduesimo Convegno di studi sulla Magna Grecia, Taranto-Sibari, 7–12 ottobre 1992*, 103–136. Taranto, Istituto per la storia e l'archeologia della Magna Grecia.

Peroni, R. (1994) *Introduzione alla protostoria Italiana*. Roma, Laterza.

Peroni, R. (1996) *L'Italia alle soglie della storia*. Roma, Laterza.

Peroni, R. and Trucco, F., eds (1994) *Enotri e Micenei nella Sibaritide*. Taranto, Istituto per la storia e l'archeologia della Magna Grecia.

Peruzzi, E. (1992) Cultura Greca a Gabii nel secolo VIII. *PP* 47, 459–468.

Pfister, R. (1934) *Textiles de Palmyre*. Paris, Les Éditions d'art et d'histoire.

Pfister, R. (1937) *Nouveaux Textiles de Palmyre*. Paris, Les Éditions d'art et d'histoire.

Pfister, R. (1940) *Textiles de Palmyre III*. Paris, Les Éditions d'art et d'histoire.

Pfister, R. and Bellinger, L. (1945) Part II. The Textiles. In M. I. Rostovtzeff *et al.* (eds), *The Excavations at Dura-Europos. Final Report IV*. New Haven, Yale University Press.

Phillips, K. M. Jr. (1993) *In the Hills of Tuscany: Recent Excavations at the Etruscan Site of Poggio Civitate (Murlo, Siena)*. Philadelphia, University Museum.

Phillips, K. M. Jr. (1994) Stamped Impasto Pottery Manufactured at Poggio Civitate. In R. D. De Puma and J. P. Small (eds), *Murlo and the Etruscans: Art and Society in Ancient Etruria*, 29–46. Madison, The University of Wisconsin Press.

Pincelli, R. and Morigi Govi, C. (1975) *La necropoli villanoviana di San Vitale*. Bologna, Museo Civico Archeologico.

Pinza, G. (1905) Monumenti primitivi di Roma e del Lazio antico. *MonAnt* 15.

Pionati Shams, G. (1987) Some Minor Textiles in Antiquity. Göteborg, Paul Åströms Förlag.

Pisani Santorio, G. (1977) L'Area sacra di S. Omobono. Esame preliminare dei materiali archeologici – Parte III. *PP* 32, 55–61.

Pizzano, N. (2005) *Scheda 10*. In Longola di Poggiomarino, folder pubblicato a cura della Soprintendenza archeologica di Pompei in occasione della XL Riunione Scientifica Italiana.

Pizzo, M. (1986) Fuseruole. In M. Castoldi (ed.), *I Greci sul Basento. Mostra degli scavi archeologici all'Incoronata di Metaponto 1971–1984*, 179–180. Como, Edizioni New Press.

Pizzo, M. (1999) Vassallaggi (S. Cataldo, Caltanissetta). La necropoli meridionale, scavi 1956. *NSc* 9–10, 207–398.

Pohl, I. (1972) *Iron Age Necropolis of Sorbo at Cerveteri*. Stockholm, Swedish Institute in Rome.

Pohl, I. (1977) *San Giovenale III.3. The Iron Age Habitations in Area E*. Stockholm, Swedish Institute in Rome.

Pohl, I. (1981) Part II. In *San Giovenale II.4. The Semi-subterranean Building in Area B*. Stockholm, Swedish Institute in Rome.

Polosmak, N. V. and Barkova, L. L. (2005) *Костюм и текстиль пазырыкцев Алтая. (IV–III вв. до н. э.)*. Novosibirsk, Infolio.

Pontrandolfo, A. and Rouveret, A. (1992) *Le Tombe dipinte di Paestum*. Modena, Franco Cosimo Panini Editore.

Potter, T. W. (1976) *A Faliscan Town in South Etruria. Excavations at Narce 1966–71*. London, The British School at Rome.

Potthast, I. and Riens, R. (2003) Conservazione di reperti bagnati. In M. Bazzanella *et al.* (eds), *Textiles. Intrecci e tessuti dalla preistoria europea*, 31–40. Riva del Garda (Trento), Provincia Autonoma di Trento.

Procopio, G. (1962) La necropoli preellenica di Calanna. *Klearchos* 4, 21–31.

Prosdocimi, A. (1887) Avanzi di antichissime abitazioni nell'agro atestino. *BPI* 13, 156–167, 185–201.

Protostoria tra Sile e Tagliamento (1996) *Antiche genti tra Veneto e Friuli (Mostra)*. Padova, Esedra Editrice.

Pugliese Carratelli, G., ed. (1985a) *Magna Grecia. Il Mediterraneo, le metropoleis e la fondazione delle colonie*. Milano, Electa.

Pugliese Carratelli, G., ed. (1985b) *Sicanie. Storia e civiltà della Sicilia greca*. Milano, Libri Scheiwiller.

Pugliese Carratelli, G., ed. (1986a) *Megale Hellas. Storia e civiltà della Magna Grecia*. Milano, Libri Scheiwiller.

Pugliese Carratelli, G., ed. (1986b) *Rasenna. Storia e civiltà degli etruschi*. Milano, Libri Scheiwiller.

Pugliese Carratelli, G., ed. (1996) *The Western Greeks*. Milano, Bompiani.

Pyrgi. Scavi del santuario etrusco (1959–1967). (1970) *NSc Suppl.* 24.

Quilici, L. (1983) Palestrina: luogi di ritrovamento di materiale votivo. *QArchEtr* 7, 88–103.

Quilici, L. and Quilici Gigli, S. (1980) *Crustumerium*. Roma, Consiglio Nazionale delle Ricerche.

Quagliati, Q. (1904) Regione III (Lucania et Bruttii) – Pisticci: Tombe lucane con ceramiche greche. *NSc* 1, 196–208.

Ræder Knudsen, L. (2002) La tessitura con le tavolette nella tomba 89. In P. von Eles (ed.), *Guerriero e sacerdote. Autorità e comunità nell'età del ferro a Verucchio. La Tomba del Trono*, 220–234. Firenze, All'Insegna del Giglio.

Rageth, J. (1974) Der Lago di Ledro im Trentino und seine Beziehungen zu den alpinen und mitteleuropäischen Kulturen. *BerRGK* 55 (no. 1), 73–259.

Raheel, M. (1994) History, Identification, and Characterization of Old World Fibers and Dyes. In S. U. Wisseman and W. S. Williams (eds), *Ancient Technologies and Archaeological Materials*. Langhorne, Gordon and Breach Science Publishers.

Rahmstorf, L. (2004) Clay spools from Tiryns and other contemporary sites. An indication of foreign influence in LH III C?, in N. Kyparissi-Apostolika and M. Papakonstantinou (eds), Η Περιφέρεια του Μυκηναϊκού Κόσμου. Β΄Διεθνές Διεπιστημονικό Συμπόσιο/*The Periphery of the Mycenaean World. 2nd International Interdisciplinary Colloquium. Lamia 26–30.9.1999*, (Lamia 2003), 397–415. Athens, Ministry of Culture.

Rallo, A., ed. (1989a) *Le donne in Etruria*. Roma, L'Erma di Bretschneider.

Rallo, A. (1989b) Fonti. In A. Rallo (ed.), *Le donne in Etruria*, 15–33. Roma, L'Erma di Bretschneider.

Rallo, A. (1989c) Classi sociali e mano d'opera femminile. In A. Rallo (ed.), *Le donne in Etruria*, 147–156. Roma, L'Erma di Bretschneider.

Randal-McIver, D. (1924) *Villanovans and Early Etruscans*. Oxford, Clarendon.

Randsborg, K. (1991) Historical Implications. Chronological Studies in European Archaeology c. 2000–500 B.C. *ActaArch* 62, 89–108.

Rast-Eicher, A. (1997) Tessuti dell'età del bronzo in Europa. In M. Barnabò Brea, A. Cardarelli and M. Cremaschi (eds), *Le Terremare. La più antica civiltà padana*, 545–553. Modena, Electa.

Rast-Eicher, A. (2000b) Die eisenzeitlichen Gewebe im Tessin. In R.C. de Marinis and S. Biaggio Simona (eds), *I Leponti tra mito e realtà. Raccolta di saggi in occasione della mostra*, 415–419. Locarno, Armando Dadò.

Rast-Eicher, A. (2003) Determinazione delle fibre. In M. Bazzanella *et al.* (eds), *Textiles. Intrecci e tessuti dalla preistoria europea*, 55–60. Riva del Garda (Trento), Provincia Autonoma di Trento.

Rast-Eicher, A. (2005) Bast before wool: the first textiles, in P. Bichler *et al.* (eds), *Hallstatt Textiles: Technical Analysis, Scientific Investigation and Experiment on Iron Age Textiles*, 117–131. BAR-IS 1351. Oxford.

Rast-Eicher, A. (forthcoming) Textilien, Leder, Fell aus eisenzeitlichen Gräbern der Schweiz und die Entwicklung der Wolle von der Hallstattzeit bis zum römischen

Epoche. (Diss. Bern 2006).

Rast-Eicher, A. and Thijsse, S. 2001. Anbau und Verarbeitung von Lein: Experiment und archäologisches Material. *ZSchwArch* 58, 47–56.

Rebecchi, F., ed. (1994) *Spina e il delta Padano. Riflessioni sul catalogo e sulla mostra ferrarese.* Roma, l'Erma di Bretschneider.

Reese D. S. (2005) Whale bones and shell purple-dye at Motya (Western Sicily, Italy). *OJA* 24(2), 107–114.

Renfrew, C. (1982) *Towards an Archaeology of Mind: An Inaugural Lecture Delivered before the University of Cambridge on 30 November 1982.* Cambridge, Cambridge University Press.

Renfrew, C. and Bahn, P. (1996) *Archaeology. Theories Methods and Practice.* 2nd Edition. London and New York, Thames and Hudson.

Renfrew, J. M. (1973) *Palaeoethnobotany: the prehistoric food plants of the Near East and Europe.* London, Methuen.

Renfrew, J. M., ed. (1991) *New Light on Early Farming: Recent Developments in Palaeoethnobotany.* Edinburgh, Edinburgh University Press.

Ridgway, D. (1988a) Italy from the Bronze Age to the Iron Age. In *CAH*, 2nd ed., vol. 4, 623–633.

Ridgway, D. (1988b) The Etruscans. In *CAH*, 2nd ed., vol. 4, 634–675.

Ridgway, D. (1992) *The First Western Greeks.* 2nd ed. Cambridge, Cambridge University Press.

Ridgway, D. (1994) Phoenicians and Greeks in the West: A View from Pithekoussai. In G. R. Tsetskhladze and F. de Angelis (eds), *The Archaeology of Greek Colonisation*, 35–46. Oxford, Oxford Committee for Archaeology.

Ridgway, D. (1996a) Greek Letters at Osteria dell'Osa. *OpRom* 20, 87–97.

Ridgway, D. (1996b) Relations between Cyprus and the West in the Precolonial Period. In G. Pugliese Carratelli (ed.), *The Western Greeks*, 117–120. Venezia, Bompiani.

Ridgway, D. (2002) Archaeology in Sardinia and South Italy 1995–2001. *AR* 2001–2002, 117–138.

Riedel, A. and Tecchiati, U. (2003) La capra e la pecora in Italia tra Neolitico e l'età del Bronzo. In Bazzanella *et al.* (eds), *Textiles. Intrecci e tessuti dalla preistoria europea*, 73–78. Riva del Garda (Trento), Provincia Autonoma di Trento.

Rilli, N. (1964) *Gli Etruschi a Sesto Fiorentino.* Firenze, Topografia Giuntina.

Rittatore Vonwiller, F. (1966) *La necropoli preromana della Ca'Morta.* Como.

Rittatore Vonwiller, F. (1975) La Civiltà del ferro in Lombardia, Piemonte, Liguria. In *PCIA 4*. Roma, Biblioteca Di Storia Patria.

Robertson, R. H. S. (1949) The Fuller's Earth of the Elder Pliny. *CR* 63, 51–52.

Robinson, D. M. (1930) *Excavations at Olynthus. Part II. Architecture and Sculpture: Houses and Other Buildings.* Baltimore, The Johns Hopkins Press.

Robinson, D. M. (1941) *Excavations at Olynthus. Part X. Metal and Minor Miscellaneous Finds.* Baltimore, The Johns Hopkins Press.

Roche-Bernard, G. (1993) *Costumes et Textiles en Gaule Romaine.* Paris, Editions Errance.

Roncalli, F. (1980) '*Carbasinis Voluminibus Implicati Libri*' Osservazioni sul Liber Linteus di Zagabria. *JdI* 95, 227–264.

Roncalli, F. (1998) Una imagine femminile di culto dalla "Tomba d'Iside" di Vulci. *AnnFaina* 5, 15–40.

Roncalli, F. and Bonfante, L., eds (1991) *Gens antiquissima Italiae. Antichità dall'Umbria a New York.* Perugia, Electa Editori Umbri.

Rosada, G. (2004) Altino e la via della transumanza nella *Venetia* centrale. In B. S. Frizell (ed.), *PECUS. Man and animal in antiquity. Proceedings of the conference at the Swedish Institute in Rome, September 9–12, 2002*, 67–79. Rome, The Swedish Institute in Rome.

Rossoni, G. (2002) Pesi da telaio. In M. L. Famà (ed.), *Mozia. Gli scavi nella "Zona A" dell'abitato*, 315–320. Bari, Edipuglia.

Rottoli, M. (1993) "La Marmotta", Anguillara Sabazia (RM). Scavi 1989. Analisi paletnobotaniche: prime risultanze. Appendice 1 in M. A. Fugazzola Delpino *et al.*, "La Marmotta" (Anguillara Sabazia, RM). Scavi 1989. Un abitato perilacustre di età neolitica. *BPI* 84, 305–315.

Rottoli, M. (1999) I resti vegetali di Sammardenchia – Cûeis (Udine), insediamento del Neolitico antico. In A. Ferrari and A. Pessina (eds), *Sammardenchia – Cûeis. Contributi per la conoscenza di una comunità del primo neolitico. Edizioni del Museo Friuliano di Storia Naturale 41*, 307–326. Udine, Edizioni del Museo Friulano di Storia Naturale.

Rottoli, M. (2003) Il lino. In M. Bazzanella *et al.* (eds), *Textiles. Intrecci e tessuti dalla preistoria europea*, 65–72. Riva del Garda (Trento), Provincia Autonoma di Trento.

Rottoli, M. (2005) Tessuti e intrecci dalla preistoria al medioevo: recupero, conservazione e analisi. Le esperienze del laboratorio di archeologia dei Musei Civici di Como. In *Intrecci vegetali e fibre tessili da ambiente umido. Analisi Conservazione e Restauro. Atti Trento 28–30 maggio 2003. Incontri di Restauro 4*, 62–92. Riva del Garda (Trento), Provincia Autonoma di Trento.

Ruby, P. (1995) *Le crépuscule des marges: le premier Âge du Fer à Sala Consilina.* Rome/Naples, Centre Jean Bérard.

Rudenko, S. I. (1970) *Frozen Tombs of Siberia: The Pazyryk Burials of the Iron Age Horsemen.* Berkley and Los Angeles, UCLA Press.

Russo Tagliente, A., ed. (1995) Armento. Archeologia di un centro indigeno. *BolArch* 35–36.

Russo, A., ed. (2006) Con il fuso e la conocchia. La fattoria lucana di Montemurro e l'edilizia domestica nel IV sec. a.C. Milano, Soprintendenza archeologica della Basilicata.

Rutschmann, H. (1985) Webgewichte als Bildträger. *AntW* 19, 46–55.

Ryder, M. L. (1962) The Origin of Felt-Making and Spinning. *Antiquity* 36, 304.

Ryder, M. L. (1965) Report of Textiles from Çatal Hüyük. *AnatSt* 15, 176.

Ryder, M. L. (1968) The Origin of Spinning. In *Textile History*, 73–82. Guilford, Butterworth.

Ryder, M. L. (1974) Wools from Antiquity. *Textile History* 5, 100–110.

Ryder, M. L. (1992) The interaction between biological and technological change during the development of different fleece types in sheep. *Anthropozoologica* 16, 131–140.

Ryder, M. L. (1993a) Sheep and goat husbandry with particular reference to textile and milk production. *Bulletin on Sumerian Agriculture 7: Domestic Animals in Mesopotamia, Part I*, 9–32.

Ryder, M. L. (1993b) The use of goat hair. An introductory historical review. *Anthropozoologica* 17, 37–46.

Sackett, L. H. and Popham, R. M. (1965) Excavations at Palaikastro VI. *BSA* 60, 248–315.

Sagona, C. (1999) Silo or vat? Observations on the ancient textile industry in Malta and early Phoenician interests in the island. *OJA* 18 (1), 23–60.

Salmon, E. T. (1985) *Sannio e i Sanniti.* Torino, Einaudi.

Salmon, E. T. (1988) The Iron Age: The Peoples of Italy. In *CAH* 4, 676–719.

Salzani, L. (1987) La Tomba 4 della necropoli di Arquà Petrarca (Padova). In D. Vitali, ed., *Celti ed Etruschi nell'Italia centro-settentrionale dal V sec. A. C. alla Romanizzazione*, 271–280. Bologna, Bononia University Press.

Santillo Frizell, B., ed. (2004a) *PECUS. Man and animal in antiquity. Proceedings of the conference at the Swedish Institute in Rome, September 9–12, 2002.* Rome, The Swedish Institute in Rome.

Santillo Frizell, B. (2004b) Curing the flock. The use of healing waters in Roman pastoral economy. In B. S. Frizell (ed.), *PECUS. Man and animal in antiquity. Proceedings of the conference at the Swedish Institute in Rome, September 9–12, 2002*, 80–93. Rome, The Swedish Institute in Rome.

Santoro, P. (1979) *I Galli e l'Italia.* Roma, De Luca.

Santoro, P. (1983) Colle del Forno (Roma). Loc. Montelibretti. Relazione preliminare di scavo della campagna settembre-ottobre 1979 nella necropoli. *NSc* 37, 105–140.

Sartori, A. (2001) Produzioni in impasto. Pesi da telaio. In M. Bonghi Jovino (ed.), *Tarquinia. Scavi sistematici nell'abitato. Campagne 1982–1988. I materiali 2*, 137–146. Tarchna III. Rome, Bretschneider.

Sartoris, A. (1997) I pesi da telaio. In M. Barra Bagnasco (ed.), *Pomarico Vecchio I. Abitato mura necropoli materiali*, 227–230. Galatina, Congedo Editore.

Sassaman, K. E. (1992) Gender and Technology at the Archaic-Woodland "Transition". In C. Claassen, ed., *Exploring Gender Through Archaeology. Selected Papers from the 1991 Boone Conference*, 71–79. Madison, Wisconsin.

Scarfì, B. M. (1962) Gioia del Colle. *NSc* 16, 1–286.

Scarpa, G. (1988) La fauna. In R. De Marinis (ed.), *Gli Etruschi al nord del Po*, 184–192. Mantova, Campanotto Editore.

Scheid, J. and Svenbro, J. (1996) *The Craft of Zeus: Myths of Weaving and Fabric*. (Transl. C. Volk). Cambridge, Harvard University Press.

Schick, T. (1988) Nahal Hemar Cave: Cordage, Basketry and Fabrics. *'Atiqot* 18, 31–43.

Schiffer, M. B. and Skibo, J. M. (1987) Theory and Experiment in the Study of Technological Change. *CurrAnthr* 28 (no. 5), 595–622.

Schlanger, N. (1994) Mindful Technology: Unleashing the chaîne opératoire for an Archaeology of Mind. In C. Renfrew and E. B. W. Zubrow (eds), *The Ancient Mind. Elements of Cognitive Archaeology*, 143–151. Cambridge, Cambridge University Press.

Schmidt Colinet, A., A. Stauffer and al-As'ad, K. (2000) *Die Textilien aus Palmyra*. Mainz, Philip von Zabern.

Seiler-Baldinger, A. (1994) *Textiles, A Classification of Techniques*. Bathurst, Crawford House.

Serra Ridgway, F. R. (2000) The Tomb of the Anina Family. Some motifs in late Tarquinian painting. In D. Ridgway *et al.* (ed.), *Ancient Italy in its Mediterranean Setting. Studies in Honour of Ellen Macnamara*, 301–316. London, Accordia Research Institute.

Serritella, A. (1990) Pesi da telaio. In G. Greco and A. Pontrandolfo (eds), *Fratte. Un insediamento etrusco-campano*, 159–167. Modena, Franco Cosimo Panini.

Shennan, S. (1989) Introduction: archaeological approaches to cultural identity. In S. Shennan (ed.), *Archaeological Approaches to Cultural Identity*, 1–32. London, Unwin Hyman.

Shepherd, G. (2004) Fibulae and Females: Intermarriage in the Western Greek Colonies and the Evidence from Cemeteries. In G. R. Tsetskhladze and F. de Angelis (eds), *The Archaeology of Greek Colonisation*, 267–300. Oxford, Oxford University Committee for Archaeology.

Sherratt, S. and Sherratt, A. (1993) The growth of the Mediterranean economy in the early first millennium BC. *WorldArch* 24 (no. 3), 361–378.

Shishlina, N. I., Orfinskaya, O. V. and Golikov, V. P. (2002) Bronze Age Textiles from North Caucasus: problems of origin. In *Steppe of Eurasia in Ancient Times and Middle Ages, Proceedings of International Conference*, 253–259. Saint Petersburg, The Hermitage.

Shishlina, N. I., Orfinskaya, O. V. and Golikov, V. P. (2003) Bronze Age Textiles from North Caucasus: new evidence of fourth millennium BC fibers and fabrics. *OJA* 22 (4), 331–344.

Sibari (1969) *NSc* 23 Suppl. 1.

Sibari (1970) Scavi al Parco del Cavallo (1960–1962, 1969–1970) e agli Stombi (1969–1970). *NSc* 24 Suppl. 3.

Silvestrini Lavagnoli, M. and Cazzella, A. (1981) L'insediamento piceno di Cartofaro (Ascoli Piceno). *Picus* 1, 151–162.

Sjöqvist, E. (1960) Excavations at Morgantina (Serra Orlando) 1959. Preliminary Report IV. *AJA* 64, 125–135.

Smith, C. J. (1998) Traders and artisans in archaic central Italy. In H. Parkins and C. J. Smith (eds), *Trade, Traders and the Ancient City*, 31–51. London and New York, Routledge.

Smith, C. J. (1999) Medea in Italy: Barter and Exchange. In G. Tsetskhladze (ed.), *Ancient Greeks West and East*, 179–206. Leiden, Brill.

Smith, J. S. (2002) Changes in Workplace: Women and Textile Production on Late Bronze Age Cyprus. In D. Bolger and N. Serwint (eds), *Engendering Aphrodite: Women and Society in Ancient Cyprus* (CAARI International Conference, 19–23 March 1998), 281–312. Boston, ASOR.

Smolnik, R. (1994) *Der Burgstallkogel bei Kleinklein II. Keramik der vorgeschichtlichen*

Siedlung. Veröffentlichungen des vorgeschichtlichen Seminars Marburg. Sonderband 9. Marburg.

Soranzo, F. (1885) *Scavi e scoperte nei poderi Nazaro di Este*. Roma, R. Accademia dei Lincei.

Speiser, N. (1991) On Classification of Primary Structures and Techniques. *Bulletin CIETA* 69, 13–15.

Spindler, K. *et al.*, eds (1995) *Der Mann im Eis. Neue Funde und Ergebnisse* 2. Veröffentlichung des Forschungsinstituts für Alpine Vorzeit der Universität Innsbruck 2. Wien/New York, Springer-Verlag.

Stærmose Nielsen, K.-H. (1999) *Kirkes Væv. Opstadvævens historie og nutidige brug*. Forsøg med fortiden 6. Roskilde, Historisk-Arkæologisk Forsøgcenter Lejre.

Stage, J. C. (1985) *Etruscan Textile Arts: The Literary and Archaeological Evidence*. MA thesis, Tallahassee, The Florida State University (unpublished).

Stage, J. C. (1991) Extant Etruscan Textiles: A Working List. *StMat* 6, 146–150.

Stager, L. (1991) When did the Philistines arrive in Canaan? Multiple clues help unravel the mystery. *Biblical Archaeology Review* 17, 10–19.

Stager, L. (1995) The impact of the Sea Peoples in Canaan (1185–1050). In T. E. Levy (ed.), *The Archaeology of Society in the Holy Land*, 332–348. New York, Facts on File.

Stauffer, A. (2000) The textiles from Palmyra: technical analyses and their evidence for archaeological research. In D. Cardon and M. Feugère (eds), *Archéologie des textiles des origines au Ve siècle, Actes du colloque de Lattes, oct. 1999*, 247–252. Monographies Instrumentum 14. Montpellier, Éditions Monique Mergoil.

Stauffer, A. (2002) I tessuti. In P. von Eles (ed.), *Guerriero e sacerdote. Autorità e comunità nell'età del ferro a Verucchio. La Tomba del Trono*, 192–219. Firenze, All'Insegna del Giglio.

Stauffer, A. (2003) Ein Gewebe mit Schnurapplikation aus der 'Tomba del Trono' in Verucchio (700 v. Chr.). In L. Bender Jørgensen, J. Banck-Burgess, and A. Rast-Eicher (eds), *Textilien aus Archäologie und Geschichte. Festschrift für Klaus Tidow*, 205–208. Neumünster, Wachholtz Verlag.

Stauffer, A. (2004) Early Etruscan Garments from Verucchio. *Bulletin du CIETA* 81, 14–20.

Stauffer, A. (2005) Tessuti archaeologici consolidati con sostanze sintetiche: problemi legati allo studio e alla conservazione. In *Intrecci vegetali e fibre tessili da ambiente umido. Incontri di restauro 4*, 174–187. Trento, Provincia Autonoma di Trento.

Stea, G. (1997) Oggetti diversi. In *Ricerche archeologiche all'Incoronata di Metaponto. 5. L'oikos greco del saggio H. Lo scavo e reperti*, 87–90. Milano, Edizioni ET.

Stefani, E. (1911) Nazzano. Nuove scoperte nel territorio capenate. Oggetti di suppellettile funebre provenienti da tombe di un antico sepolcro in contrada S. Lucia. *NSc*, 431–442.

Stefani, E. (1916) Terni. Scoperta di antichi sepolcri nella contrada di "S. Pietro in Campo", presso la stazione ferroviaria di Terni. *NSc*, 191–226.

Stefani, E. (1928) Veio. – Scoperta di antichi sepolcri nella tenuta di Monte Oliviero, presso Prima Porta. *NSc*, 95–105.

Stefani, E. (1958) Capena. Scoperte archeologiche nell'agro capenate. Ricerche archeologiche nella contrada "Le Saliere." *MonAnt* 44, 1–203.

Stegmann-Rajtár, S. (1998) Spinnen und Weben in Smolenice-Molpír. Ein Beitrag zum wirtschaftlichen und religiös-kultischen Leben der Bewohner des Hallstattzeitlichen "Fürstensitzes." *SlovArch* 46–2, 263–287.

Steigerwald, G. (1986) Die antike Purpurfärberei nach den Berichten Plinius des Älteren in seiner *Naturalis Historia. Traditio: Studies in Ancient and Medieval History, Thought and Religion* 42, 1–57.

Stein, G. J. (1999) *Rethinking World-systems: Diasporas, Colonies, and Interaction in Uruk Mesopotamia*. Tucson, University of Arizona Press.

Steinberg, A. R. (1966) *The Bronzes of the Bernardini Tomb*. Ph.D. diss., University of Pennsylvania (microfilm P38: 1966).

Steingräber, S. (1985) *Catalogue Raisonné of Etruscan Tomb Paintings*. New York, Harcuort Brace Jovanovich.

Steinhauser-Zimmermann, R.-A. (1989) *Der Montlingeberg im Kanton St. Gallen. Funde und Grabungen von 1898–1960*. St. Gallen.

Stevens, G. P. (1936) The Periclean Entrance Court of the Acropolis of Athens. *Hesperia* 5 (no. 4), 443–520.

Stoop, M. W. (1972) II. Terrecotte e ceramiche. In Francavilla Marittima: Varia. B. Santuario sul Timpone della Motta. *AttiMGrecia* 1970–1971, 7–81.

Stopponi, S. (1968) Παραπετασματα Etruschi. *BA* 53, 60–62.

Stopponi, S. (1983) *La Tomba della "Scrofa Nera"*. Materiali del Museo Archeologico Nazionale di Tarquinia, VIII. Roma, Bretschneider.

Stopponi, S. (1985) *Case e palazzi d'Etruria*. Regione Toscana, Electa.

Strömberg, A. (1993) *Male or Female: a methodological study of grave gifts as sex indicators in Iron Age burials from Athens*. Jonsered, P. Åstrom Förlag.

Tagliente, M. (1986) La necropoli di Alianello. In *Siris-Polieion*, 167–170. Galatina, Congedo.

Taglione, C. (1999) *L'abitato etrusco di Bologna*. Bologna, Bononia University Press.

Tamburini, P. (1995) *Un abitato villanoviano perilacustre. Il "Gran Carro" sul lago di Bolsena (1959–1985)*. Roma, Bretschneider.

Tatton-Brown, V. (1992) Loom-weights, Bobbins, and Spindle-whorls. In A. M. Small (ed.), *An Iron Age and Roman Republican Settlement on Botromagno, Gravina in Puglia: Excavations of 1965–1974*. Archaeological Monographs of the British School at Rome No. 5. London, British School at Rome.

Terranova, F. and Lo Campo, P. (1999) Indagini bioarcheologiche sui reperti di Colle Madore. In S. Vassallo (ed.), *Colle Madore. Un caso di ellenizzazione in terra sicana*, 283–284. Entella, Giuseppe Nenci.

Thompson, W. (1982) Weaving: A Man's Work? *CW* 75, 217–222.

Tiboni, F. (2005) Weaving and Ancient Sails: Structural Changes to Ships as a Consequence of New Weaving Technology in the Mediterranean Late Bronze Age. *IJNA* 34, 127–130.

Tinè, S. and Vagnetti, L. (1967) *I Micenei in Italia*. Fasano, Arti grafiche N. Schena.

Tissage, Corderie, Vannerie. IXe Rencontres Internationales d'Archéologie et d'Histoire, Antibes, Octobre 1988 (1989) Juan-les-Pins, Éditions Association pour la Promotion et la Diffusion des Connaissances Archéologiques.

Tinè Bertocchi, F. (1985) *Le necropoli daunie di Ascoli Satriano e Arpi*. Genova, Sagep Editrice.

Tizzoni, M. (1981) La Cultura tardo La Tène in Lombardia. *Studi Archeologici I*, 21–25. Bergamo.

Tocco Sciarelli, G. (1985) La fondazione di Cuma. In *Napoli Antica*, 87–99. Napoli, Gaetano Macchiaroli.

Tomedi, G. (2000) *Italische Panzerplatten und Panzerscheiben. PBF III, 3*. Stuttgart, Steiner.

Tomlinson, R. A. (1969) Perachora: the remains outside the two sanctuaries. *BSA* 64, 155–258.

Toms, J. (1986) The relative chronology of the Villanovan cemetery of Quattro Fontanili at Veii. *AION* 8, 41–97.

Toms, J. (1998) The Construction of Gender in Early Iron Age Etruria, in R. D. Whitehouse (ed.), *Gender and Italian Archaeology: Challenging the Stereotypes*, 157–179. Accordia Specialist Studies in Italy 7. London, Accordia Research Centre and Institute of Archaeology.

Tonini, A. (1970) La Tomba Tarquiniese del Cacciatore. *StEtr* 38, 45–65.

Torelli, M. (1984) *Lavinio e Roma*. Roma, Edizioni Quasar.

Torelli, M. (1986) History: Land and People. In L. Bonfante (ed.), *Etruscan Life and Afterlife. A Handbook of Etruscan Studies*, 47–65. Detroit, Wayne State University Press.

Torelli, M. (1997) "Domiseda, lanifica, univira". Il trono di Verucchio e il ruolo e l'imagine della donna tra arcaismo e repubblica. In M. Torelli, *Il rango, il mito e l'immagine. Alle origini della representazione storica romana*. Milano, Electa.

Torelli, M., ed. (2000) *Gli Etruschi*. Milano, Bompiani.

Törnkvist, S. (1969) Note on Linen Corslets. *OpRom* 7, 81–82.

Tosi, G., ed. (1992) *Este antica dalla preistoria all'età romana*. Este, Zielo.

Toti, O. (1964) Allumiere. – L'Abitato "protovillanoviano" di Monte Rovello. *NSc* 18, 12–28.

Tovoli, S. (1989) *Il sepolcreto villanoviano Benacci Caprara di Bologna*. Bologna, Gratis Edizioni.

Tovoli, S. and Vitali, D. (1979) *La necropoli villanoviana di Ca' dell'Orbo a Villanova di Castenaso. Problemi del popolamento dal IX al VI secolo a.C.* Bologna, Museo Civico Archeologico.

Trever, C. (1932) *Excavations in Northern Mongolia*. Leningrad, Russkoe geograficheskoe obshchestvo.

Trinkl, E. (2007) Artifacts Related to Preparation of Wool and Textile Processing Found Inside the Terrace Houses of Ephesus, Turkey. In M.-L. Nosch and C. Gillis (eds), *Ancient Textiles: Production, Craft and Society*, 81–86. Oxford, Oxbow Books.

Trucco, F. (1994) Le necropoli di Sala Consilina: analisi dei corredi, configurazioni planimetriche, gruppi sociali. In *La presenza etrusca in Campania meridionale, Atti Salerno-Pontecagnano 1990*, 135–152. Firenze, Olschki.

Trucco, F. (2006) Considerazioni sul rituale funerario in Etruria meridionale all'inizio dell'età del Ferro alla luce delle nuove ricerche a Tarquinia. In P. von Eles (ed.), *La ritualità funeraria tra età del Ferro e Orientalizzante in Italia, Atti del convegno, Verucchio, 26–27 giugno 2002*, 95–102. Pisa and Roma, Fabrizio Serra Editore.

Tuck, A. S. (1994) The Etruscan Seated Banquet: Villanovan Ritual and Etruscan Iconography. *AJA* 98, 617–628.

Tuck, A. S. (1996) *Burials from Poggio Aguzzo: The Necropolis of Poggio Civitate (Murlo)*. Ph.D. diss., Brown University.

Tunzi Sisto, A. M. (1988) *Pesi fittili da Adelfia. Dalla protostoria all'età classica. Tipologia di forme e decorazioni*. Manduria, Piero Lacaita Editore.

Turfa, J. M. (1982) The Etruscan and Italic Collection in the Manchester Museum. *PBSR* 50, 166–193.

Turfa, J. M. (1986) International Contacts: Commerce, Trade, and Foreign Affairs. In L. Bonfante (ed.), *Etruscan Life and Afterlife*, 69–72. Detroit, Wayne State University Press.

Turfa, J. M. (2004) Rummaging in the basement. *DIG* 6 (No. 3), 22–25.

Turfa, J. M. (2005) *Catalogue: Etruscan Gallery of the University of Pennsylvania Museum of Anthropology and Archaeology*. Philadelphia, University Museum.

Turfa, J. M. (2006a) Votive Offerings in Etruscan Religion. In N. T. De Grummond and E. Simon (eds), *The Religion of the Etruscans*, 90–115. Austin, The University of Texas Press.

Turfa, J. M. (2006b) Review of Gilda Bartoloni, Filippo Delpino, Oriente e Occidente: metodi e discipline a confronto. Riflessioni sulla cronologia dell'età del ferro in Italia, Atti dell'Incontro di studi, Roma, 30–31 ottobre 2003. Mediterranea 1 (2004), Pisa, Istituti Editoriali e Poligrafici Internazionali, 2005. *BMCR* 2006.08.10.

Turfa, J. M. and Gleba, M. (2005) Book review of Larissa Bonfante, *Etruscan Dress*, Updated Edition, 2003, The Johns Hopkins University Press. *BMCR* 2005.04.60.

Turfa, J. M. and Steinmayer, A. G. Jr. (1999a) The *Syracusia* as a giant cargo vessel. *IJNA* 28.2, 105–125.

Turfa, J. M. and Steinmayer, A. G. Jr. (1999b) The Earliest Foresail, on Another Etruscan Vase. *IJNA* 28.3, 292–296.

Tusa, S. (1999) *Sicilia Preistorica*. Palermo, Dario Flaccovio Editore.

Tusa Cutroni, A. (1971) Lilibeo (Marsala). – Nuovi scavi nella necropoli punica (1969–1970). *NSc* 25, 662–769.

Tzachili, I. (2007) Weaving at Akrotiri, Thera. Defining Cloth-making Activities as a Social Process in a Late Bronze Age Aegean Town. In M.-L. Nosch and C. Gillis (eds), *Ancient Textiles: Production, Craft and Society*, 190–196. Oxford, Oxbow Books.

Ucelli, G. (1950) *Le Navi di Nemi*. Roma, Libreria dello Stato.

Ugolini, L. M. (1927) Le antichità primitive dell'Albania. *BPI* 47, 159–180.

Vagnetti, L. (1971). *Il deposito votivo di Campetti*

a Veio (Materiale degli scavi 1937–1938). Firenze, Sansoni Editore.

Vagnetti, L., ed. (1985) *Magna Grecia e Mondo Miceneo. Atti del XXII Convegno di studi sulla Magna Grecia (Taranto 1982)*. Taranto, Istituto per la storia e l'archeologia della Magna Grecia.

Vagnetti, L. (1996) The First Contacts between the Minoan-Mycenaean and the Western Mediterranean Worlds. In G. Pugliese Carratelli (ed.), *The Western Greeks*, 109–116. Venezia, Bompiani.

Vagnetti, L. (2000) Western Mediterranean Overview: Peninsular Italy, Sicily and Sardinia at the Time of the Sea Peoples. In E. D. Oren (ed.), *The Sea Peoples and Their World: A Reassessment,* 305–326. Philadelphia, the University Museum.

Valentino, M. (1997) La Montagnola di Marineo. Pesi da telaio e oscilla. In C. A. Di Stefano (ed.), *Archeologia e territorio*, 197–207. Palermo, G. B. Palumbo.

Vanden Berghe, I. (2002) Appendice 1: analisi del colore. In P. von Eles (ed.), *Guerriero e sacerdote. Autorità e comunità nell'età del ferro a Verucchio. La Tomba del Trono*, 220. Firenze, All'Insegna del Giglio.

van der Meer, B. (2007) *The Liber linteus zagrabiensis. The Linen Book of Zagreb. A Study of the Longest Etruscan Text.* Leuven, Peeters.

van Driel-Murray, C. (1990) The Roman army tent. *Exercitus* 2.8, 137–144.

Vassallo, S., ed. (1999) *Colle Madore. Un caso di ellenizzazione in terra sicana*. Entella. Giuseppe Nenci.

Vay, I. (1988) I pesi da telaio della casa R18. In R. de Marinis (ed.), *Gli Etruschi al nord del Po*. Vol. I, 170–174. Mantova, Campanotto Editore.

Vellani, S. (2005) Resti di tessuti mineralizzati dalla necropoli di Monte Bibele. In *Intrecci vegetali e fibre tessili da ambiente umido. Analisi Conservazione e Restauro. Atti Trento 28–30 maggio 2003. Incontri di Restauro 4*, 370–375. Riva del Garda (Trento), Provincia Autonoma di Trento.

Vianello Cordoba, A. P. (1967) Una tomba "protovillanoviana" a Veio. *StEtr* 35, 295–306.

Vicari, F. (1994) Economia della Cispadana Romana: la produzione tessile. *RivStorAnt* 24, 239–260.

Vicari, F. (2001) *Produzione e commercio dei tessuti nell'Occidente romano*. BAR-IS 916. Oxford, John and Erica Hedges.

Vighi, R. (1955) Il sepolcreto arcaico del Sorbo. *MonAnt* 42, 25–200.

Višniauskaitė, A. and Laniauskaitė, J. (1977) *Valstiečių lininінkystė ir transportas*. Vilnius, Mokslas.

Vitali, D., ed. (1983) *Monterenzio e la Valle dell' Idice*. Bologna, Banca del Monte di Bologna e Ravenna.

Vitali, D., ed. (1987a) *Celti ed Etruschi nell'Italia centro-settentrionale dal V sec. a.C. alla Romanizzazione*. Bologna, Bononia University Press.

Vitali, D. (1987b) Monte Bibele tra Etruschi e Celti: dati archeologici e interpretazione storica. In D. Vitali (ed.), *Celti ed Etruschi nell'Italia centro-settentrionale dal V sec. a.C. alla Romanizzazione*, 309–380. Bologna, Bononia University Press.

Vitali, D. (1991a) The Celts in Italy. In S. Moscati (ed.), *The Celts*, 220–235. Milano, Bompiani.

Vitali, D. (1991b) The Settlement and Cemetery at Monte Bibele. In S. Moscati (ed.), *The Celts*, 288–289. Milano, Bompiani.

Vitali, D. (1992) *Tombe e necropoli galliche di Bologna e del territorio*. Bologna, Istituto per la Storia di Bologna.

Vitali, D. (ed.) (2003) *La necropoli di Monte Tamburino a Monte Bibele*. Bologna, Gedit Edizioni.

Vitri, S., *et al.* (1991) Gli impianti produttivi seminterrati di Pozzuolo del Friuli. In *Tipologia di insediamento e distribuzione antropica nell'area Veneto-Istriana dalla protostoria all'alto medioevo. Atti di seminario di studio, Asolo, 3–5 Novembre 1989*, 17–32. Padova, Edizioni delle Lagune.

Vogelsang-Eastwood, G. M. (1987) A Re-Examination of the Fibres from the Çatal Hüyük Textiles. *Oriental Carpet and Textile Studies* 3, part 1, 15–19.

Vogt, E. (1937) *Geflechte und Gewebe der Steinzeit*. Basel, Birkhäuser.

Waarsenburg, D. J. (1995) *The Northwest Necropolis of Satricum. An Iron Age Cemetery in Latium Vetus*. Amsterdam, Thesis Publishers.

Wace, A. J. B. (1948) Weaving or Embroidery? *AJA* 52, 51–55.

Wace, A. J. B. and Thompson, M. S. (1912) *Prehistoric Thessaly*. Cambridge, Cambridge University Press.

Waldstein, C. (1905) *The Argive Heraeum*. Vol. II. Boston and New York, Houghton, Mifflin and Company.

Walton, P. and Eastwood, G. (1983) *A brief guide to the cataloguing of archaeological textiles*. York, Wordplex Wordprocessing Bureau.

Walton Rogers, P. (1999) Dyes in the Hochdorf Textiles. In J. Banck-Burgess (ed.), *Hochdorf IV, Die Textilfunde aus dem späthallstattischen Fürstengrab von Eberdingen-Hochdorf (Kreis Ludwigsburg) und weitere Grabtextilien aus Hallstatt- und Latenezeitlichen Kulturgruppen*, 240–245. Stuttgart, Theiss.

Walton Rogers, P. (2003) Dyes and dyeing. In D. Jenkins (ed.), *The Cambridge History of Western Textiles*, Vol. I, 25–29. Cambridge, Cambridge University Press.

Walton Rogers, P., Bender Jørgensen, L. and Rast-Eicher, A., eds (2001) *The Roman Textile Industry and its Influence. A Birthday Tribute to John Peter Wild*. Exeter, Oxbow Books.

Warden, P. G. (2003) The Anatomy of an Etruscan Tomb Forgery: Case Unresolved. *IFAR Journal* 5 no. 4, 36–42.

Warden, P. G. and Kane, S. (1997) Excavations at Poggio Colla (Vicchio) 1995–1996. *EtrStud* 4, 159–186.

Watson, J. (1992) The mammals. In A. Small (ed.), *An Iron Age and Roman Republican Settlement on Botromagno, Gravina di Puglia. Excavations of 1965–1974. Volume I. The Site*, 93–120. London, British School at Rome.

Webster, G. S. (1996) *A prehistory of Sardinia 2300–500 BC*. Monographs in Mediterranean Archaeology 5. Sheffield, Sheffield Academic Press.

Weinberg, G. D. and Weinberg, S. S. (1956) Arachne of Lydia at Corinth. In S. S. Weinberg (ed.), *The Aegean and the Near East: Studies Presented to Hetty Goldman on the Occasion of her 75th Birthday*. Locust Valley, J. J. Augustin.

Wells, P. S. (1980) *Culture Contact and Culture Change: Early Iron Age Central Europe and the Mediterranean World*. Cambridge, Cambridge University Press.

White, K. D. (1967) *Agricultural Implements of the Roman World*. Cambridge, At the University Press.

White, K. D. (1970) *Roman Farming*. Ithaca, Cornell University Press.

Whitehead, J. (1996) New Researches at La Piana. *EtrStud* 3, 105–46.

Whitehouse, R., ed. (1998) *Gender in Italian Archaeology*. Accordia Specialist Studies in Italy 7. London.

Whitehouse, R. D. and Wilkins, J. B. (1985) Magna Graecia before the Greeks: towards a reconciliation of the evidence. In C. Malone and S. Stoddart (eds), *Papers in Italian Archaeology IV. Part iii. Patterns in Protohistory*, 89–109. BAR-IS 245. Oxford.

Whitehouse, R. D., Wilkins, J. B. and Herring, E. (2000) *Botromagno. Excavation and Survey at Gravina in Puglia 1979–1985*. Accordia Specialist Studies on Italy Vol. 9. London.

Wikander, C. (1986) La vita quotidiana nell'ambiente domestico. In *Architettura Etrusca nel Viterbese. Ricerche svedesi a San Giovenale e Acquarossa 1956–1986*, 109–128. Roma, De Luca Editore.

Wild, J. P. (1970) *Textile Manufacture in the Northern Roman Provinces*. Cambridge, Cambridge University Press.

Wild, J. P. (1976) Textiles. In D. E. Strong and D. Brown (eds), *Roman Crafts*, 167–177. New York, New York University Press.

Wild, J. P. (1988) *Textiles in Archaeology*. UK, Shire Publications.

Wild, J. P. (2000) Textile Production and Trade in Roman Literature and Written Sources. In D. Cardon and M. Feugère (eds), *Archéologie des textiles des origines au Ve siècle, Actes du colloque de Lattes, oct. 1999*, 209–213. Monographies Instrumentum 14. Montpellier, Éditions Monique Mergoil.

Wild, J. P. (2002) The Textile Industries of Roman Britain. *Brittania* 33, 1–42.

Wild, J. P. (2003) Introduction, in D. Jenkins (ed.), *The Cambridge History of Western*

Textiles, Vol. I, 9–29. Cambridge, Cambridge University Press.

Wild, J. P., *et al.* (1998) Vindolanda: Some Results of the Leverhulme Trust Programme. In L. Bender Jørgensen and C. Rinaldo (eds), *Textiles in European Archaeology. Report from the 6th NESAT Symposium, 7–11th May 1996 in Borås*, 85–98. GOTARC Series A, Vol. 1. Göteborg, Göteborg University.

Wilkens, B. (1992) Primi dati sui resti faunistici provenienti da alcuni abitati piceni (Ancona, Osimo, Moie di Pollenza). In *La civiltà picena nelle Marche. Studi in onore di Giovanni Annibaldi. Atti del convegno. Ancona 1988*, 38. Ripatransone, Maroni.

Wilson, A. (2004) Archaeological evidence for textile production and dyeing in Roman Africa. In C. Alfaro, J. P. Wild, B. Costa (eds), *Purpurae vestes: Actas del I Symposium Internacional sobre Textiles y Tintes del Mediterráneo en época romana (Ibiza, 8 al 10 de noviembre, 2002)*, 155–164. València, Universitat de València.

Wilson, J-P. (2000) Ethnic and state identities in Greek settlements in southern Italy in the eighth and seventh century BC. In E. Herring and K. Lomas (eds), *The Emergence of State Identities in Italy in the First Millennium BC*, 31–43. Accordia Specialist Studies on Italy 8. London.

Wilson, L. (1930) Loom weights. In D. M. Robinson (ed.), *Excavations at Olinthus. Part II. Architecture and Sculpture: Houses and Other Buildings*, 118–128. Baltimore, The Johns Hopkins Press.

Wilson, V. (1977) The Loomweights. In J. Du Plat Taylor *et al.*, Gravina di Puglia III (Part two). *PBSR* 45, 132–136.

Wincott Heckett, E. (1998) A Late Bronze Age Horsehair Ornament from Cromaghs, Armoy in Ireland. In L. Bender Jørgensen and C. Rinaldo (eds), *Textiles in European Archaeology. Report from the 6th NESAT Symposium, 7th–11th May 1996 in Borås*, 29–37. GOTARC Series A, Vol. 1. Göteborg, Göteborg University.

Winiger, J. (1995) Die Bekleidung des Eismannes und die Anfänge der Weberei nördlich der Alpen. In K. Spindler *et al.* (eds), *Der Mann im Eis. Neue Funde und Ergebnisse* 2, 119–187. Veröffentlichung des Forschungsinstituts für Alpine Vorzeit der Universität Innsbruck 2. Wien/New York, Springer-Verlag.

Wuilleumier, P. 1932. Les disques de Tarente. *RA* 35, 26–64.

Zanasi, C. and Gianferrari, A. (1990) *Nonantola Antiquarium*. Nonantola, Tracce del Tempo.

Zancani Montuoro, P. (1966) Francavilla Marittima – Edificio Quadrato. *AttiMGrecia* 6–7, 1–214.

Zancani Montuoro, P. (1972) Francavilla Marittima: Varia. B. Santuario sul Timpone della Motta. III Statuetta dedalica incompleta. *AttiMGrecia* 11–12 (1970–71), 67–74.

Zancani Montuoro, P. (1975) I labirinti di Francavilla Marittima ed il culto di Athena. *RendNap* 50, 125–140.

Zanini, A. (1997) *Dal Bronzo al Ferro. Il II millennio a.C. nella Toscana centro occidentale.* Pisa, Pacini Editore.

Zannoni, A. (1876) *Gli scavi della Certosa.* Bologna, Regia Tipografia.

Zevi, F. and Bedini, A. (1973) La necropoli arcaica di Castel di Decima. *StEtr* 41, 27–44.

Zhao, F., and Yu, Z. (2002) *The Legacy of the Desert King*. Hong Kong, ISAT/Costume Squad.

Zifferero, A. (1995) Rituale funerario e formazione delle aristocrazie nell'Etruria protostorica: osservazioni sui corredi femminili e infantili di Tarquinia. In N. Negroni Catacchio (ed.), *Preistoria e Protostoria in Etruria II. Tipologia delle necropoli e rituali di deposizione. Richerche e scavi.* Milano, Centro studi di preistoria e archeologia.

Zuffa, M. (1976) La civiltà villanoviana. In V. Cianfarani, G. D. Lollini and M. Zuffa (eds), *PCIA* 5, 199–363. Roma, Biblioteca Di Storia Patria.

3000 anni fa a Verona. (1976) Verona, Museo Civico.

Index